NATO Science Series

A Series presenting the results of scientific meetings supported under the NATO Science Programme.

The Series is published by IOS Press, Amsterdam, and Kluwer Academic Publishers in conjunction with the NATO Scientific Affairs Division

Sub-Series

I.	**Life and Behavioural Sciences**	IOS Press
II.	**Mathematics, Physics and Chemistry**	Kluwer Academic Publishers
III.	**Computer and Systems Science**	IOS Press
IV.	**Earth and Environmental Sciences**	Kluwer Academic Publishers
V.	**Science and Technology Policy**	IOS Press

The NATO Science Series continues the series of books published formerly as the NATO ASI Series.

The NATO Science Programme offers support for collaboration in civil science between scientists of countries of the Euro-Atlantic Partnership Council. The types of scientific meeting generally supported are "Advanced Study Institutes" and "Advanced Research Workshops", although other types of meeting are supported from time to time. The NATO Science Series collects together the results of these meetings. The meetings are co-organized bij scientists from NATO countries and scientists from NATO's Partner countries – countries of the CIS and Central and Eastern Europe.

Advanced Study Institutes are high-level tutorial courses offering in-depth study of latest advances in a field.
Advanced Research Workshops are expert meetings aimed at critical assessment of a field, and identification of directions for future action.

As a consequence of the restructuring of the NATO Science Programme in 1999, the NATO Science Series has been re-organised and there are currently five sub-series as noted above. Please consult the following web sites for information on previous volumes published in the Series, as well as details of earlier sub-series.

http://www.nato.int/science
http://www.wkap.nl
http://www.iospress.nl
http://www.wtv-books.de/nato-pco.htm

Series IV: Earth and Environmental Sciences – Vol. 13

Lake Issyk-Kul:
Its Natural Environment

Lake Issyk-Kul: Its Natural Environment

edited by

Jean Klerkx
International Bureau for Environmental Studies,
Brussels, Belgium

and

Beishen Imanackunov
National Academy of Sciences of the Kyrgyz Republic,
Bishkek, Kyrgyzstan

Kluwer Academic Publishers

Dordrecht / Boston / London

Published in cooperation with NATO Scientific Affairs Division

Proceedings of the NATO Advanced Research Workshop on
The Issyk-Kul Lake: Evaluation of the Environmental State and its Remediation
Cholpon-Ata, Kyrgyz Republic
25–28 September 2000

A C.I.P. Catalogue record for this book is available from the Library of Congress.

ISBN 1-4020-0899-6

Published by Kluwer Academic Publishers,
P.O. Box 17, 3300 AA Dordrecht, The Netherlands.

Sold and distributed in North, Central and South America
by Kluwer Academic Publishers,
101 Philip Drive, Norwell, MA 02061, U.S.A.

In all other countries, sold and distributed
by Kluwer Academic Publishers,
P.O. Box 322, 3300 AH Dordrecht, The Netherlands.

Printed on acid-free paper

Printed in the Netherlands.

CONTENTS

PREFACE

Lake Issyk-Kul is a closed lake, considered as monomictic, located in the Tien Shan mountain belt of the Republic of Kyrgyzstan. It is the world's fifth deepest lake (668 m.) and the second largest high-altitude lake in the world (1607 m above sea level). The lake basin is surrounded by high mountain ranges rising 2000-3500 m above the lake. Between 1973 and 1999, the average monthly temperature in the lake basin increased by more than 4°C, resulting in an acceleration of glaciers melting in the surrounding mountains and an increased input of river during the summer period. Surprisingly the lake level is declining by about 13 m since the middle of the 19th century, while 3 m of lake level decline has been recorded from 1927 to 1966.

The lake is affected by several environmental threats, as well of natural as of anthropogenic origin: decline of the lake level resulting in a progressive increase in salinity of the water, incomplete vertical water exchange which results in eutrophication of the deep waters, risk of contamination by former or present industrial activities and uncontrolled agriculture

The region of the Issyk Kul Lake is one of the most densely populated areas of Kyrgyzstan. Due to its particularly mild climate, agriculture is intensively developed around the entire lake basin. It still is a popular tourist destination during the summer months and numerous touristic centres are located along the shorelines of the lake. The economic development of the region is consequently largely dependent from the environmental state of the lake.

A NATO Advanced Research Workshop entitled *"The Issyk-Kul Lake: evaluation of the environmental state and its remediation"* has been organised in Cholpon Ata (Kyrgyzstan) from September 25 to 28, 2000 with the aim to make a multidisciplinary evaluation of the present state of knowledge about Lake Issyk-Kul, with particular emphasis on the environmental aspects; to define and to evaluate the environmental threats affecting the lake and its catchment area; to define essential gaps in the present knowledge about Lake Issyk-Kul, particularly for what concerns the environmental situation; to establish a strategy for further investigation of the environmental problems concerning the lake basin; finally to inform the Kyrgyz authorities about the present environmental situation of the lake basin and to propose plans for future actions.

Since many years the lake has been monitored at several permanent observatories located on the lake shores; Kyrgyz scientific institutes have concentrated their efforts on the study of Issyk Kul from an environmental point of view. During the last years, several international scientific projects have been initiated in the Issyk Kul region. Scientific teams from different countries are investigating problems related to the environment within several disciplines: seismology, tectonics and earth movements, chemical contamination of the lake waters, hydrobiology, health problems. The

Workshop was bringing together local scientists involved in the study of the lake, foreign scientists active in projects on Lake Issyk-Kul and specialists involved in similar investigations in other places of the world.
The results of the Workshop are presented in this volume. In a first section is assembled the information collected on the lake during many years of observation, as well as interpretations of scientific investigations by local scientists and some concepts about the environmental protection of the Lake that are prevailing in Kyrgyzstan. This information is often only available in unpublished reports or published in local scientific journals that are not easily accessible. For the first time this unique information is presented for an international audience. It is completed by recent results of international projects focusing on Lake Issyk Kul. A second section assembles the results of case-studies on similar problems in other parts of the world. In a last section, ethical aspects concerning the environmental protection of Lake Issyk-Kul are presented.
The Statement of the Workshop is included. It addresses priority fields of research for which complementary knowledge is required. It stresses on the urgent need for assembling all available data relevant to the lake basin and its catchment area in an electronic database, in order to define the present-day baseline of the lake, its recent evolution and its possible future trends. Finally it formulates recommendations to the authorities.

The organizers gratefully acknowledge the financial support of the Scientific Affairs Division of NATO for the organization of the Workshop.
They express their most sincere thanks to the persons who have participated in the preparation of this volume: Mrs. Svetlana Kolobova for carefully editing the contributions, Mrs. Tatiana Peripelova for accurate translations.

Address by the President of the National Academy of Sciences of the Kyrghyz Republic
Dr. Janybeck JEYENBAEV

GENERAL TENDENCIES OF THE ISSYK-KUL LAKE ENVIRONMENTAL STATE AND MECHANISMS OF ITS ENVIRONMENTAL SAFETY MAINTENANCE

Dear Colleagues,
Ladies and Gentlemen,

Let me heartily greet you in the blessed land of Kyrghyzstan, on the shore of its pearl – the largest high altitude Lake of Issyk-Kul.

This Workshop has noble goals – to unite a scientific potential of foreign scientists and scientists of Kyrghyzstan, specialists and scientists of the world, who have already been involved or acquainted with the ecological problems of the Issyk-Kul Lake, and those experts also, who are experienced with such problems in other parts of the world, to make a multidisciplinary evaluation of the present conditions of knowledge about the Issyk-Kul lake, making an accent to, especially, ecological aspects. To evaluate the degree of the present ecological threat to the lake and its basin, define the priorities of further researches, define the strategy of joint multidisciplinary approach to the problems of the Issyk-Kul basin.

You are apparently acquainted with brief historical information and brief characteristics of the Issyk-Kul lake, that have been described in the booklet. But, nevertheless, I will stop on some of the general tendencies of its environmental state and mechanisms of its safety maintenance.

The Issyk-Kul Lake is one of the largest little salted lakes of the world, the 4th deep continental reservoir of the world, the largest high altitude lake in Central Asia. The lake and its basin are characterized with own original natural conditions, which refer this region to the category of the most unique objects of our planet. The originality is displayed in everything – its morphometry and morphology, relief structure, climate conditions and vegetation, mineral and recreational resources etc.

Caused by big heights of the mountain range, framing the lake, and their nearness to the basin, there occurs a sharp change of landscape zones – from semi-desert and steppe up to forest meadow alpine and glacial-cornfield ones. Such diversity of the landscape in the distance of 40-60 km from the coast attracts researchers, tourists and lets develop various branches of agriculture here.

The main wealth of the Issyk-Kul region and adjoining to it high altitude lands is a non-freezing lake Issyk-Kul itself, deposit of mineral water, gold, tin and other non-ferrous metals, various materials for structure, water-energetic resources, huge reserve of medical mud, a favourable climate for cure and recreation, a large extent of natural sandy beaches, agricultural lands (pasture and area under crops) etc. In order to preserve natural beauty of this land, protect the flora and fauna a number of national preserves and reserves was established on the shore of the Issyk-Kul Lake. It should be noted that the Issyk-Kul Lake is one of places for the largest wintering of water-fowl. They are swans, wild ducks, geese.

The economy of the region is specialized in production of non-ferrous metals, electro-technical goods, woods, meat, potato; development of sheep-breeding, meat and milk, cattle-breeding, pedigree horse-breeding, etc. The big height and climate differentiation caused difference in character and intensity of farming in the basin, particularly, degrees of influence upon the environment. The most influenced is the valley near the Lake. Here there is a maximal density of population, there develop actively an agriculture and not large industrial enterprises. In the foot hill of broad development are the irrigation farming and cattle winter keeping. The mountain zone is mainly used as a base of distant pasture stock rising, forestry, as a region of active tourism and alpinism.

The large part of the lake is not frozen in winter and because of the significant depth in summer seasons, it accumulates a large quantity of heat (till 10^{13} ton-calorie), which is used in winter for increasing the temperature in coastal zones. It is distinctive for the Issyk-Kul Lake to have a stable circling movement of water with sufficiently high speeds, that promote a good water mixture and oxygen enrichment. A high content of oxygen in winter favors the active life of water organisms and active running of bio-chemical processes.

In the coastal zones of the Issyk-Kul Lake are located more than 30 mineral sources, with different chemical compositions and various medical properties. Part of them is thermal, constituting the base for specialized resort centers.

As the most important and valuable natural resource of the Issyk-Kul basin is the climate. Caused by its position inside the mountains many synoptic processes are not manifested here, a large volume of water mass to keep down summer heat and to increase winter temperatures. Except the eastern part of the Lake, the climate of the coastal zone is similar to the sea climate. The intensity of precipitation falls is not high.

Especially favorable for the lake and its near zones is a regime of solar radiation and solar radiance. This is indeed a Land of sunshine. Unsunny days especially in summer, are very few, which is favorable for helio-energetics. In order to use non - traditional sources of energy it is worth to pay attention also to the wind. It is significant for the western and eastern parts to have local winds called as "Ulan", "Santash", "Gorniak" reaching speeds of up to 25-40 m/sec. Their clashes over the surface of the Lake form vortical formations of the water spout type. A study of this phenomenon in a small space could play a significant role for the explanation of the mechanism of genesis and development of such a natural phenomenon like typhoon.

Glaciers that occupy an area of 650,4 km^2 are one of the specifics of the Issyk-Kul depression. Their main value is ability to maintain high water content in the period of maximum water use, even in years with humidity of 30-40 % lower than usual. Now glaciers are decreasing. This process has started in the middle of the 19th century, but especially activated during the last 30 years due to the global warming.

Global warming and change in water content determine one of the modern problems of the lake - decrease of its level. Dynamics of Issyk-Kul level's change and its prognosis are complex problems. It should be mentioned that in modern history there were periods of level increase and stabilization. During the last 30 years, this took place in 1970-71, 1987-89 and 1999. Reasons of level fluctuation in current century are connected with climatic, tectonic, hydrogeological, anthropogenic factors. It is naturally, that in the process of interaction of a man with the nature there appear some negative eco-situations.

Today there exists a real threat to biosphere, which may cause changes of the formed biological systems. For example, conditions of forest and mountain meadows, water meadow and fir forests etc. arouse a great anxiety.

Unfortunately, Issyk-Kul periodically experiences an anthropogenic pressing on the unique biogeocenosis, in forms of cultivation of foreign ichthyofauna - antagonistic population (Sevan trout, pike-perch); irrational irrigation; consequences of uranium and other polluting industries; "wild" recreation and bad condition of recreational centers, sanatoriums, health resorts; rules and norms of agricultural works are being violated (ineffective irrigation, use of mineral fertilizers without control, violation of norms of cattle-breeding, dates of pasture, etc.).

Increase of rates of water area assimilation without any nature-protecting arrangements causes the intensive technogenic pollution of this region, and this leads to the real threat to human health. According to the data of Kyrghyz scientists, immunity decreasing and a great number of diseases are the consequences of this phenomenon. Factor of high altitude and simultaneous effect of unfavorable ecology on human organism is an unstudied problem and of a great interest not only for Kyrghyzstan.

Scientists of the National Academy of Sciences of the Kyrghyz Republic have studied the Issyk-Kul Lake during many years regularly. Particularly, the Tien-Shan physico-geographic station has started systematic researches of temperature regime of glaciers, dynamics of ice melting, changing of river water, flowing to the lake, firstly in Central Asia. The National Academy of Sciences has carried out a great volume of works on regular monitoring of environment condition of the region.

The problems of Issyk-Kul have also drawn the attention of scientists from different countries of the world. At present, several international scientific projects are being carried out for the investigation of ecological problems of this region from the point of view of seismology, tectonics and land movement, chemical pollution of lake water, hydrobiology, health problems, and these are important not only for Issyk-Kul, but also for other regions of the planet. Therefore, we hope that the Issyk-Kul region, because of its exclusive natural specifics, may become an international natural experimental training ground for global ecological studies.

Thank you for your attention!

NATO Advanced Research Workshop

THE ISSYK-KUL LAKE : EVALUATION OF THE ENVIRONMENTAL STATE AND ITS REMEDIATION

Cholpon Ata (Kyrgyzstan), September 25-28, 2000

STATEMENT

The NATO Advanced Research Workshop: *The Issyk-Kul Lake: Evaluation of the Environmental State and its Remediation* was attended by 44 scientists from Belgium, Canada, Finland, France, Italy, Kazakhstan, Kyrgyzstan, Russia, Spain, Switzerland, Tajikistan, Turkey, the United States of America, Uzbekistan. Representatives of the NATO Science and Environment Division, the European Commission and INTAS also attended the workshop.

The first part of the Workshop was devoted to various aspects of the environmental state of the lake and its basin :

◊ the earthquake risk and its consequences, such as landslides, rockfalls, ...
◊ the environmental changes related to earth's movements
◊ the lacustrine physico-chemicalpresent-day status of the lake.

Subsequently, the environmental threats affecting the lake were discussed:

- the risk related to the presence of radioactive wastes due to the Kadji Sai uranium mining activities along the lake shore was mentioned,
- the 1998 accident at the Kumtor gold mine in the SW part of the basin was discussed along different aspects. Alternative ore processing methods were proposed,
- the lake level variations were discussed. The progressive decrease since the beginning of the 20th century is attributed mainly to water consumption by irrigation for agricultural purposes,
- the potential pollution of the lake resulting from human activities along the lake shores and particularly due to touristic activity concentrated on the northern shore was mentioned.

The second part of the Workshop focused on case studies related to similar problems in other parts of the world.

Finally some ethical aspects related to the environment were considered.

It appears from the presented information that important gaps exist in the knowledge of basic physical and chemical processes triggering the lake water mass:

- an overall hydrophysical model based on a complete set of factual data for the entire water basin is missing;
- the chemical structure of the water mass is insufficiently known;
- the water balance of the basin and its catchment basin is insufficiently supported by data.
- the trophic chain and its relation with the potential pollution of the lake.

Fundamental studies on these topics are required. They should be based on a complete set of data collected by modern methods.

The urgent need exists for a database assembling all available data relevant to the lake basin and its catchment area, in order to define the present-day baseline of the lake, its recent evolution and its possible future trends.

It also appears that the lake is potentially at risk by several environmental threats, but that their importance is insufficiently ascertained :

- the potential risk resulting from the radioactive wastes of the Kadji Sai mining activities has to be evaluated;
- the risk from the mining activity in the Kumtor gold mine has to be strictly controlled, in order to prevent possible ecological disasters;
- the lake is at risk from human activities along its shores. Water quality has to be systematically controlled particularly in the areas of intense human occupation and turism activity. Discharge of waste waters coming from these activities into the lake has to be controlled. The implementation of a monitoring system for water quality, particularly in the areas of intense human activity, has to be considered.
- the irrigation system for agricultural purposes around the lake is assumed to be responsible for the lake water decrease. Its impact of the lake water balance has to be better ascertained.

Recommendations

From the information considered at the Workshop and the discussions among the participants, the following recommendations are formulated to the authorities :

1. to collect existing environmental data into universal access database in a timely and efficient manner
2. to analyse existing data to determine:
   (i) baseline values for significant environmental variables
   (ii) long term trends of environmental concern
   (iii) a prioritized list of areas where significant environmental data is lacking
3. to seek support for socio-economic studies to:
   (i) enhance understanding of current economic organization
   (ii) identify future economic trends of likely environmental significance

(iii) identify opportunities for economic sustainable development, taking into account natural capital, human capital, social capital

4. to seek support for evaluating the need of a monitoring system of the lake basin at least in the areas of intense human activity
5. to restart at least a minimal programme of monitoring of the lake in order to overcome the gap of information during the last 10 years
6. to seek support for training of scientists and technicians in modern techniques of environmental monitoring
7. to start a programme of environmental education of the population, particularly of scholars, students and local officials.

Part I

PHYSICAL AND CHEMICAL CHARACTERISTICS

CLIMATE AND PHYSICAL PROPERTIES OF WATER IN LAKE ISSYK-KUL

G.D. SHABUNIN
Issyk-Kul Lake Observatory
Cholpon Ata, Kyrgyzstan

A.G. SHABUNIN
Instiute of Water Problems and Hydropower
533, Frunze str., Bishkek, Kirghyzstan

1. General background

Lake Issyk-Kul is a unique natural water basin of tectonic origin located in the Tien Shan mountain belt of Central Asia, at an altitude of 1606 m a.s.l. Its drainage basin occupies a surface area of 22 080 km^2 and is surrounded by 4 000 – 5 200 m high ridges of the Northern Tien-Shan (Fig. 1). All major tributaries of Issyk-Kul originate from glaciers in the high mountains, where 834 glaciers occupy an area of 650 km^2 and a volume of 48 km^3.

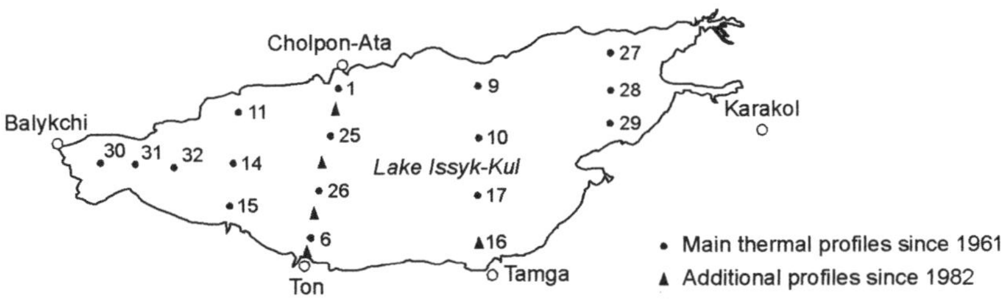

Figure 1. Location map of the Issyk-Kul basin and locations of measurements.

Precise measurements of the lake dimensions has been made by Kodyaev [1]: 178 km length, 60.1 km width, 668 m maximum depth, 278.4 m average depth, 623 km^2 water surface area, 1738 km^3 water volume (Table 1). Three decades have elapsed since these measurements were made, and in the meantime the lake level has dropped by nearly 1.3 m; as a result, the water surface area became 60 km^2 smaller, and the water volume was reduced by 7.3 km^3, etc. Over the entire period of instrumental monitoring – from 1928

J. Klerkx and B. Imanackunov (eds.), Lake Issyk-Kul: Its Natural Environment, 3–11.

up to now - the level of Issyk-Kul decreased for 3.3 m, at an average yearly drop rate of 4.5 cm.

TABLE 1. Variations of area and volume of Lake Issyk-Kul at different depths [1]

Isobath, m	Area		Volume	
	km^2	%	km^3	%
0	6236.0	100.00	1738.0	100.0
10	5797.4	92.96	1679.1	96.6
20	5393.3	86.48	1623.2	93.4
50	4481.6	72.02	1465.3	84.3
100	3842.7	61.63	1267.4	72.8
150	3498.2	56.09	1083.9	62.3
200	3150.7	50.71	917.8	52.7
250	2741.9	43.96	770.6	44.3
300	2431.8	38.99	641.3	36.8
350	2260.8	36.27	524.0	30.1
400	2129.7	34.14	414.3	23.8
450	2008.8	32.37	310.9	17.9
500	1840.9	29.69	214.7	12.4
550	1621.4	26.00	128.2	7.4
600	1357.1	21.76	53.8	3.1
650	674.3	10.81	4.0	0.2
668	0.0	0.0	0.0	0.0

The age of Issyk-Kul is estimated at about 10 m.y. The Issyk-Kul water has a high salinity (up to 5.968 g/kg) due to the closed basinal character of the lake that exists at present, and, possibly, repeatedly in the past. The water chemistry shows a high alkalinity and a Na-Mg chloride-sulfate composition [2]. The temperature of maximum density of the water is 2.75° C and the freezing point 0.3°C.
The water physics is mainly controlled by temperature and salinity and their ratios. The salinity is constant in about 95% of the lake [2]; the variations of the physical properties of water are due to temperature dynamics and to external effects, primarily the wind.

2. Wind

The predominant wind directions estimated as monthly means over a long period of observations are shown in Table 2.

TABLE 2 . Frequency of predominant wind directions (%) at different stations in Issyk-Kul basin, averaged over many years of observation.

Station	N	NE	E	SE	S	SW	W	NW	Still
Balykchi	12	7	20	10	2	2	41	6	24
Cholpon-Ata	20	17	17	16	11	5	6	8	23
Karakol	6	9	29	5	17	8	17	9	49
Tamga	12	6	4	10	41	5	8	14	19

On the background of strongly variable wind dynamics, the general predominant direction is seaward. Breezes constitute an important component of the wind regime on Issyk-Kul. However, although very frequent, they are rather weak (3-4 m/s in average, maximum 6-7 m/s) and cannot have a significant influence on the hydrodynamics of the deep lake.

Table 3 contains data on strong winds recorded by weather stations in the Issyk-Kul basin. For instance, at the Balykchi station wind strength may attain 8 – 15 m/s for at least 171 days in a year, from 15 to 20 m/s for 83 days, from 20 to 30 m/s for 60 days, and at least once a year wind is as strong as 30 m/s.

TABLE 3 . Wind strength (m/s), monthly and annual mean, averaged over many years of observation.

Station	I	II	III	IV	V	VI	VII	VIII	XI	X	XI	XI I	Annual mean
Balykchi	3.7	4.3	4.5	4.8	4.3	3.6	3.3	3.6	4.0	4.4	4.3	3.6	4.1
Cholpon -Ata	1.9	2.0	1.7	1.8	1.8	1.7	1.6	1.5	1.7	1.9	2.1	2.0	1.8
Karakol	1.4	1.6	1.7	2.0	2.0	1.6	1.4	1.4	1.6	1.6	1.7	1.4	1.6
Tamga	1.9	1.7	1.4	1.5	1.5	1.5	1.4	1.4	1.6	1.7	1.8	1.9	1.6

Strong winds related to invasion of cold air into the Kyrghizstan territory exert a major control on the thermodynamic processes in Issyk-Kul. Winds invade through the lows in the west and in the east of the basin surroundings. Because of the daily Earth's rotation, the wind from the west (locally called "ulan") turns toward the southern basin side and the wind from the east (coming through the San-Tash pass and thus called "santash") turns toward the northern side. Both winds often blow simultaneously. Their strength reaches 35 – 40 m/s and 25 m/s, respectively, and the duration is from several hours to several days.

The "hydrodynamically significant" wind that can change the currents and temperature patterns and can produce considerable waves in Issyk-Kul should attain a strength at least 7 m/s and last at least six hours. These limits were estimated [3] on the basis of wind-induced water-level changes in the western part of the lake and applied to the data of three years of observations (1982 – 1984) at three weather stations in the west and in the east of the lake: Balykchi (B), Cholpon-Ata (Ch), and Karakol (K). As a result, nine types of wind patterns were distinguished (Table 5) as a function of various combinations of "ulan" (U) and "santash" (S) and the absence of an hydrodynamically significant wind (N). During a storm, several wind patterns (up to 11) as a rule follow one another without still-weather breaks. For the purpose of classification of storms, all wind patterns were grouped into three classes: (1) dominated by "ulan" (types 1 – 3); (2) dominated by joint effect of "ulan" and "santash" (types 4 – 7); (3) dominated by "santash" (types 8,9), which yielded 7 types of storms with different seasonal frequencies (Table 6).

TABLE 4. Average number of days with wind equal or exceeding a certain strength

Station	Wind strength, m/s	I	II	III	IV	V	VI	VII	VIII	IX	X	XI	XII	Annual mean
Balykchi	≥8	11.3	12.2	16.0	17.4	16.7	14.5	14.0	14.3	15.4	15.6	13.0	10.9	171.3
	≥15	5.9	6.3	8.6	8.9	8.2	5.7	5.7	6.5	7.2	8.4	6.5	5.5	83.4
	≥20	5.4	5.6	5.0	7.3	6.5	4.0	2.6	4.5	5.2	5.7	4.7	3.1	59.6
	≥30	0.2		0.3		0.2	0.09					0.3		1.1
Chol pon-Ata	≥8	1.8	2.1	2.0	3.2	3.5	3.5	2.4	2.4	3.0	3.3	3.0	2.5	32.7
	≥15	0.7	1.1	1.0	1.8	2.2	2.0	1.5	1.3	1.7	2.0	1.7	1.5	18.5
	≥20	0.5	0.8	0.2	0.6	0.7	0.7	0.7	0.3	0.3	0.5	0.7	0.6	6.6
Kara kol	≥8	3.2	3.3	4.1	4.8	5.1	3.3	2.8	3.6	3.5	3.5	3.7	2.9	43.8
	≥15	1.3	1.2	1.2	1.7	2.0	1.3	1.2	1.2	1.3	1.3	1.6	1.3	16.6
	≥20		0.1	0.07		0.2	0.07			0.07		0.1	0.07	0.7
Tamga	≥8	0.5	0.4	0.8	1.9	2.1	2.5	2.3	1.8	1.4	1.0	1.2	0.6	16.5
	≥15	0.1	0.1	0.3	0.8	1.0	1.2	1.4	1.0	0.6	0.5	0.4	0.2	7.6
	≥20	0.1			0.1		0.2	0.5	0.1		0.1	0.07		1.2

TABLE 5 . Wind patterns and their parameters over 1982-1984

Type	Station			Number of cases	Duration, (hrs)			Frequency (%)	
	B	Ch	K		Total	Average	Max.	From number of cases	From duration
1	U	U	U	73	786	10.8	45	14	12
2	U	U	N	43	390	9.1	21	8	6
3	U	N	N	17	204	12.0	30	4	3
4	U	N	S	22	321	15.0	48	5	5
5	U	S	N	64	816	12.8	60	12	13
6	U	S	S	164	2562	15.6	84	31	39
7	U	U	S	73	747	10.2	45	14	12
8	N	S	S	18	162	9.0	18	3	2
9	S	S	S	48	465	9.7	27	9	8

Thus, the most frequent is the pattern defined by a joint effect of the western and eastern winds (type 6 in Table 5), and the storm conditions are most often produced by their alternation or simultaneous action (types D and B in Table 6). These features of the wind dynamics control the circulation and the physical properties of the water.

TABLE 6 . Seasonal frequency of storms of various types (1982-1984)

Storm type	Wind pattern class	Number of cases for three years					
		Winter	Spring	Summer	Autumn	Total	%
A	I	4	8	3	7	22	10
B	II	20	21	18	26	85	39
C	III	-	3	1	1	5	2
D	I,II	18	15	5	18	56	26
E	I,III	-	1	4	1	6	3
F	II,III	1	7	10	3	21	10
G	I,II,III	2	12	4	5	23	10
Total		45	67	45	61	218	100

3. Temperature regime and water physics in Issyk-Kul

Monitoring of water temperature in the open part of Issyk-Kul (Fig. 1) was conducted from 1961 to 1991. Before 1978, measurements were taken monthly and were restricted to the upper 100 m of water. In 1978 – 1981 the number of vertical measurement profiles increased to 35, but measurements on additional profiles were irregular. From 1982 to 1991, measurements were carried out weekly, to depths till 200 m, along the vertical profiles shown in Fig. 1 plus 4 more profiles along the line Cholpon-Ata – Ton. At depths below 200 m, only a few special measurements were taken.

First the seasonal dynamics of the temperature field in the upper 100 m of water will be considered. The heat budget of the lake water is the lowest in late February – early March, when the main portion of water (95%) is in a nearly isothermal state with temperatures 4-5° C (Fig. 2). Lower temperatures were observed in shallow basins in the western and eastern lake extremities, and the ice covers less than 1% of the water surface. The temperature pattern during this transition period is shown in Fig. 2. In February, there exist two zones of cold near-surface water in the west and in the east and one zone in the central part of the lake. This distribution holds down to 20 m (Fig. 2b). The lower temperatures in the west and in the east of the lake are accounted for by shallow depths and more intense cooling, and the low-temperature zone in the lake center is related to cyclonic circulation of water that causes upwelling of cold water from greater depths.

The central low-temperature zone is encircled by a warmer periphery. The origin of the latter is due to two main causes: (1) great depths (100 – 300 m) allow unlimited accumulation of heat during the period of lake warming; (2) the upwelling of cold pelagial waters in the center is accompanied by outflow of warm near-surface waters towards the periphery. As a result, by the end of the warming period, the heat budget of the upper 100 m in the peripheral zone is 15 – 20% higher than that in the central zone. Having accumulated such a storage of heat during the warming period, the peripheral zone keeps warmest through the period of cooling.

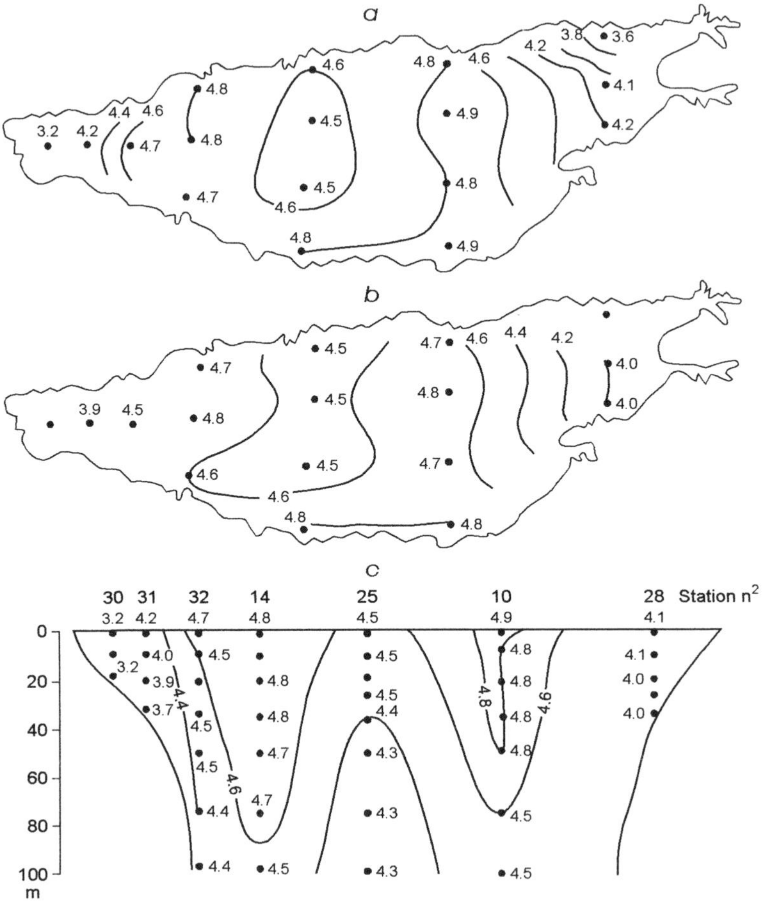

Figure 2. Mean monthly water temperatures in February at depths 0 m (a), 20 m (b), and along a west-east cross section of Issyk-Kul (c)

The spring warming of water in March is evident only from temperature growth in the shallow zone and in the upper 10 m section in the central part of the lake. In April and May, warming involves almost the entire 100 m section. At 100 m, the temperature increase is within 0.1 – 0.2° C, and at 75 m the increment may attain 0.7° C. The temperature of surface water increases to 6.5 – 8° C by mid-April and to 10 – 12° C by mid-May. In the thermocline that forms then, the vertical temperature gradient may reach 1 deg/m. The average thickness of the epilimnion is 4 – 5 m and up to 25 – 30 m in the littoral zone.

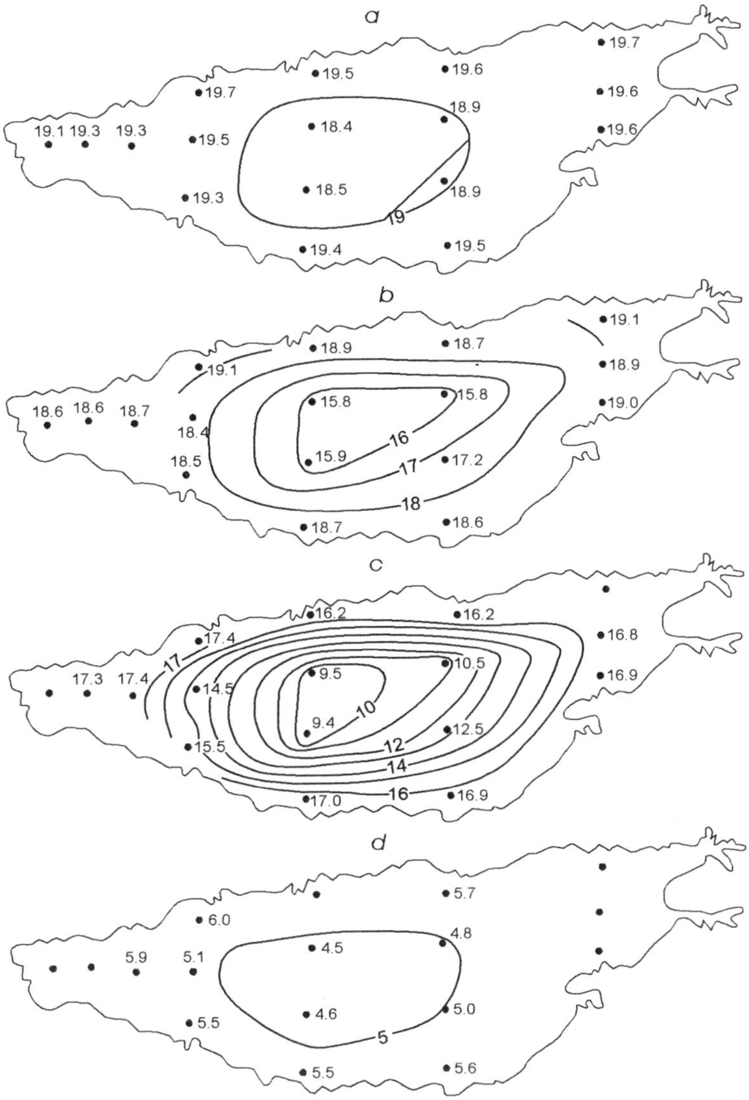

Figure 3. Mean monthly water temperatures in April at depths 0 m (a), 10 m (b), 20 m (c), and 75 m (d). At some points no values are shown because the section did not reach the mapped depth level.

The growth of horizontal temperature (density) gradients between the colder central and warmer peripheral zones causes more intense density cyclonic circulation, with an average velocity of 3 cm/s, as estimated from the mean temperature distribution averaged over many years of observations. The instrumentally measured velocity of currents in that period is 2-4 times higher, i.e., density circulation is strengthened by the circulation produced by strong winds. The integral water transport in cyclonic circulation is divergent and directed toward the periphery of vortices. That is why warm near-surface waters are pushed shoreward and give way to the upwelling of colder waters from deeper layers. As a result, the thermal structure of the water in July-August acquires a pronounced dome-like shape (Fig. 3).

On the background of general warming and thickening of the epilimnion from July to August, its thickness reduces notably in the central part of the lake. Especially strong winds observed in some years caused such an intense upwelling that the 6-8° C colder metalimnion waters were involved.

The period of warming in Issyk-Kul ends in August, when the epilimnion attains 15 – 20 m in average thickness and 20° C in temperature. The highest vertical temperature gradient in the thermocline reaches 3 deg/m, the horizontal gradient between the cold central part of the lake and its periphery is up to 1 deg/km, and the velocity of currents in cyclonic density circulation increases to 20 cm/s.

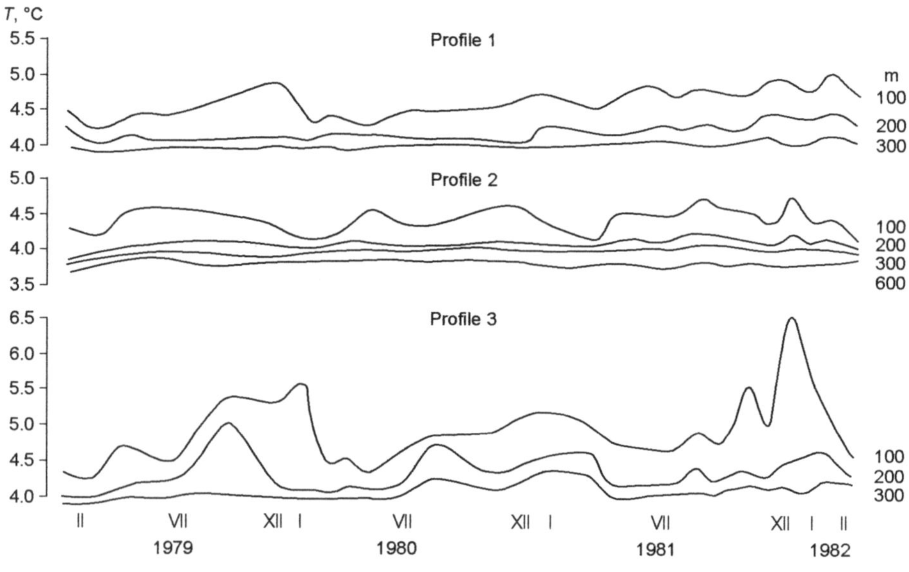

Figure 4. Water temperature variations along the vertical profiles on the line Cholpon-Ata – Ton, at depths from 100 to 650 m.

In autumn, under the effect of convective mixing, the epilimnion thickens rapidly in the littoral zone (up to 40 – 50 m in November), and in the central part of the lake its thickness either remains invariable or decreases. As a result, the cold-water dome becomes more convex, though the horizontal temperature gradient decreases. The vertical temperature gradient in the thermocline becomes 4-5 times lower from August to November, and the water temperature on the lake surface drops from 20° C to 10° C.

The cyclonic density currents (calculated from the water temperature field averaged over many years) are up to 10 – 12 cm/s.
In winter, as the lake cools down, its thermal structure gradually arrives at the February pattern. Note that the cold-water dome in the central lake part exists in Issyk-Kul all year around, only changing its configuration and horizontal and vertical temperature gradients. The cyclonic water circulation also exists through all seasons.
Temperature measurements at depths below 100 m were carried out by Shabunin in 1979 – 1982, along three vertical profiles across the central zone of the lake (line Cholpon-Ata – Ton, Fig. 1): (1) in the north of the lake to a depth of 320 m, (2) in the center to a depth of 600 m; (3) in the south of the lake, to a depth of 320 m. The measurements were taken monthly (from February 1979 to February 1982) by deep-water thermometers, at depths 0, 5, 15, 25, 50, 100, 200, 300, 600, and 650 m; in addition, bathythermograph sounding was carried out in the upper 200 m of water.
Fig. 4 shows the water temperature variations in the depth range 100 to 650 m. Maximum contrasts are observed on the 3d vertical profile: 2.4° C at 100 m, 1.1° C at 200 m, and 0.4° C at 300 m. At the depth levels of 600 and 650 m, the temperature contrasts remained within 0.2° C for the three years of measurements.

4. References

1. Kodyaev, G.V. (1973) Morphometry of Lake Issyk Kul [in Russian], *Izv. VGO* **105**,
2. Kadyrov, V.K. (1986) *Water Chemistry in Lake Issyk Kul and its Catchment Basin* [in Russian], Ilim, Frunze.
3. Revyakin, V.I. and Shabunin, G.D. (1988) Classification of strong winds in Lake Issyk Kul [in Russian], in Dickich, A.N (ed), *Physiographic Studies of Lake Issyk Kul and its Surroundings*, Ilim, Frunze, 79-87.
4. Shabunin, G.D. (1982) Long-term dynamics of the temperature regime in Lake Issyk Kul [in Russian], *Izv. Kirg. AN SSR* **3**, 39-46.

WATER CHEMISTRY AND ECOLOGY OF LAKE ISSYK-KUL

A.S. KARMANCHUK
Institute of Irrigation of Kyrgyzstan
4a, Toktonalieva street, 720055 Bishkek, Kyrgyzstan

In recent years, Lake Issyk-Kul became a subject of special concern because of continuous water level decrease and contamination related to human activity (recreational facilities, constructions, irrigational land use, cattle breeding, etc.). The development of agriculture is associated with increasing water withdrawal from lake tributaries, from 9% of total runoff in 1966, 34% in 1982, to 43% in 1998.

The ecology of the Issyk-Kul drainage basin is consequently affected by natural and man-induced factors. The natural factors (rocks, soils, climate, surface topography) that control the chemistry of the 60 inlets remain practically invariable within the recent period. At the same time, man-induced factors (industrial, farming, and life waste waters) disturb the natural ecological balance in river systems and in Lake Issyk-Kul and input contamination with organic compounds (petroleum products, detergents, phenols, pesticides) and heavy metals. Moreover, the lake and the water systems in its catchement drain an extensive land-use irrigation area contaminated with fertilizers and pesticides. The man-induced threats to ecology are aggravated by the physiographic conditions of Issyk-Kul, which is a closed lake surrounded by mountains and thus has a limited self-cleaning capacity.

1. Water chemistry in the littoral and pelagial zones of the lake

The recent data on the littoral and pelagial water chemistry of Issyk-Kul were obtained in 1998-2000, for five distinct segments of the lake on the basis of their specific geomorphology and hydrochemistry:

1. the western extremity - the gulf of Rybachiy (Fisher's gulf) - , relatively shallow (0.5 to 30 m), with a smoothly cut shoreline. Frequent winds provide continuous water mixing and roiling of bottom sediments which is a cause of enhanced concentration of suspended matter. The water has a salinity of 5801.2 – 6019.7 mg/l, and a Na-sulfate-chloride composition.
2. the northern coast, open to the lake, with numerous bays resulting in a strongly dissected shoreline, depths within 40 – 50 m. The gulf of Cholpon-Ata is the best studied. The water has a salinity of 5519.6 – 5817.7 mg/l, and a Na-sulfate-chloride composition.

J. Klerkx and B. Imanackunov (eds.), Lake Issyk-Kul: Its Natural Environment, 13–26.

3. the southern coast, closed, with large gulfs that are submerged river mouths. Bays are deep (>50 m), separated from the main water. Water in the gulfs has a salinity of 5466.1 – 5809.9 mg/l and a Na-sulfate-chloride composition.
4. the open-water central part of the lake, more than 95% of surface area, depths from 50 to 668 m. Salinity is 5842.1 mg/l at a depth of 50 m and 5889.5 mg/l at a depth of 650 m, measured along the profile Cholpon-Ata – Koltsovka; 5990.9 mg/l at a depth of 50 m and 5888.2 mg/l at a depth of 290 m, measured along the profile Grigorievka – Tamga. The water chemistry is of Na-sulfate-chloride composition.
5. freshwater gulfs of the eastern coast, far advanced landward (for 10 – 12 km), with strongly dissected shoreline and narrow backwaters. Depths are <35 m in the Tyup gulf and <45 m in the Djergalan gulf. Salinity is 2572.9 – 2777.8 mg/l as far as 3 km away from the Tyup river mouth; at 9 km away from the Tyup mouth it is 4323.3 mg/l at the surface, 5616.8 mg/l at a depth of 10 m, and 5773.5 mg/l at the bottom; in the Djergalan gulf it is within 3078.6 – 5913.7 mg/l.

1.1. THE LITTORAL ZONE

The water chemistry data are summarized in Tables 1 – 5. Water was sampled at river mouths and at 57 stations on the lake, including 19 stations inshore and 38 in the littoral zone of the lake (Fig. 1).

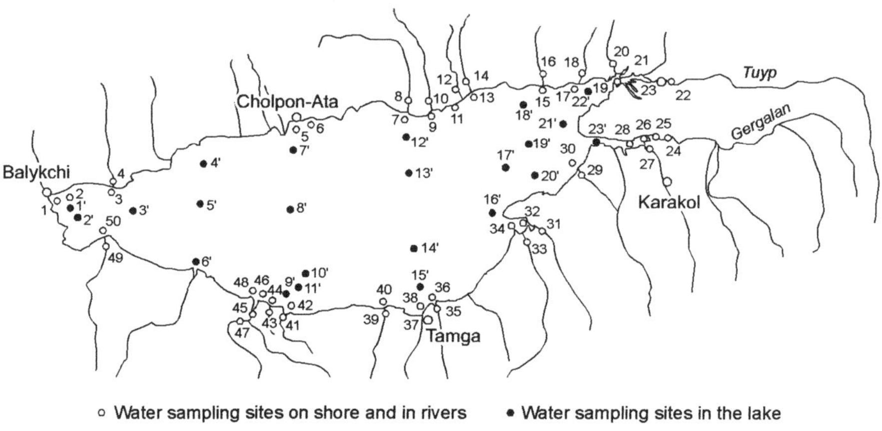

Figure 1. Location of water sampling sites

The water chemistry of the littoral zone is inhomogeneous. The salinity and major ion composition of the water is controlled by the river runoff, though its effect becomes evident at a certain distance offshore and at certain depths and decreases toward the open lake.

Gradual mixing of the river and lake waters produces zones of lower mineralisation distributed over the lake by vertical circulation and strong winds [1].

TABLE 1 . Water chemistry in Rybachiy gulf

Sampling site	*Depth m*	*mg/dm³ / mg-equ.*						*Salinity mg/l*
		Ca^{2+}	Mg^{2+}	$Na^{+}+K^{+}$	HCO_3^-	SO_4^{2-}	Cl^-	
0.1 km away from city beach	0.5	112.5	282.5	1340.4	249.1	1924.0	1524.0	5432.5
		5.61	23.23	58.28	4.08	40.06	42.98	
0.5 km away from city beach	0.5	117.2	284.4	1527.9	251.0	2247.0	1588.0	6015.5
		5.85	23.39	66.43	4.11	46.78	44.78	
1.0 km away from city beach	0.5	117.2	283.4	1492.5	248.0	2209.0	1560.0	5910.1
		5.85	23.31	64.89	4.07	45.99	43.99	
2.0 km away from city beach	0.5	115.6	280.6	1536.6	250.1	2247.0	1588.0	6017.9
		5.77	23.08	66.81	4.10	46.78	44.78	
3.0 km away from city beach	0.5	117.2	279.7	1499.6	248.2	2209.0	1560.0	5913.7
		5.85	23.00	65.20	4.07	45.99	43.99	
	7.0	117.2	283.4	1461.7	250.0	2133.0	1567.0	5812.8
		5.85	23.31	63.55	4.11	44.41	44.19	
0.1 km offshore, region of meat factory	0.5	117.2	282.5	1486.7	251.9	2190.0	1560.0	5888.3
		5.85	23.23	64.64	4.13	45.60	43.99	
0.5 km offshore, region of meat factory	0.5	104.8	284.4	1518.9	248.6	2228.0	1567.0	5951.7
		5.23	23.39	66.04	4.08	46.39	44.19	
	7.0	120.2	278.7	1461.4	247.7	2133.0	1560.0	5801.0
		6.00	22.92	63.54	4.06	44.41	43.99	
1.0 km offshore, region of meat factory	0.5	106.4	285.3	1510.0	246.3	2228.0	1560.0	5836.0
		5.31	23.46	65.65	4.04	46.39	43.99	
	10.0	109.4	288.1	1451.3	246.3	2095.0	1581.0	5771.1
		5.46	23.69	63.10	4.04	43.62	44.59	
11.0 km away from Rybachiy light beacon	0.5	117.2	284.4	1499.4	247.2	2228.0	1560.0	5936.2
		5.85	23.39	65.19	4.05	46.39	43.99	
	10.0	117.2	288.1	1497.5	248.7	2190.0	1595.0	5936.0
		5.85	23.69	65.11	4.07	45.60	44.98	
	20.0	117.2	293.7	1487.4	249.6	2190.0	1595.0	5932.9
		5.85	24.15	64.67	4.09	45.60	44.98	

33.6 km away from Rybachiy light beacon	0.5	117.2	294.6	1452.2	245.3	2190.0	1546.0	5845.3
		5.85	24.23	63.14	4.02	45.60	43.60	
	10.0	117.2	297.4	1462.8	245.3	2171.0	1581.0	5874.7
		5.85	24.46	63.60	4.02	45.20	44.59	
	25.0	117.2	296.5	1471.8	245.8	2190.0	1581.0	5902.3
		5.85	24.38	63.99	4.03	45.60	44.59	
	50.0	117.2	283.4	1496.2	245.8	2171.0	1595.0	5908.6
		5.85	23.31	65.05	4.03	45.20	44.98	
	100.0	117.2	291.8	1494.3	246.7	2190.0	1602.0	5942.0
		5.85	24.0	64.97	4.04	45.60	45.18	

TABLE 2 . Water chemistry in Cholpon-Ata gulf

Sampling site	*Depth*	$\frac{mg / dm^3}{mg - equ.}$						*Salinity mg/l*
		Ca^{2+}	Mg^{2+}	$Na^{+}+K^{+}$	HCO_3^-	SO_4^{2-}	Cl^-	
0.2 km offshore, region of "Goluboi Issyk-Kul" spa	0.5	115.6	290.0	1479.1	228.3	2190.0	1581.0	5884.0
		5.77	23.85	64.31	3.74	45.60	44.59	
0.5 km offshore, region of "Goluboi Issyk-Kul" spa	0.5	114.1	281.6	1490.4	218.8	2171.0	1588.0	5863.9
		5.69	23.16	64.80	3.59	45.28	44.78	
	47.0	115.6	280.6	1494.1	233.5	2171.0	1588.0	5882.8
		5.77	23.08	64.96	3.83	45.20	44.78	
1.0 km offshore, region of "Goluboi Issyk-Kul" spa	0.5	115.6	279.2	1496.4	238.7	2209.0	1560.0	5898.9
		5.77	22.96	65.06	3.91	45.89	43.99	
	9.0	115.6	281.6	1518.7	242.5	2228.0	1581.0	5967.4
		5.77	23.16	66.03	3.98	46.39	44.59	
2.0 km offshore, region of "Goluboi Issyk-Kul" spa	0.5	115.6	285.3	1498.9	245.3	2228.0	1560.0	5833.1
		5.77	23.46	65.17	4.02	46.39	43.99	
	16.0	115.6	285.3	1496.2	249.6	2190.0	1581.0	5917.7
		5.77	23.46	65.05	4.09	45.60	44.59	
0.5 km offshore, region of hippodrome	0.5	115.6	278.7	1485.3	246.7	2190.0	1547.0	5863.3
		5.77	22.92	64.58	4.04	45.60	43.63	
	9.0	117.2	277.8	1471.5	248.1	2152.0	1553.0	5819.6
		5.85	22.85	63.98	4.07	44.81	43.80	
1.0 km offshore,	0.5	120.2	287.2	1477.5	248.1	2209.0	1553.0	5895.0
		6.00	23.62	64.24	4.07	45.99	43.80	

region of hippodrome	67.0	123.3	285.3	1459.6	248.1	2171.0	1553.0	5840.3
		6.15	23.46	63.46	4.07	45.20	43.80	
2.0 km offshore, region of hippodrome	0.5	114.1	283.4	1471.3	246.3	2152.0	1574.0	5841.1
		5.69	23.31	63.97	4.04	44.81	44.39	
	22.0	114.1	290.9	1464.2	248.1	2152.0	1574.0	5843.3
		5.69	23.92	63.66	4.07	44.81	44.39	

TABLE 3 . Water chemistry in Tyup gulf

Sampling site	*Depth*	$\frac{mg/dm^3}{mg-equ.}$						*Salinity mg/l*
		Ca^{2+}	Mg^{2+}	$Na^{+}+K^{+}$	HCO_3^{-}	SO_4^{2-}	Cl^{-}	
0.1 km away from Tyup mouth	0.5	83.2	102.9	629.3	169.5	1028.0	560.0	2572.9
		4.15	8.46	27.36	2.78	21.40	15.79	
3.0 km away from Tyup mouth	0.5	90.9	124.4	662.9	180.9	1067.0	652.0	2777.8
		4.54	10.23	28.81	2.97	22.22	18.39	
	13.0	120.0	273.1	1285.7	240.6	1847.0	1488.0	5254.6
		6.00	22.46	55.90	3.94	38.46	41.96	
6.0 km away from Tyup mouth	0.5	97.1	144.5	829.4	185.6	1333.0	780.0	3369.5
		4.85	11.88	36.06	3.04	27.75	22.00	
	10.0	120.2	273.6	1380.9	240.1	1999.0	1524.0	5537.8
		6.00	22.50	60.04	3.94	41.62	42.98	
	22.0	123.3	279.7	1459.8	246.3	2190.0	1524.0	5823.1
		6.15	23.00	63.47	4.04	45.60	42.98	
9.0 km away from Tyup mouth	0.5	100.2	209.0	1065.1	215.0	1733.0	1007.0	4323.3
		5.00	16.69	46.31	3.52	36.08	28.40	
	10.0	114.1	273.1	1413.1	242.5	2057.0	1517.0	5616.8
		5.69	22.46	61.44	3.98	42.83	42.78	
	25.0	117.2	283.4	1446.7	247.2	2133.0	1546.0	5773.5
		5.85	23.31	62.90	4.05	44.41	43.60	
	30.0	118.7	284.4	1435.7	277.2	2133.0	1517.0	5766.0
		5.92	23.39	62.42	4.54	44.41	42.78	
15.5 km away from Tyup mouth	0.5	114.1	260.0	1286.4	233.0	1828.0	1461.0	5183.5
		5.69	21.46	55.93	3.82	38.06	41.20	
	10.0	115.6	275.9	1400.5	240.1	2076.0	1496.0	5604.0
		5.77	22.69	60.89	3.94	43.22	42.19	
	25.0	121.8	286.2	1408.1	245.3	2076.0	1546.0	5683.4
		6.08	23.54	61.22	4.02	43.22	43.60	
	40.0	123.3	290.0	1428.8	251.0	2133.0	1546.0	5772.1
		6.15	23.85	62.12	4.11	44.41	43.60	

TABLE 4 . Water chemistry in Djergalan gulf

Sampling site	*Depth m*	*mg / dm³* / *mg – equ.* Ca^{2+}	Mg^{2+}	$Na^{+}+K^{+}$	HCO_3^{-}	SO_4^{2-}	Cl^{-}	*Salinity mg/l*
0.3 km away from Djergalan mouth	0.5	84.8	182.4	710.7	189.4	933.3	97.8	3078.6
		4.23	15.0	30.90	3.10	19.43	27.6	
	17.0	115.6	281.6	1037.1	244.8	2038.0	978.0	4695.1
		5.77	23.16	45.09	4.01	42.43	27.58	
14.3 km away from Djergalan mouth	0.5	110.0	244.1	1353.8	230.2	1981.0	1397.0	5316.1
		5.49	20.07	58.86	3.77	41.25	39.40	
	10.0	115.6	266.6	1410.1	243.4	2038.0	1510.0	5583.7
		5.77	21.92	61.31	3.99	42.43	42.58	
	25.0	143.4	266.6	1435.0	249.1	2152.0	1510.0	5756.0
		7.16	21.92	62.39	4.08	44.81	42.58	
	50.0	117.2	276.9	1506.0	250.5	2190.0	1574.0	5913.7
		5.85	22.77	65.48	4.11	45.60	44.39	

TABLE 5 . Water chemistry in gulfs of southern lake side

Sampling site	*Depth m*	*mg / dm³* / *mg – equ.* Ca^{2+}	Mg^{2+}	$Na^{+}+K^{+}$	HCO_3^{-}	SO_4^{2-}	Cl^{-}	*Salinity mg/l*
Pokrovsky gulf, 3.8 km away from Chon-Kyzylsu mouth	0.5	114.1	263.8	1375.6	241.1	1981.0	1489.0	5464.6
		5.69	21.69	59.81	3.95	41.25	41.99	
	36.0	117.2	279.7	1472.5	248.2	2133.0	1574.0	5823.9
		5.85	23.0	64.02	4.07	44.41	44.39	
Ton gulf, 0.5 km away from Ton mouth	0.5	117.2	275.0	1355.4	241.5	1981.0	1496.0	5466.1
		5.85	22.62	58.93	3.96	41.25	42.19	
	10.0	117.2	287.2	1425.8	246.3	2057.0	1581.0	5714.5
		5.85	23.62	61.88	4.04	42.83	44.59	
	40.0	117.2	298.4	1409.2	247.2	2057.0	1588.0	5716.6
		5.85	24.54	61.27	4.05	42.83	44.78	
Gulf Tamga, 1.0 km away from Tamga mouth	0.5	112.5	291.8	1450.4	248.2	2133.0	1567.0	5802.9
		5.61	24.0	63.06	4.07	44.41	44.19	
	10.0	111.0	275.0	1469.9	249.1	2152.0	1531.0	5788.0
		5.54	22.62	63.91	4.08	44.81	43.18	
	25.0	109.4	279.7	1471.3	247.2	2171.0	1531.0	5809.6
		5.46	23.00	63.97	4.05	45.20	43.18	
	47.0	115.6	277.8	1486.7	248.2	2171.0	1560.0	5859.3
		5.77	22.85	64.64	4.07	45.20	43.99	

Water in the gulfs is fresher in the upper layer (<10 m in most gulfs); mineralisation and composition change bottomward, and in the lower layer it is close to the lake water. In the Rybachiy gulf, where the river runoff is low, water salinity and chemistry correspond to that in the open lake. Water in the Cholpon-Ata gulf is also similar to that in the open lake due to active exchange with the pelagial. The gulfs of the southern coast (Tamga, Ton, Akterek) are well protected against the wind and show a direct salinity stratification of water with lower mineralized uppermost layers. Water in the Tyup and Djergalan gulfs experiences the strongest influence of the river discharge, which is 9.6 m^3/s in the Tyup river (measured in Tyup village) and 17.1 m^3/s in the Djergalan (measured in Mikhailovka village). As a result, the water is diluted and salinity increases seaward, though the salt and ion composition is affected only in the uppermost layer, and in the lower layers, the composition is dominated by sulfates, chlorides, and Na, as in the open lake.
The effect of the river discharge consists primarily in modifying the concentrations of bicarbonate and calcium ions related to precipitation of $CaCO_3$. The carbonate balance in the littoral water shows a high alkalinity of 2.49 – 2.65 mg/equ.
The lake water is undersaturated in $MgCO_3$, as $Mg^{2+} \times CaCO_3^{2-} = 0.22 \cdot 10^{-4}$ is below the $MgCO_3$ solubility ($3.1 \cdot 10^{-4}$), which inhibits its precipitation.
Water in the gulfs is supersaturated in $CaCO_3$, 2.95 – 3.77 times in the upper layer and 3.08 – 4.19 times at a depth of 50 m. A major role in $CaCO_3$ supersaturation belongs to the riverine water, which is mostly of calcium-bicarbonate composition and is 1.13 to 4.11 times supersaturated in $CaCO_3$. $CaCO_3$ precipitation is important for the stability of water chemistry, as it evacuates the greatest portion of calcium and bicarbonate ions brought by rivers. Carbonate concretions deposit as crusts on rocks and constitute a specific feature of Issyk-Kul. The calcareous deposits occur at depths from 0.1-1.5 m to 15 m throughout the lake except for places with a muddy and sandy bottom.
The composition of the littoral water is predominantly Na-Mg sulfate; locally it is sulfate-chloride, and mineralisation varies from 2572.9 mg/l (Tyup gulf) to 6017.9 mg/l (Rybachiy gulf), but the proportions of major ions remain invariable.

1.2. THE PELAGIAL ZONE

The water chemistry in the open lake is homogeneous both laterally and vertically [2], which is confirmed by our measurements along four profiles (Choktal village – Akterek bay; Cholpon-Ata – Koltsovka bay; Grigorievka – Tamga bay; Orto-Yryukty – Pokrovka bay). Depth-dependent variations of water chemistry in the open lake at depths from 0.5 to 650 m are shown in Table 6 .

TABLE 6 . Water chemistry in open part of Lake Issyk-Kul (1998-2000)

Depth, m	*Ion composition* $\frac{mg/dm^3}{mg-equ.}$						*Salinity mg/l*
	Ca^{2+}	Mg^{2+}	$Na^{+}+K^{+}$	HCO_3^{-}	SO_4^{2-}	Cl^{-}	
0.5	115.1 / 5.74	279.9 / 23.02	1465.1 / 63.70	247.8 / 4.06	2138.4 / 44.50	1556.4 / 43.90	5801.7
10.0	115.5 / 5.76	286.4 / 23.55	1471.1 / 63.96	247.9 / 4.06	2145.0 / 44.60	1579.3 / 44.55	5845.2
25.0	116.5 / 5.81	290.3 / 23.87	1446.2 / 62.88	248.0 / 4.07	2118.1 / 44.10	1573.4 / 44.39	5828.5
50.0	116.0 / 5.79	283.3 / 23.30	1461.9 / 63.56	248.8 / 4.08	2121.3 / 44.17	1573.9 / 44.40	5805.2
100.0	116.1 / 5.79	287.7 / 23.66	1475.5 / 64.15	249.2 / 4.09	2148.9 / 44.73	1587.4 / 44.78	5864.8
200.0	116.1 / 5.79	284.4 / 23.39	1463.0 / 63.61	249.7 / 4.09	2127.7 / 44.30	1574.0 / 44.40	5814.9
300.0	115.1 / 5.59	284.7 / 23.41	1462.6 / 63.59	249.3 / 4.09	2121.2 / 44.16	1571.7 / 44.34	5804.6
400.0	116.1 / 5.79	288.6 / 23.73	1458.0 / 63.39	249.1 / 4.08	2129.5 / 44.34	1577.5 / 44.49	5818.8
500.0	116.4 / 5.81	291.4 / 23.96	1448.1 / 62.96	248.2 / 4.07	2121.3 / 44.17	1577.5 / 44.49	5802.7
650.0	115.6 / 5.77	285.3 / 23.46	1463.4 / 63.62	249.6 / 4.09	2121.0 / 44.16	1581.0 / 44.60	5815.9

Both vertical and lateral salinity variations are insignificant, which attests for active lateral exchange and vertical circulation of water.

The ion composition is dominated by Na and Mg cations and chloride and sulfate anions. Chlorides and sulfates are nearly equal in equivalent amounts but sulfates predominate in absolute concentrations.

The sulfate-to-chloride ratio (in mg/equ) approaches 1. The Cl index (salinity to Cl ion ratio) is 3.69 in the open lake, which strongly distinguishes Lake Issyk-Kul from other large water basins (cf. 1.807 for the oceanic water, 1.813 for the Black Sea, 1.844 for Azov Sa, 1.843 for Aral Sea, 2.39 for the Caspian Sea, and 5.102 for Lake Balkhash).

The total alkalinity of the Issyk-Kul water determined by the bicarbonate and carbonate ion content is 5.21 mg-equ, which is higher than in other water basins.

Average water chemistry is shown in Table 7.

By its salinity (5.9 mg/l), Lake Issyk-Kul is a brackish-water basin, and occupies an intermediate position between Aral (10.5 – 12.5 g/l) and Balkhash (3.0 g/l).

TABLE 7 : Average chemical composition of Issyk-Kul water (1998-2000)

	mg/l	*mg-equ*	*% equ*
Ca^{2+}	115.9	5.78	6.2
Mg^{2+}	285.9	23.51	25.3
$Na^{+}+K^{+}$	1461.7	63.55	68.5
HCO_3^-	248.8	4.08	4.4
SO_4^{2-}	2129.1	44.33	47.7
Cl^-	1575.2	44.43	47.9
Salinity 5815.6 mg/l			

1.2.1. *Dissolved oxygen*

The presence of dissolved oxygen in the water is important for water-living organisms. Its concentration depends on both physical (water-air exchange and turbulence) and chemical (photosynthesis and oxidation) factors. Solubility of oxygen in the Issyk-Kul water is controlled by temperature and atmospheric pressure and shows pronounced seasonal dynamics.

Water in Issyk-Kul is highly oxygenated, with O concentration attaining 6.98 – 9.74 mg/l and a saturation of 94.6 – 121.2%. The littoral water is the richest in oxygen to depths <50 m, and the uppermost water layer till 0.5 m below the surface is the least oxygenated.

Most of oxygen concentration (on average 9.15 mg/l or 112%) is due to photosynthesis, which occurs in the 0 to 10 m section in summer and in autumn (106 – 108%). The absence of photosynthesis in winter reduces oxygen saturation of the near-surface water to 98.8%. In spring the concentration of oxygen grows (up to 107 – 109%) due to water warming and development of phytoplankton.

The vertical distribution of oxygen is documented for site 28 located 20 km away from the northern shore, on the profile Cholpon-Ata – Koltsovka bay (measurements taken on June 28, 1998):

Depth, m	Oxygen concentration, mg/l	Depth, m	Oxygen concentration, mg/l
0.5	7.67	200	9.05
10	7.87	300	9.24
25	9.05	400	9.05
50	9.34	500	8.85
100	9.34	650	8.65

The high oxygen concentration at great depths evidences the good vertical circulation of water.

1.2.2 *Biogenic elements*

By the concentration of biogenic elements and bioproductivity, Lake Issyk-Kul is a low-productive oligotrophic basin. Biogenic elements participate in photosynthesis and are mineralised during the decay of organisms.

The concentration of biogenic elements in the near-surface water of the littoral zone is shown in Table 8.

TABLE 8 : Concentrations of biogenic elements in the water of littoral zone, mg/l (May 2000)

Sampling site	$N_{NO_2^-}$	$N_{NO_3^-}$	$N_{NH_4^+}$	P_{tot}
Rybachiy gulf, western extremity	0.025	0.26	0.10	0.000
Rybachiy gulf, docks	0.082	0.19	0.12	0.000
Rybachiy gulf, grain mill	0.020	0.22	0.11	0.000
Cholpon-Ata gulf, children spa	0.003	0.13	0.05	0.020
Cholpon-Ata gulf, "Goluboi Issyk-Kul spa"	0.002	0.10	0.00	0.000
Cholpon-Ata gulf, city beach	0.000	0.18	0.00	0.000
Cholpon-Ata gulf, region of purification structures	0.000	0.10	0.08	0.007
Camping, v. Ananyevo	0.007	0.27	0.08	0.024
Pier, v. Kurmenty	0.006	0.13	0.00	0.030
Tyup gulf	0.012	0.09	0.00	0.017
Djergalan gulf, pier	0.010	0.24	0.00	0.000
Pokrovsky gulf	0.010	0.06	0.00	0.000
Chon-Djargylchak mouth	0.009	0.00	0.00	0.007
Tamga gulf	0.003	0.13	0.00	0.000
Kajisai mouth	0.004	0.09	0.08	0.033
Ton gulf	0.006	0.10	0.05	0.046

The vertical distribution of biogenic elements, measured in July 1998 in site 28 of the profile Cholpon-Ata – Koltsovka bay, is given in Table 8a.

TABLE 8a

Depth, m	$N_{NH_4^+}$	$N_{NO_2^-}$	$N_{NO_3^-}$	P_{tot}
0.5	0.00	0.00	0.10	0.003
10	0.03	0.00	0.11	0.001
25	0.00	0.00	0.11	0.00
50	0.00	0.00	0.10	0.00
100	0.00	0.00	0.11	0.00
200	0.03	0.00	0.10	0.00

300	0.04	0.00	0.10	0.00
400	0.04	0.00	0.11	0.00
500	0.05	0.00	0.15	0.00
650	0.08	0.00	0.16	0.00

The main portion of biogenic elements is brought by rivers, and the low concentrations of biogenic elements in Issyk-Kul are due to the low proportion of riverine water per unit volume (0.2% in Issyk-Kul against 13.4% in Balkhash and 5.3% in Aral [3]

1.2.3. *Concentrations of contaminants*

The degree of contamination of the Issyk-Kul water has been established from data collected by the Hydrometeorological Department through 1975 – 1982 and by the Institute of Irrigation through 1985 – 1992, as well as from measurements in the course of the last years (1998 – 2000). Concentrations of contaminants (variation ranges and average values) are shown in Table 9.

TABLE 9 : Concentrations of chemical elements and physical properties of water in Issyk-Kul

	1975-1982		1985-1992	
	Range	Average	Range	Average
Water temperature, ^{0}C	2.8-22.2	8.6	2.6-23.0	8.7
	2.4-22.2	8.6	3.9-21.4	8.9
pH	6.50-9.98	-	8.05-8.78	-
	8.13-9.68	-	8.57-8.95	-
Transparency, m	0.0-23.0	7.5	0.3-19.0	5.6
	1.5-34.0	13.6	8.0-35.0	9.4
Suspended matter, mg/l	0.0-51.2	21.7	-	-
	0.0-17.7	9.8	-	-
Dissolved oxygen, mg/l	4.74-15.02	8.93	6.25-10.18	7.4
	6.24-10.50	8.50	7.08-9.78	8.7
Dissolved solids,‰	2.2-6.7	5.5	2.6-5.9	5.3
	4.4-7.3	6.0	5.7-6.0	5.9
BPK_5, mgO_2/l	0.0-10.54	0.85	0.29-21.4	1.51
	0.02-5.38	0.99	0.29-21.4	1.97
C_{org}, mg/l	0.5-7.2	2.0	-	-
	0.3-13.0	2.0	-	-
Petroleum products, mg/l	0.0-0.69	0.10	0.0-0.02	0.01
	0.0-0.37	0.09	0.0-0.03	0.01
Phenols, mcg/l	1-3	1	0-2	1
	1-7	1	0-2	1
SAS, detergents, mg/l	0.0-0.08	0.03	0.0-0.04	0.01
	0.01-0.04	0.02	0.0-0.02	0.01
Zn, mcg/l	0-40	10	0-77	12

	0-30	10	0-46	2
Cu, mcg/l	0-9	1	0-14	2
	0-26	4	0-4	1
Hg, mcg/l	0.05-12.0	0.18	-	-
	0.02-2.5	0.16	-	-

Note: first line is for littoral, second line is for pelagial zone.

Most parameters of water chemistry remain constant during the period of measurements (pH, transparency, dissolved oxygen), but contaminants show a complex behaviour.

Most contamination is due to petroleum products and heavy metals. The most strongly polluted water is in the littoral zone of the lake, where the average concentration of contaminants varied from 0.00 to 0.69 mg/l through 1975 – 1982, with yearly variations within 0.00 to 1.87 mg/l. The pelagial zone is less polluted, with the concentrations of contaminants varying from 0.05 to 0.2 (annual mean). Most strongly contaminated is the Rybachiy gulf and the Eastern bay near Cholpon-Ata town (in the northern lake side), the surroundings of the village of Kadji Sai and the Ton gulf, as well as Koltsovka bay in summer (on the southern lake side), and the Tyup and Dejrgalan gulfs (on the eastern lake side), in which various contaminants highly exceed maximum admissible concentrations (MAC): 1.4 – 37 MAC in Rybachiy gulf, 1.2 – 9.6 MAC in Tyup gulf, 1.4 – 15 MAC in the eastern bay of Cholpon-Ata, and 1.4 – 6.4 MAC in Kadji Sai (Table 10).

The content in petroleum products decreased in 1985 – 1992, as new purification structures were put into operation, specially protected recreation zones were established, transport of petroleum by water was prohibited, and petroleum storages were moved farther away from the shore. In the recent years (1998 – 2000), the concentration of petroleum products in the Issyk-Kul water has continued to decrease in relation to the economical decline accompanied by reduced industrial input of waste waters and lower recreational activity.

The concentrations in heavy metals (Cu, Zn, Pb, Hg, Cd) in the uppermost 0.5 m of water also show a decreasing tendency. At the same time, the concentrations of copper in the littoral water near Kadji Sai village and in the eastern bay near Cholpon-Ata amount to 0.009 and 0.008 mg/l, respectively, and are still higher in the pelagial (0.026 mg/l maximum, 0.004 mg/l average). Phenols that exist in the littoral water in concentrations from 0 to 0.014 mg/l (2 MAC in Rybachiy gulf and Cholpon-Ata) have not been found in the pelagial.

Therefore, the water chemistry in the littoral zone suffers a significant man-induced contamination evident in high concentrations of petroleum products, heavy metals, detergents, and phenols. If contamination remains within the littoral, where the water chemistry changes are restricted to a depth of 50 m, we may expect that the self-cleaning capacity of the lake will maintain, for some uncertain time, the chemical water balance. In the central part of the lake, the water chemistry remains almost invariable. However, the lake ecology may be strongly disturbed if the control over waste water discharge becomes less strict.

TABLE 10 : Concentrations of contaminants in the littoral zone of lake Issyk-Kul, mg/l (May 2000)

Sampling site	*Fe*	*Cu*	*Zn*	*Pb*	*Cd*	*SSAM*	*Petro-leum products*	*Phenols*
Rybachiy gulf, western extremity	0.00	0.002	0.008	0.001	0.000	0.005	0.002	0.002
Rybachiy gulf, docks	0.00	0.003	0.007	0.001	0.000	0.009	0.005	0.000
Rybachiy gulf, grain mill	0.00	0.001	0.001	0.001	0.000	0.013	0.001	0.000
Cholpon-Ata gulf, children spa	0.01	0.000	0.000	0.001	0.000	0.005	0.003	0.001
Cholpon-Ata gulf, "Goluboi Issyk-Kul spa"	0.00	0.002	0.002	0.004	0.000	0.009	0.002	0.002
Cholpon-Ata gulf, city beach	0.02	0.000	0.000	0.000	0.000	0.018	0.001	0.000
Cholpon-Ata gulf, region of purification structures	0.00	0.000	0.001	0.001	0.000	0.000	0.000	0.000
Camping, v. Ananyevo	0.00	0.006	0.001	0.003	0.000	0.000	0.000	0.000
Pier, v. Kurmenty	0.01	0.006	0.002	0.017	0.000	0.005	0.002	0.000
Tyup gulf	0.01	0.020	0.004	0.001	0.000	0.000	0.002	0.000
Djergalan gulf, pier	0.00	0.002	0.001	0.000	0.000	0.000	0.000	0.000
Pokrovsky gulf	0.03	0.002	0.005	0.001	0.000	0.000	0.000	0.001
Chon-Djargylchak mouth	0.00	0.000	0.000	0.000	0.000	0.000	0.001	0.000
Tamga gulf	0.01	0.001	0.001	0.002	0.000	0.000	0.001	0.000
Kajisai mouth	0.00	0.000	0.000	0.000	0.000	0.018	0.000	0.000
Ton gulf	0.02	0.000	0.000	0.000	0.000	0.009	0.003	0.000

1.2.4. *Self-cleaning capacity of the lake*

Self-cleaning in Lake Issyk-Kul that occurs under the effect of natural physical and chemical factors was investigated by experimental modeling of oxidation of petroleum products.

During the experiment, diesel fuel (MP 14298-89) was added to water samples from the pelagial zone and the Cholpon-Ata gulf. Water was poured into 20 l water bottles, 10 cm^3 of fuel was put into each bottle and mixed for an hour. In order to monitor fuel decomposition, concentrations of O_2, BPK_5, NO_2, NO_3, NH_4, PO_4, and petroleum products, as well as temperature and pH, were measured for the following 15 days.

It was observed that oil saturation occurred during the first four days up to the maximum of 1.0 mg/l. During the following 12 days the concentration of petroleum products decreased to 0.5 mg/l. Therefore, the half-life of petroleum products in the lake water of salinity 5.85 g/l and a temperature of 14-16° C is 288 hours. The decomposition occurs by oxidation and its rate is thus a function of oxygen concentration. In the experiment we observed a decrease in oxygen concentration from 7.20 to 7.01 mg/l. Thus, oxidation of 0.5 mg of diesel fuel in Issyk-Kul takes 0.19 mg/l and oxidation of 1 mg requires 0.38 mg of dissolved oxygen.

The experiment showed a linear relationship between logarithmic concentrations of petroleum products and their life time, which indicates that oil oxidation is a first-order reaction, i.e., the rate of decomposition of diesel fuel is expressed as

$$K_t = \frac{2.3}{t} \lg \frac{c_0}{c_t},$$

where t – time of reaction;
c_0 – initial concentration, mg/l;
c_t – final concentration, mg/l.

Proceeding from this formula, the decomposition rate is $2.4 \cdot 10^{-3}$ hr^{-1}.

The man-induced effect on the Issyk-Kul water is to reduce the concentration of dissolved oxygen which is spent on oxidation of petroleum products and other organic and mineral contaminants. This oxygen loss in some areas of the lake is compensated by photosynthetic oxygen production. Thus, the self-cleaning mechanism of the lake is based on oxidation of contaminants and oxygen regeneration by photosynthesis.

2. References

1. Stavissky, Ya.S. (1977) Some regularities of temperature exchange in the water of Lake Issyk Kul [in Russian], *Bull. Central Asian Inst. Hydrometeorol.* **50**, 75-80.
2. Kadyrov, V.K. (1986) *Water Chemistry in Lake Issyk Kul and its Catchment Basin* [in Russian], Ilim, Frunze.
3. Tarasov, M.N. (1961) Studies of carbonate formation in closed water basins of the arid zone [in Russian], *Hydrochem. Materials* **31**, 78-87.
4. Alekin, O.A. (1946) Physics and chemistry of water in Lake Issyk Kul [in Russian], in Molchanov, I.V. (ed), *Lake Issyk Kul*, Gidrometeoizdat, Leningrad, pp. 80-102.
5. Gribovskaya, I.F., Vorotnitskaya, I.E., and Letunova, S.V. (1974) Migration of chemicals in Lake Issyk Kul [in Russian], *Proc. Vernadsky Inst. Geochem. and Analyt. Chem.* **13**, 224-234.
6. Kadyrov, V.K. and Karmanchuk, A.S. (1964) Concentrations of some trace elements in the water of Lake Issyk Kul [in Russian], in Slonim, O.A. (ed), *Trace Elements in the Water of Lake Issyk Kul*, Ilim, Frunze, pp. 101-106.
7. Karmanchuk, A.S. and Kadyrov, V.K. (1964) Self-cleaning capacity of water in Lake Issyk Kul [in Russian], *Proc. IV Congr. Geogr. Soc. Kirg.*, Ilim, Frunze, pp. 98-103.
8. Karmanchuk, A.S. (1990) *Cultural Effects and Protection of Water Resources in the Issyk Kul Basin* [in Russian], Ilim, Bishkek.

HYDROBIOLOGY OF LAKE ISSYK-KUL

V.V. ROMANOVSKY
Institute of Water Problems and Hydropower
533, Frunze str., Bishkek, Kyrgyzstan

1. Macrophytes

The most complete study of the specific composition of macrophyte algae in Issyk-Kul belongs to T. Matyukova [1] who described vegetation assemblages in the Tyup gulf, in which the indented coastline and strongly uneven hydrochemical conditions have favoured proliferous development of various aquatic vegetation species.

Aquatic plants in Issyk-Kul are either above-water or immersed. The former group includes, primarily, Phragmites communis Trin, Typha latifolia L., Scripus Tabernaemonatni Cmel, Bolboschoenus affinis Drob, and Bolboschoenus compactus Drob. Immersed plants include higher and lower spore algae: charophytes and periphyton. Higher phanerogams of Issyk-Kul include Potamogeton pectinatus L., P.otusifolius M.K., P.pussilis L., P.perfoliatus L., Najas marina L., Myriophyllum spicatum, Ceratophyllum demestrum, Urticularja vulgaris L., Ranunculus natans C.A.M., Zannichellia palustris L., and Ruppia maritima L.

The specific composition of charophytes in Issyk-Kul was studied by Matyukova [1], Mambetalieva [2], and Gollerbakh (references in [3]). They identified 17 species of charophytes: Nitellopsis obtusa (Desv.) Groves, Tolypella nidifica (Mill) Leonh, Chara altaica A.Br., Ch. aspera (Lethard) Willd, Ch. aspera (Dethard) Willd var brevispina Mig, Ch. tomentosa L., Ch. connivens Salzm., Ch. contraria A.Br., Ch. contraria A. Br. hispidula A. Br., Ch. crinita Wallr, Ch. delicotula Ag. var. bulblliflora A. Br., Ch. Fischeri Mig, Ch. intermedia A. Br., Ch. Schaffneri (A. Br.) Robinson, Ch. crinitoides Hollerb. sp. nov., Ch. vulgaris L., and Ch. fragifera Durun.

Chara tomentosa, the most abundant and the largest (up to 2 m high) alga in the lake [3], grows both in desalinated shallow waters of gulfs and bays and in the open lake. As a rule, this charophyte forms thick close thickets in the open lake at depths from 6-8 to 25-30 m, which cover 90-100% of the bottom and reach 35 kg/m^2 in phytomass. At greater depths (to 50 m) Ch.tomentosa is less abundant and grows in patches.

Charophytes inhabiting the littoral zone of the lake (35-40 m deep) constitute a major component of aquatic vegetation in Issyk-Kul. They occupy the largest areas, have the greatest phytomass, and enter almost all assemblages of aquatic plants in which they most often predominate. Thus, according to the dominant species of aquatic plants, Issyk-Kul can be considered charophyte lake. The thickets of charophytes extend along

J. Klerkx and B. Imanackunov (eds.), Lake Issyk-Kul: Its Natural Environment, 27–43.

the shoreline in strips (charophyte belts), which are wide or narrow in function of disturbance and transparency of water and bottom topography [1]. So, in the shallow-sloping bottom of the Rybachiy gulf, the charophyte belt is spread as far as 15 km offshore but narrows to 200 m over the steeply dipping bottom at Mount Prishib and near Tossor village. Vegetation is absent to the depths of 5-10 m on the sandy-bars of the Dolinka, Bozteri, Sukhoy Ridge, and Kara-Bulun peninsulas exposed to active stormy turbulence. Because of steep bottom slopes, the charophyte belt is much narrower at the southern than at the northern shore. In the turbid and steeply downsloping Tyup gulf, charophytes grow at depths of 6-8 m and their belt is as narrow as 20-30 m. In the upper weirs of the gulf, where the inflow of biogenic elements is high and the water transparency is as low as 0.5 m, charophytes give way to higher hydromacrophytes, and these are undergrown by Vaucheria from the depth of 2.5 m.

Periphyton (fouling algae) occurs in the littoral (to 5 m) as continuous fouling on stones and stalks of macrophytes or as concentrations on the bottom and on the water surface. Light-green filamentous alga Spirogira sp. occurs as floating mucous foam covering the littoral macrophytes with a kind of thick felt. Cladophora exists in two species (Cladophora fracta and C. Vulgaris) and Enteromorpha in three species (Enteromorpha compessa, E. Salina and E. Prolifera).

The distribution of annual productivity of macrophytes at different depths is constrained by the upper disturbance and lower transparency limits (Table 1) [3].

TABLE 1. Macrophyte productivity at different depths

Depth, m	Productivity, g/m^2
0-5	1020
5-10	1580
10-15	2360
15-20	2660
20-25	1310
25-30	580
30-35	300
35-40	30

Annual production of macrophytes in Issyk-Kul is 1,730,000 tons of green weight [3], in which the contribution of charophytes is 96% and 2.5% belong to Potamogeton. The highest productivity of macrophytes (950,000 tons or 54.9%) is at the depth range of 10 to 20 m (which can thus be considered the ecological optimum), and the average annual production is 277 g/m^2.

The specific composition of immersed macrophytes in Issyk-Kul is controlled by the balance of carbon dioxide, which is rarely free but mostly bound in acid carbonates. As a result, the predominant species are those that have their vegetative shoots in contact with water. This is achieved either due to extremely narrow leaves (Myriophylllum spicatum, Potamogeton pectinatus L. and Najas marina L.) or thick branching with numerous side spiny shoots, as is typical of charophytes. This morphology of macrophytes allows them to adapt to the specific conditions of the biotope by assimilation of both free and bound (in bicarbonates) carbon dioxide.

Thus, the specific composition of Issyk-Kul macrophytes is controlled by the ability of algae to assimilate bound CO_2, while the predominance of charophytes indicate high calcium enrichment of the water.

2. Phytoplankton

The first detailed study of Issyk-Kul phytoplankton belongs to I. Kiselev in 1932 [4], who studied samples of plankton in fishermen's nets and showed that the floristic composition of algae in Issyk-Kul was diverse and more specific than elsewhere. He identified 309 species of algae, generally diatoms inhabiting the lake bottom, most of which occurred in the littoral zone. The phytoplankton from the open lake showed a much less diversity limited to three leading species: Botryococcus braumi, Peredinium borgei and Amphiprora paludosa var. Issykkulentis, the latter two being more frequent. Botryococcus is adapted to surface layers and Amphiprora inhabits greater depths.

Matyukova [1] studied bottom algae of the Tyup gulf in the late 1950s and identified 240 species and forms of diatoms, among which only 13 species belonging to plankton.

Mambetalieva [2] investigated the floristic composition of bottom algae of the northern shore in the early 1960s and made a list of 346 species and forms of algae.

Kulumbaeva [4] studied phytoplankton in the open lake and identified 111 species, varieties and forms, including 44 species of blue-green, 2 species of yellow-green, 19 species of diatom, 6 species of pyrophyte, and 400 species of green algae. The pelagic phytoplankton is dominated by planktonic algae species (84%), as is typical of deep lakes.

Diatoms are the most diverse and include bottom-living and littoral fouling species. The abundance of phytoplankton in the open lake is 166-488 thousand cells per liter, and the biomass is 73-153 mg/l [4], attaining 1513-4126 thousand cells/l and 207-398 mg/l, respectively, in the Tyup gulf. Blue-green and green algae constitute 44% of species of which blue-green Merismopedia punctata is the most abundant (12356 thousand cell/l).

Peaks of abundance and biomass of phytoplankton in Issyk-Kul are observed at depths of 15-50 m, and, what is specific of Issyk-Kul, significant algae concentrations occur at great depths as well (100-150 m). Issyk-Kul shares this feature of vertical distribution of phytoplankton with Lake Baikal, probably due to high water transparency in both lakes.

Unlike many other large lakes, in which the specific composition of phytoplankton is annually variable, the main algal assemblage in Issyk-Kul remains more or less the same. The greatest portion of phytoplankton production belongs to small blue-green and green algae (nannoplankton) with cells of 2-3 microns, which is about the sizes of bacteria plankton. Note that though abundant, phytoplankton in Issyk-Kul has a low biomass. The development of small algae in the lake might be adaptive to low concentrations of biogenic nutrients. The concentrations of phytoplankton are the lowest in January-February and in September, and two peaks of growth are in May and

October-December [4]. The winter (February) and autumn (September) maxima of water transparency in Issyk-Kul coincide with the minimum growth of phytoplankton. At the same time, the spring burst of phytoplankton has no relation to water transparency.

Davydova [5] analyzed diatoms from a sedimentary core sampled at the depth of 330 m and revealed paradoxic accumulation of diatoms in fouling and on the bottom of the littoral. Of 120 diatoms, only 9 were plankton species. This may be related to past water level fluctuations in Issyk-Kul. However, any inference from just one core would be preliminary.

3. Zooplankton

The list of zooplankton species in Issyk-Kul includes 154 species and varieties, among which there are 76 protozoa, 59 rotifers, 11 cladocera, and 8 copepods. Most species inhabit shallow littoral waters and 3 species of rotifers live in the open lake [6].

The most abundant are 15 zooplankton species, including 9 species of rotifers, 2 cladocera and 4 copepods. The open lake community consists of only 5 species: 3 rotifers and 2 copepods (A.viridis and Arctodiaptomus salinus, a predominant species in all regions of the lake). Diaptomus makes up 75-95% of the number and 95-99% of the biomass of the lake zooplankton. The abundance and biomass of diaptomus in the upper water layer (< 100 m) are the highest in Late August-Early September (6-8 thousand specimens per cubic meter of water and 0.17-0.22 g/m^3, respectively) and the lowest in Late February-Early March (to 2-2.5 thousand specimens/m^3 and 0.12-0.15 g/m^3, respectively).

The analysis of stomachs of crustaceous diaptomus, the dominant representative of zooplankton, showed that it feeds mostly with mass species of phytoplankton.

4. Benthos

The Issyk-Kul benthos includes 224 species and forms: 3 species of molluscs, 35 species of chironomides, 2 species of two-winged larvae (excluding chironomides), 2 species of aquatic beetles, 5 species of aquatic bugs, 7 species of caddisworms, 1 species of mayfly larvae, 4 species of dragon-fly larvae, 5 species of ticks, 5 species of amphipods, 12 species of ostracods, 5 species of leeches, 5 species of oligochaets, and 55 species of protozoa, plus turbillaria and nematodes worms. Not all species are well studied. Earlier studies revealed 12 species of ostracods in Issyk-Kul but later [7], 28 species were found in a sedimentary core from the central part of the lake and 40 species in cores from the Rybachiy gulf.

The mean biomass of zoobenthos in bays and gulfs, the most productive localities of the lake, is 104.2 kg/ha [8] (cf. 220 kg/ha in the littoral of Lake Baikal). The benthos biomass is the highest in spring and decreases by autumn being eaten out by fishes and

hatched by chironomides through summer: in the Pokrovsky gulf it is 22.14 g/m^2 in spring, 8.8 g/m^2 in summer, 5.5 g/m^2 in autumn, and 11.9 g/m^2 in winter.
The littoral zone is abundant in Chironomus genus with the most frequent species of Strictochironomus pictulus and Microspectra sp.; orthocladines in charophyte thickets are of broad occurrence, and Radix auricularia is the predominant mollusc species. Gammarus bergi M. and Gammarus ocellatus M. are the most abundant gammarids. The former is typical of stony biotopes and the latter lives among the thickets of charophyte algae. The mean biomass of zoobenthos in the littoral charophyte belt is 8-10 g/m^2. Outside this belt, to depths about 60-70 m, it is 2 or 3 times as low (2.5-3.5 g/m^2), with the predominant species of Chironomis anthracinus, Ch. plumosus, Strictochironomus pictulus, Micropsectra, and mollusc Radix auricularia. In the pelagial (deeper than 70 m), zoobenthos is minor, both in abundance and biomass (0.2-0.3 g/m^2), and is mostly represented by just two endemic oligochaets (Enchytraeus przewalskii Hrabe and E.issykulensis Hrabe) and one species of gammarids (Issykkogammarus hamatus Chevreux).
The main portion of zoobenthos biomass (up to 70%) in the bays and gulfs of the lake, is made by molluscs and their egg laying, except for the Tyup gulf in which it is dominated by chironomide larvae [8]. In 1965-1968, three species of Paramisis were introduced into the lake and got acclimatized well in macrophyte thickets at depths 1.5-8 m, where the mean annual biomass is 1.5-2.5 g/m^2.
The main foodstuff for fish in Issyk-Kul is provided by molluscs, especially Radix auricularia, the most abundant and frequent species in the littoral zone of the lake (< 60 m). In the Tyup gulf, it is chironomide larvae, which comprise about 60% of benthos biomass and are among highest calorie groups of invertebrates. The food base of Issyk-Kul fish had been considered poor before quantitative estimates obtained in the 1960s showed that the benthos upon which Issyk-Kul fish fed was sufficient to feed a population twice as great as the existing one [9]. This was the reason for acclimatization of carp and pike-perch in Issyk-Kul (see below).
An analysis of ostracods in three sedimentary cores from the western, eastern, and central parts of the lake [7] showed that the number of ostracods, especially halophilic species, in the Rybachiy gulf has increased through the two last centuries. In the Tyup gulf, on the contrary, their number reduced while a halophilic species of Cyprideis torosa has appeared.
Studies of ostracods in the bottom sediments of the Tyup gulf indicate greater salinity of water and unfavourable conditions for ostracods related to the water level drop since the beginning of the 20th century. The unfavourable conditions may be also associated with eutrophication and related productivity increase of autotrophic organisms and worse oxygenation of the bottom water, which is confirmed by the predominance of larvae of two-winged chironomides in the present-day benthos of the gulf.

5. Pisces (Fish)

Fishes in Issyk-Kul currently include 21 species of 5 families:

- Family Salmonidae (salmons) is represented by two genera (Salmo and Coregonus) and 5 species.
  Genus of salmons: Salmo ischchan issykogegarkuni Lushin, Salmo gairdneri irideus Gibb; whitefish: Coregonus lavaretus L., C. Autumalis migratorius, Coregonus peled Gmelin.
- Family Cyprinidae (carps): 11 genera and 14 species.
  Genus Leuciscus: L.schmidti Herzenstein and L.bergi Kaschkarov.
  Genus Phoxinus: Ph.issykkulensis Berg.
  Genus Alburnoides: A.taeniatus Kessler.
  Genus Tinca: T.tinca L.
  Genus Gobio: G.gobia latus Anikin.
  Genus Pseudorasbora: P.parva (Schleg.).
  Genus Abramis: A.brama orientalis.
  Genus Schizothorax: Sch.issykkulensis Berg.
  Genus Diptychus: D.dibovskii lansdelli Gunter and D.dibovskii turdakovi Tjurin.
  Genus Carasius: C.auratus Gibelio.
  Genus Ciprinus: Ciprinus carpio L.
- Family Cobitidae includes one genus Noemacheilus (Arctic salmons) and 6 species:
  Noemacheilus dorsalis Kessler.
  Noemacheilus strauchi (Kessl.) Herz.
  Noemacheilus ulacholicus Anikin.
  Noemacheilus strauchi dorsaloides Turd.
  Noemacheilus strauchi ulacholicus var. Pedaschenko.
  Noemacheilus labiatus Kessler.
- Family Percidae (perch) is represented by one genus Stizostedion (pike-perch) and one species Stizostedion lucioperca (Linne).
- Family Eleotridae is represented by one genus Hypseleotris Gill and one species Hypseleotris cintis (Dabri de Thiersant).

Among 21 species of fish inhabiting the lake, 14 species are native and the others were introduced from elsewhere. Three introducers have failed to acclimatize.

In the early 1930s, the ichthyofauna of Issyk-Kul and its tributaries was represented by the following species and forms [Lindberg, 1935]:

Leuciscus schmidti Herzenstein
Leuciscus bergi Kashkarov
Diptychus dybowskii Kessler;
Diptychus maculatus Steindachner
Schizothorax pseudaksaensis issykkuli Berg
Cyprinus carpio L
Phoxinus issykkulensis Berg
Gobio gobio latus Anikin
Noemacheilus stoliczkai (Stein)

Noemachtilus ulacholicus (Anirin) and its subspecies N.strauchi ulacholicus var. pedaschenkoi Berg

Noemachtilus dorsalis (Kessler).

There were no predatory fishes and phytophagans, except for large individuals of Diptychus and Schizothorax which were optional predators. Leuciscus bergi Kashkarov made up over 90% of the total fish biomass.

There are 8 endemic species of fish in Issyk-Kul: 2 species of dash, Phoxinus issykkulensis Berg, Gobio gobio latus Anikin (gudgeon), Schizothorax, Diptychus dybowskii Kessler, Noemacheilus ulacholicus Anikin, Noemacheilus strauchi dorsaloides Turd. Endemic fish comprise 57% of native, and only 27% of the contemporary content of ichthyofauna. The fish endemism in Issyk-Kul is at the level of species. All fich endemics are of relic forms.

According to their type of feeding, Issyk-Kul fishes are of two main groups: (1) fishes feeding upon invertebrates and plants: Leuciscus schmidti Herzenstein, Leuciscus bergi Kashkarov, Phoxinus issykkulensis Berg, Gobio gobio latus Anikin, Schizothorax pseudaksaensis issykkuli Berg, Diptychus, wild carp, carp, Noemacheilus, bream, Tinca tinca L., Prussian carp, and Pseudorasbora parva (Schleg.) and (2) predatory fishes: salmon, trout, and pike-perch.

All native fishes eat aquatic plants anyway but none feeds uniquely upon vegetation. Diptychus dybowskii Kessler feeds upon charophyte algae. Schizothorax, Leuciscus, Cyprinus eat both living organisms and algae: stomachs of some fish contain 75-80% lower algae and fragments of higher plants. The ration of an adult Leuciscus includes 40-50% filamentous algae, young shoots of charophyte and cladocera.

Many species of fish have seasonal migrations. In late autumn-early winter, adult trouts migrate from deep waters to the spawning rivers, while the juveniles born in the previous year roll into the lake. The species that migrate within the littoral include Leuciscus schmidti Herzenstein and Leuciscus bergi Kaschkarov: from deep waters to shallow-water spawning and fattening sites in spring, and back to wintering sites at depths about 100 m in autumn. Moreover, Leuciscus bergi Kaschkarov makes summer diurnal migrations stipulated by migrations of zooplankton, its main foodstuff: It moves into the open lake at night and is concentrated near the shores at the day time. Diptychus dybowskii Kessler has spawning migration twice a year: winter Diptychus spawns in the lake in winter, while summer Diptychus spawns in the rivers in summer.

Leo S. Berg, well-known natural scientist visited Issyk-Kul in 1928 and found the living conditions there favorable for trout and whitefish [10], i.e. for salmon. Following his recommendations, the valuable species of Salmo ischchan infraspecies gegarcuni Kessler biotipus Fortunatov from Sevan was introduced in 1932 and 1936; Stizostedion lucioperca (L) and bream were introduced in 1954-1957 and have acclimatized well in the eastern part of the lake in the Tyup and Djergalan gulfs. In the same period carp and Varicorhinus capoeta were acclimatizing, and tench, crucian, Albunoides bipunctatus, Pseudorasbora parva (Schleg), and Hypseleotris cintis (Dabri de Thiersant) penetrated occasionally. The acclimatization of Varicorhinus capoeta was a failure. Some scientists proposed to replace Pseudorasbora by Coregonus albula, of a higher food quality.

In the late 1960s, Coregonus peled Gmelin, Baikal omul Coregonus autumnalis migratorius Georgi, Sevan whitefish Coregonus lavaretus L., and omul were introduced with the aim to create a flock of whitefish [11], and the lower-value pike-perch was introduced as its food. Trout Salmo gairdneriirideus Gubb., penetrated into the lake from fish breeding cages near Baetov village and has settled there since the latest 1980s. Sevan trout became an active predator under new conditions. Thus, the fish assemblage of Issyk-Kul was added with 3 new predatory species.

Trout was introduced into Issyk-Kul from Lake Sevan. The character of feeding of adult individuals of Issyk-Kul trout is quite different from that in Sevan: there it feeds mainly on benthic organisms (96.6% gammarids) and in Issyk-Kul mostly on fish (81.8%).

The ration of trout includes about 50% Leuciscus bergi Kaschkarov and juveniles of Leuciscus schmidti Herzenstein and about 20% benthic organisms (gammarids, larvae of air insects, misids) [12]. Trout has a narrow throat and can swallow rather small fish only. Therefore, the greatest portion of two-year-old Leuciscus schmidti Herzenstein fall out of its ration, while Leuciscus bergi Kaschkarov, which does not exceed 13 cm, remain chased all their life cycle through. Like all trouts, Sevan and Issyk-Kul trout stop feeding during the spawning period.

As it got into Issyk-Kul, the Sevan trout has changed the colour of adult individuals, as well as the size and length ratio of some parts of the body. Of 21 parameters, 9 changed at males and 10 at females. The mean weight of Issyk-Kul salmon is 3,550 g. A catch of a 17 kg and 95 cm long specimen was reported in the literature; poachers claim that fishes weighing over 20 kg are quite frequent; the largest specimen of 28 kg was caught in Djergalan gulf.

The native Sevan species rarely are 4 kg and 60 cm long, being mostly about 400-500 g. The rate of growth of Issyk-Kul male trout is 1.5 times and female trout is twice as high as that of their Sevan counterparts. During the short acclimatization period, the Sevan trout became a new subspecies and was first described in 1951 as Salmo ischchan issykogegarcuni Lushin [13].

To create and maintain the economic flock, trout was reproduced artificially, with an annual output of 14-15 mln. incubated eggs and about 3 mln. juveniles let out into the lake rivers. The economic catch of Issyk-Kul trout was set up in 1971 and attained 17.6 tons. Further on trout catches increased from year to year and reached a maximum of 53.8 tons in 1979.

For the reconstruction of the Issyk-Kul ichthyofauna, various whitefish species were introduced into the lake in the 1960s: Coregonus peled, Baikal omul, Sevan whitefish, etc. The rate of linear and weight growth, fat content, terms of sexual maturity, and absolute reproduction of the introduced whitefish increased in Issyk-Kul relative to the initial parameters. E.g., the maximum weight of six-year-old whitefish in Lake Ladoga is 950 g, while its eight-year old counterparts in Issyk-Kul weights 2,364 g [14].

Experiments showed that the saline water of Issyk-Kul was destructive for whitefish eggs, and since the 1970s whitefish had been bred at the Karakolsky Fish-Breeding Farm. Later on the breeding ceased, but catches have increased significantly in the recent years. Therefore, whitefish did find suitable conditions for natural reproduction, which was proved only in 1988.

The first information on catches of mature individuals of Baikal omul appeared in the early 1970s. The number of omul in the lake increased gradually, and its catches made up 20-25% of the whitefish catches in the second half of the 1970s. However, in 1983 the catches dropped to 3%, and only few individuals are caught at present.
Considerable concentrations of omul were observed in the late 1970s (in September-October) in the littoral zone of the Tyup and Djergalan gulfs [15]. The fish had mature sexual products and high individual reproduction (from 40 to 99 thousand eggs). According to fishermen's witness, they caught omul juveniles at the western shore of the Rybachiy gulf, where the lake water is desalinated by the influx of ground waters. It was expected that the Baikal omul would find conditions for natural reproduction in Issyk-Kul, but the sharp drop in catches in the 1980s showed that the expectations failed.
In 1954 it was recommended [9] to introduce pike-perch into Issyk-Kul in anticipation that it would act as a "biological meliorator" by reducing the population of non-economic fishes, such as Noemacheilus, Phoxinus and others. In the 1950s there were about 13,000 tons of Leuciscus and other small fishes [9], which was sufficient to feed 600-1000 tons of predator fishes annually.
Pike-perch juveniles feed upon zooplankton from birthtime to autumn and then, at the age of 2 months, start feeding on chironomide larvae and small (2-3 cm) fish. Adult pike-perch is an active predator whose ration consists of Leuciscus schmidti Herzenstein, Leuciscus bergi Kaschkarov, gudgeon, Schizothorax, Diptychus, Noemacheilus, and pike-perch juveniles. Thus the menu which was initially "prescribed" to pike-perch when introducing it into Issyk-Kul has actually become much more diverse.
Endemic Leuciscus schmidti Herzenstein exists in 2 forms: fast growing "limnic" and slowly growing "weir" fishes (fishermen call it "yellow belly"). Limnic Leuciscus lives for 11 years the longest and reaches 31 cm in length and 650 g in weight; in the spawning period its weight varies from 157 to 226 g and the length is 21-23 cm. The limnic Leuciscus is among the largest daces in Eurasia [14] being second large after chub and ide. The slowly growing "weir" Leuciscus reaches a length of 23 cm, a mass of 220 g, and a maximum age of 13 years. Spawning females are mostly 14-17 cm long.
It has been shown experimentally [16] that Leuciscus eggs grow faster in the saline water of Issyk-Kul: in fresh water eggs develop for 252.7 degrees-days; in mixed limnic-fresh water this duration is 237.7, and in limnic water with normal salinity it is 196.3 degrees-days. Moreover, the Leuciscus embryo hatched in limnic water are larger. In fresh water, the egg development is the longest and the hatched embryos are the smallest. Thus, although Leuciscus is a typical freshwater species, it has well adapted to the conditions of Issyk-Kul saline water.
Till the beginning of the 1970s, Leuciscus bergi Kaschkarov was the most abundant fish of Issyk-Kul. It lived in all regions of the lake to depths below 100 m, being especially abundant at the eastern shore. Within the natural habitat, there were several races of L.bergi Kaschkarov: semi-transit race, which went spawning into rivers, slowly growing weir race, and the most abundant limnic race. However, the number and the natural habitat of L.bergi Kaschkarov were considerably reduced through the past 20-25 years:

its eastern population was supplanted by pike-perch, and the weir and semi-transit races have disappeared. In the regions of lower abundance, a new fast growing form has appeared [17].

The Issyk-Kul L.bergi Kaschkarov is usually 16.5 cm long, weighs 60 g and lives for 7 years the longest. It reaches its sexual maturity at the age of 5-6 years when it is 11.7-13 cm long. The fast growing L.bergi Kaschkarov matures at 3-4 years old and is then 12.5-13 cm.

In the 1930-60s, the greatest portion of spawning L.bergi Kaschkarov were four-year-old individuals with a body length within 11-12 cm; since the 1970s, the portion of 5-6-year-old 13-15 cm long spawning specimens has been increasing, and the maximal life longevity increased from 5 to 7 years.

L.bergi Kaschkarov spawns on the same substrates as Leuciscus schmidti Herzenstein, at depths 1-10 m. Till the beginning of the 1970s, its spawning period was from May to October, but now it peaks in May-June at water temperatures within 12-17°C and ends in July. Juveniles of L.bergi Kaschkarov feed mostly on small organisms (rotifers, Cyclopoida, and unicell algae), and the ration of adult individuals consists of Cyclopoida, larvae and imago of insects, filamentous algae, and diatoms. The intestines of L.bergi Kaschkarov contain abundant Diaptomus salinus, planktonic algae, and imago of chironomides floating on the lake surface, which is evidence of mostly planktonic feeding. In winter, when the biomass of plankton is low, it mainly feeds upon detritus, cladocera, and filamentous algae. Such a broad range of feeding is a good adaptive feature of L.bergi Kaschkarov to the living conditions in such a low productive oligotrophic lake as Issyk-Kul.

Wild carp (some specimens reach a weight of 12 kg) exists in Issyk-Kul in two (western and eastern) flocks that inhabit two most favorable shallow-water areas of the Rybachiy and Tyup gulfs, respectively, with the highest summer water temperature and most productive higher aquatic vegetation. The ration of wild carp in Issyk-Kul includes more animal components, such as molluscs, oligochaets, and chironomides larvae, which, possibly, stimulates its fast growth.

There is evidence that a large flock of wild carp lived in Issyk-Kul in the past and was caught mostly in the weirs of the Tyup and Djergalan gulfs. Some individuals reached a weight of 12 kg [18], with the average of 3 kg. In 1938-1941, the catches of wild carp attained 23 tons; in the 1940-60s the catches reduced but still reached 3-1 tons annually. Now the flock of wild carp in Issyk-Kul is about extinction.

In mid June 1857, P. Semenov (famous Russian geographer and explorer) with his team had a rest at the Pokrovsky gulf on their way back from the Central Tien-Shan and were fascinated by its fish riches. Semenov wrote [19]: "Huge wild carps were sparkling in the sun with their nice scales; they splashed profusely on the water surface getting stuck in the thickets of water plants. ...We did not have anything to catch the fish with. But Cossacks entered the water and began chopping the fish with their cavalry swords. ... This improvised way of fishing gave us about 11 poods of fish..." (1 pood $\approx$ 16.38 kg).

This fishing may have occurred in the period of spawning of wild carp. In July 1984 we witnessed that a flock of spawning wild carp in the shallow old bed of the Djergalan

River was wiped out in a day with harpoons and pitchfork by people from neighbouring Mikhailovka village.

The population of wild carp has diminished, as it has been caught during spawning, its spawning areas have reduced by lake level drop, and its eggs are eaten by fishes as Phoxinus, loach, and gudgeon (in the spawning period of wild carp, its eggs make 90-96% of the content of loach intestines).

Diptychus, an Issyk-Kul endemic, is the most slowly growing species. Diptychus males mature at the age of 4-5 years when they are 18-22 cm long, and females reach maturity a year later. Its spawning period lasts from the second half of April to September, and during this time females spawn several times as their eggs mature. Spawning occurs among stromatolites in the region of the Tong and Ak-Terek gulfs on the southern shore and near Ornok village on the northern shore. The flock includes several age groups from 7-8 to 20 years old and older. Winter transit Diptychus, of very low number, spawns in Issyk-Kul tributaries in February. Diptychus juveniles feed upon bottom-living organisms (gammarids, chironomide larvae, and filamentous algae). 10-18 cm long juveniles feed mostly upon molluscs and young shoots of charophytes, while 20-50 cm long adults eat small fishes and Leuciscus bergi Kaschkarov. Benthic organisms are often found in their intestines together with decomposing vegetative detritus.

The catching of Diptychus and other Issyk-Kul fish was started in the earliest 20th century by first Russian settlers. The native Kirkghiz people never ate fish. Till the 1930s, Diptychus made the main portion of catches (70-80 tons annually), and in the 1940s, its mean annual catches reduced to 15-20 tons as its population diminished. As long as 1962, the economic catch of Diptychus was 10-16 tons (private fishing not included), and because of further drop in the number of this valuable fish, its catching was officially prohibited in 1965. The drop was caused by tens of years long catches in spawning sites and in the spawning period, which is especially destructive, as Diptychus grows very slowly: it reaches 2-3 kg by the age of 20-25 years only.

The dynamics of fish catches in Issyk-Kul is shown in Table 2 (by [17])

It is seen from the table that the catches of Leuciscus bergi Kaschkarov, which made up 95% of total catches, has been decreasing since 1968, but the catches of Leuciscus schmidti Herzenstein, its food competitor, and of pike-perch, its predator, have been increasing. After 1976, the total catch, which had been constant before (1,000-1,350 tons), began decreasing, and the catches of all fishes reduced. For Leuciscus bergi Kaschkarov, this reduction is more than 35 times, which shows that the flock of this main economic fish of Issyk-Kul has been damaged.

The acclimatization and related outburst of predatory species caused the decrease in the number of the natives, especially, Leuciscus bergi Kaschkarov which became the main food for pike-perch and trout. However, the introduction of predators into the lake was not the main reason for the drop in the abundance of Leuciscus bergi Kaschkarov [20]. It was rather caused by overcatching and destructive catches during spawning, as the annual catches of Leuciscus bergi Kaschkarov remained as high 1000 tons in the mid 1960s-early 1970s, when pike-perch had acclimatized, and reduced to 100-200 tons as late as in the early 1980s.

TABLE 2. Dynamics of fish catches from 1963 till 1985.

Year	Total catch, tons	Leuciscus schmidti Herzenstein, t/%	Leuciscus bergi Kaschkarov, t/%	Pike-perch, trout, t/%	Others, t/%
1963	914.2	34.1/3.8	818.7/90.7	22.7/2.5	38.7/4.2
1964	1129.3	30.0/2.7	1072.1/94.9	3.6/0.3	23.6/2.1
1965	1180.3	23.3/2.0	1134.4/96.1	15.6/1.3	7.0/0.6
1966	1350.4	32.0/2.4	1257.4/93.1	20.3/4.5	40.7/3.0
1967	1183.7	19.4/1.6	1141.0/96.4	16.9/1.4	6.4/0.5
1968	1089.6	22.7/2.1	1009.9/92.7	42.6/3.9	14.4/1.3
1969	929.9	20.8/2.2	833.7/90.3	63.6/6.8	10.8/1.2
1970	956.3	18.9/2.0	803.3/84.0	121.8/12.7	12.3/1.3
1971	1194.0	44.9/3.8	956.0/89.1	186.5/15.6	6.6/0.6
1972	1168.3	64.2/5.5	994.9/85.2	99.8/8.5	9.4/0.8
1973	1239.3	69.2/5.6	993.2/80.1	166.7/13.5	10.2/0.8
1974	1086.0	74.8/6.9	826.2/76.1	177.3/16.3	7.7/0.6
1975	927.9	77.1/8.3	685.6/73.9	159.6/17.2	5.6/0.6
1976	978.3	67.1/6.9	740.4/75.7	163.4/16.7	7.4/0.8
1977	834.3	58.6/7.0	627.7/75.2	138.4/14.1	9.6/1.2
1978	615.6	26.1/4.2	473.7/76.9	108.0/17.5	7.8/1.3
1979	474.5	19.0/4.0	340.8/71.8	106.6/22.4	8.1/1.7
1980	303.8	35.6/11.7	223.8/73.7	36.2/11.9	8.2/2.8
1981	266.6	30.1/12.3	150.0/56.3	79.2/29.7	7.3/2.7
1982	142.4	31.5/22.1	33.6/23.6	66.3/46.6	11.0/7.7
1983	158.0	34.2/21.6	71.3/45.1	36.7/23.2	15.8/10.0
1984	140.0	14.6/10.4	81.4/58.1	27.7/19.9	16.3/11.6
1985	160.0	13.4/8.3	86.0/53.7	40.0/25.0	20.8/13.0

6. Eutrophication in Issyk-Kul: hypothetic and true cultural causes

Eutrophication is an increase in bioproductivity of water basins caused by accumulation of biogenic substances in the water as a result of natural and cultural effects. Cultural eutrophication is caused by an increased input of nutrients with waste waters. Sometimes eutrophication is identified as contamination of water with biogenic substances. Eutrophication is accompanied by rapid growth of algae.

Laboratory and field experiments worldwide show that immersed and half-immersed macrophytes, together with periphyton attached to them, actively absorb nutrients and contaminants from the environment and thus serve as a natural filter for lakes. After macrophytes die off, the nutrients they accumulated are partially deposited in littoral silts, forming an internal source of biogenic material, or penetrate into the water.

The macrophyte phytomass in Issyk-Kul attains its highest concentration at depths between 10 and 20 m (Table 1). The recent water level drop has caused opening of the internal source of nutrients formed in the middle of the 19th century when the absolute level of the lake was about 1,620 m, i.e. 13 m higher than now.

The increased cultural input of biogenic substances induces an active growth of all algal species, which is however limed by light for immersed macrophytes. At the initial stage of eutrophication, the number of macrophytes reduces as they receive less light because of growing phytomass of periphyton and filamentous algae. Suppressed by periphyton, macrophytes cannot capture biogenic elements supplied from the drainage basin, which increases the productivity of phytoplankton in the open lake.

Macrophytes have recently shown evidence of suppression in the Tyup gulf, where they grow at depths of 10-20 m, at relative water transparency of 2-5 m, and the production of phytoplankton is very high. Therefore, this part of the lake is already considered mesotrophic. The decrease or complete disappearance of charophyte algae from typically charophyte basins may be caused by an increase in nutrients, especially phosphorus, which is harmful for charophytes [14](Pivnev, 1990). Thus, charophytes are a biological indicator of the cultural input of biogenic elements.

The absence of charophytes from the upper weirs of the Tyup gulf is associated with high concentrations of nutrients there [21]. Observations conducted in lakes of Canada, England, and the former Soviet Union showed that highest concentrations of biogenic elements can be withstood by Potamogeton pectinatus L., which is thus believed [22] to be an indicator of cultural disturbances of the lake regime. However, Potamogeton pectinatus L. in Issyk-Kul grows profusely along the shores of the Rybachiy and Tyup gulfs, where it is the main food for carp, but it rather occupies its natural ecological niche in the warm shallow waters.

Intense growth of periphyton, especially the filamentous algae like spirochaeta and cladophora, is another typical feature of cultural eutrophication. These algae can live on any substrate and often occur throught the water thickness in the shallow littoral. An increase in filamentous algae is evidence of an increased input of nutrients from the catchment basin. Cladophora is more sensitive to concentrations of biogenic elements than other filamentous algae [14]. Having a great productivity potential, it can develop high reproduction immediately in response to increased input of nutrients and spread very fast through the littoral. It has extremely high concentrations in many sewage-polluted basins worldwide. E.g., the fast development of cladophora was a problem in Erie and Ontario in the early 1930s.

Spirogira, entheromorpha, and cladophora live at the depth of 5 m on the surface of boulders and pebbles and on charophyte algae in many regions of Issyk-Kul, and this alarming fact is worth a special study.

Changes in specific composition and productivity of phytoplankton is the best notable indicator of ecological disturbances, as phytoplankton is highly sensitive to the environmental conditions. The prevalence of blue-green algae, which though being of a very limited specific diversity, reach mass growth and cause water "bloom", is the commonest consequence of saturation with nutrients [23]. Blue-green algae supplant diatoms, which are normally the dominant phytoplankton species in deep oligotrophic lakes like Issyk-Kul, and this change is one of typical indicators of growing cultural eutrophication.

Diatoms predominate in the littoral, where they make up 56.3% of total plankton algae [4], but are much less abundant in the open lake (17.1% against 39.7% blue-green and 36.0% green algae). Their growth in the open lake may be restrained by deficiency in soluble silica, while the bottom forms of diatoms can catch it in the littoral zone. The predominance of blue-green algae in the open lake is an alarming factor, though they do not include Anabaena flos-aquae, Aphanizomenon flos-aquae and others that cause "blooming water".

Macrophytes capture nutrient flux from the catchment basin and thus suppress the growth of phytoplankton in the open lake. A field experiment in Lake Vinera showed that active growth of phytoplankton was induced by the decomposition of Myriophyllum spicatum with release of nitrogen and phosphorus [23]. The macrophyte control over the growth of phytoplankton is typical of deepwater oligotrophic lakes such as Issyk-Kul. Macrophytes predominate in these lakes at a moderate input of nutrients and suppress phytoplankton in clear high-transparency water. Issyk-Kul macrophytes culminate between spring and autumn peaks of phytoplankton development, as they catch biogenic elements coming from the catchment basin.

Changes in biological populations is another indicator of the ecological state of water basins, and bottom dwellers are the best informative organisms. Well-balanced ecosystems normally include a great number of non-dominant species. The effect of contamination on the habitat can be estimated from changes in the ratio of resistant and vulnerable species.

Changes in the specific composition and vertical distribution of zoobenthos related to eutrophication are preceded at its initial stage by continuously greater contrasts in the vertical distribution of water chemistry, especially dissolved oxygen and hydrogen ions. Another important parameter is availability of food for benthic organisms [22]. Some parts of the bottom above the temperature jump layer are inhabited by abundant and variable fauna, including the hydrobionts that are especially sensitive to oxygen content. Beneath this layer, benthos is less abundant and less diverse, as its composition is confined to organisms adapted to lower oxygen concentrations. Progress of eutrophication is accompanied by growing oxygen deficit in hypolimnion, and the contrast in the composition and distribution of zoobenthos becomes better evident. At later stages of eutrophication, only few organisms that have adapted to anaerobic exchange can persist in the deep water.

Two mollusc subspecies (Radix auricularia (L.) var. obliquata Martens and R. Auricularia (L.) var. obliquata ampla Hartman) that were rather frequent in Issyk-Kul [24] are now about extinction. The former is inhabitant of the littoral zone (to 65 m), and is almost completely absent from the sandy bottom. It lives in charophyte thickets, where it finds food and shelter and feeds upon various diatoms and filamentous algae which form foulings on the surface of boulders, pebbles, and microbial mounds in the areas of stony bottom and carbonate sedimentation. It also eats rotifers and Harpacticidae which live in the foulings.

The second subspecies has adapted to a deep-water habitat, from 40 to 70 m. Unlike many other lung molluscs, it constantly remains at those depths, as it does not need ascending for oxygen due to good aeration of the near-bottom water. Its lung is filled with oxygenated water and functions as gills. As it lives in colder water, this subspecies of mollusc grows slower (its life cycle is as long as several years) but it is a few times bigger than the littoral variety.

Living specimens of the deep-water and R. Auricularia (L.) var. obliquata ampla Hartman were encountered in bottom samples in the late 1950s-early 1960s [24]. Now it is no longer found, though its empty shells are quite frequent. Therefore, the modern tendencies of natural lake development are dangerous for some hydrobionts, including the deep-water form of Radix auricularia. Its disappearance might be related to the deterioration of oxygen regime in the near-bottom water of the littoral. This mollusc is possibly a stenooxide organism, which is sensitive to fluctuations of dissolved oxygen. This fact, as well as the predominance of oxygen-independent chironomide larvae in the Tyup gulf, is alarming evidence of eutrophication in Issyk-Kul. Note that changes in composition, density, and distribution of benthic populations in Great Lakes were associated with the progress in features of cultural eutrophication and occurred as rapidly as in 2-3 decades.

Almost complete disappearance of endemic Diptychus dybowskii Kessler is another alarming fact, which apparently has ecological causes. Till recently Diptychus dybowskii Kessler has inhabited the littoral zone along the entire Issyk-Kul perimeter except for the mesotrophic Tyup gulf, where chironomide larvae have supplanted molluscs because of eutrophication. Therefore, the drop in the number of Diptychus dybowskii Kessler may have been caused by the disappearance of the deep-water form of Radix auricularia, which was its winter food.

In the mid 1970s, the input of phosphorus into Issyk-Kul from various sources was evidenced [3]:

253 tons (28%) from the catchment basin, without cultural landscapes;
200-300 tons (26%) from arable fertilized lands;
206 tons (22.5%) from excretion of cattle;
180-200 tons (20.5%) with municipal sewage;
28 tons (3%) with atmospheric precipitation,
with the total annual amount of 840-974 tons.

The total input of phosphorus from sources in the catchment basin is 0.13-0.15 g/m^2, whereas the annual load needed to transform an oligotrophic lake as Issyk-Kul, with average depth of 278 m, into a mesotrophic one is about 0.5 g/m^2 [22]. Therefore, it may appear that the cultural activity in the Issyk-Kul basin has not yet caused a severe damage to the lake. However, this conclusion is preliminary, as the above estimates do not include the input of nutrients from the main internal source disclosed by the lake level drop.

7. References

1. Matyukova, T.G. (1958) Charophytes in Lake Issyk-Kul [in Russian], *Proc. Kirghiz Uniersity, Biol. Department* iss. 7, Frunze, 43-49.
2. Mambetalieva, S.K. (1960) Algal vegetation of the northern Lake Issyk-Kul [in Russian], *Izv. Kirg. AN*, Ilim, iss. 3, Frunze, 14-19.
3. Romanovsky, V.V. (1978) Water vegetation [in Russian], in Koralev, V.G. (ed.), *Lake Issyk-Kul*, Ilim, Frunze, 149-156.
4. Kulumbaeva, A.A. (1982) *Phytoplankton of Lake Issyk-Kul* [in Russian], Ilim, Frunze, 106 p.
5. Davydova, N.N. (1991) Diatomaceous deposits in the bottom fill of Lake Issyk-Kul [in Russian], in *The History of Lakes Sevan, Issyk-Kul, Balkhash, Zaisan, and Aral*, Nauka, Leningrad, pp. 103-108.
6. Foliyan, L.A. (1973) Zoobenthos in Lake Issyk-Kul [in Russian], in Turdakov, A.F., *Ichthyological and Hydrobiological Studies in Kirghizia*, Ilim, Frunze, 3-11.
7. Neustrueva, I.Yu, Yurmalietis, R.A., Zubovich, S.F., and Shlyakhova N.A. (1991), Ostracods in the bottom fill of Lake Issyk-Kul [in Russian], in Treshnikov, A.F. (ed.), *The History of Lakes Sevan, Issyk-Kul, Balkhash, Zaisan, and Aral*, Nauka, Leningrad, 108-115.
8. Pavlova, M.V. (1964) *Benthos in Lake Issyk-Kul* [in Russian], Ilim, Frunze, 84 p.
9. Turdakov, F.A. (1961) Reorganization of fishing industry on Lake Issyk-Kul [in Russian], *Izv. AN SSSR* N 2, 54-56.
10. Berg, L.S. (1930) Hydrological studies in Lake Issyk-Kul in 1928 [in Russian], *Bull. Mining Institute* Vol. 28, 14-37.
11. Nikitin, A.A. (1976) *Acclimatization and farming of salmon fishes in Kirghizia* [in Russian], Ilim, Frunze, 122 p.
12. Gareev, Sh.E. (1980) Pre-spawning feeding of trout in Issyk-Kul [in Russian], in Konurbaev, A.O. (ed.), *Bioecology of Animals and Plants and Environment Protection*, Ilim, Frunze, 19-24.
13. Luzhin, B.P. (1956) *Trout in Lake Issyk-Kul* [in Russian], AN Kirg.SSR Press, Frunze, 133 p.
14. Pivnev, I.A. (1990) *Fishes in Kirghizia*, Ilim, Frunze, 138 p.
15. Alpiev, M.N. (1983) Biology of salmon species in Lake Issyk-Kul [in Russian], in Konurbaev, A.O. (ed.), *Limnological Studies in Kirghizia*, Ilim, Frunze, 75-81.

16. Kiselev, Yu.E. (1983) Some properties of eggs in Leuciscus schmidti Herzenstein in Issyk-Kul [in Russian], in Konurbaev, A.O. (ed.), *Limnological Studies in Kirghizia*, Ilim, Frunze, 54-65.
17. Zhadin, A.B. (1989) *Species Diversity of Leuciscus, Cyprinidae, Pisces in Issyk-Kul*, Abstract, Cand. Dissertation, Ilim, Frunze, 24 p.
18. Berg, L.S. (1904) *Lake Issyk-Kul* [in Russian], Zemlevedenie, St. Peresburg. Vol. 1-2.
19. Semenov, P.P. (1948) *Travel to Tien Shan in 1850-1857* [in Russian], Geograggiz, Moscow, 147 p.
20. Shukurov, E.D. (1991) Animals in the Issyk-Kul basin: ecological aspects [in Russian], in Shukurov, E.D. (ed.) *Fauna and Ecology of Land Vertebrates in Kirghizia*, Ilim, Bishkek, 37-48.
21. Kasymalieva, E.B. and Romanovsky, V.V. (1994) Protective role of macrophytes against eutrophication in Issyk-Kul [in Russian], *Izv. NAN Kirg. Rep.*, 77-80.
22. Rossolimo, L.L. (1977) *Changes in Limnic Ecosystems under Cultural Effects* [in Russian], Nauka, Moscow, 144 p.
23. Pieczynska, E. (1990) Littoral habitats and communities, *Guidelines of Lake Management*, vol.3, 39-71.
24. Pavlova, M.V. (1983) Biology and production of *Limneae (Radix auricularia (L.) var obliquata Martens* in Lake Issyk-Kul [in Russian], in Gurvich, V.F. (ed.), *Limnological Studies in Kirghizia*, Ilim, Frunze, 34-47.

WATER LEVEL VARIATIONS AND WATER BALANCE OF LAKE ISSYK-KUL

V. V. ROMANOVSKY
Institute of Water Problems and Hydropower
533, Frunze str., Bishkek, Kirghizia

Evidence of water level variations in Lake Issyk-Kul is obtained from the position of lake terraces, from archaeological findings on the lake shore and the bottom of the lake, from historical accounts, and from recent instrumental monitoring.

1. Data from lake terraces

Emerged lacustrine terraces are observed at different places around the lake shore. Lacustrine sediments outcropping at 1640 m. a.s.l. have been dated at about 26.000 yr B.P. by Aleshinskaya et al. [1]. Most of radiocarbon dating related to water level variations in Issyk-Kul has been obtained by Shnitnikov [2], Berdovskaya and Egorov [3] and Sevast'yanov [4]. According to these data, the uppermost lacustrine silts rich in $CaCO_3$, at depths up to 1 m in various parts of the Holocene terrace, have ages from 9950 ± 200 to 13540 ± 400 yr B.P., and at depths 2 and 5 m, the ages are respectively 16500 ± 700 and 18600 ± 400 yr B.P. It may consequently be concluded that Lake Issyk-Kul had been elevated to an altitude higher than now (1640 – 1610 m) for a quite long period between ca. 26 and 10 ka ago.
Younger lacustrine deposits at altitudes about 1620 m have been constrained by two ages of 1190+-160 and 1400+-120 yr B.P. These ages record a maximum transgression whose level was limited by the Karatash discharge treshold at 1622-1623 m. The geomorphic expression of this transgression on the lake shores is very poor.
A set of ten radiocarbon dates, mostly for remnants of wood sampled on the Holocene lake terrace and inside the lake to depths about 5-6 m below the water surface account for the period from 700 ± 50 to 170 ± 60 yr B.P. These ages show that 700 – 500 years ago Issyk-Kul experienced transgression and its level was 2 – 5 m lower than now; in the beginning and in the middle of the 19th century it was 13 m above the present level. During the maximum transgression, wave-cut escarpments formed on the lake shores along its perimeter, at the absolute altitude of 1620 m, many river mouths were dammed by debris, and loop-like shoals formed in the sites of shoreline bending.
The radiocarbon data show that from 10,000 to 300 yr B.P. the level of Issyk- Kul was lower than now. According to Trofimov [5], based on seismo-acoustic data, the

J. Klerkx and B. Imanackunov (eds.), Lake Issyk-Kul: Its Natural Environment, 45–57.

considerable drop of lake level may have been caused by a tectonic subsidence of the central block of the lake basin. The depth of the lake level drop is marked by submerged river valleys that end at depths 110 – 120 m. Rivers formed deeply incised valleys on land . The rise of water level, recorded by the dates from 700 to 170 yr., may have been related to the Small Ice age cooling from the 13th to the 19th century. Spore and pollen data [6] indicate that the coldest and wettest climate over the last 10 ka occurred in the Issyk Kul basin in the recent millenium.

2. Historical information

The relatively recent history of the lake can be reconstructed from ancient manuscripts, epic legends, and archeological findings on the lake shores and bottom. One oldest Early Stone Age site was discovered by archaeologists Ranov [7] and Yunusaliev [8] on the western side of the lake on Boz-Barmak mount. The deposits that hosted stone tools are Upper Pleistocene, and the age of the finding was estimated as pre-Musterian.
A primitive man's site at the southern side of Issyk-Kul, 4.5 km away from the Tossor mouth, is another ancient evidence of the Stone Age culture. The site was assigned to Late Musterian time, which corresponds to the earliest Late Pleistocene. Not far from the Tossor site of stone tools, there is another site, Kensai, dated at the Late Stone Age. Geomorphically the Boz-Barmak and Kensai sites are attributed to an Upper Pleistocene lake terrace at an altitude 1640 m. The position of the Boz-Barmak site in a dry area, far from any source of fresh water might evidence that Issyk-Kul at that time may have been an open fresh-water lake.

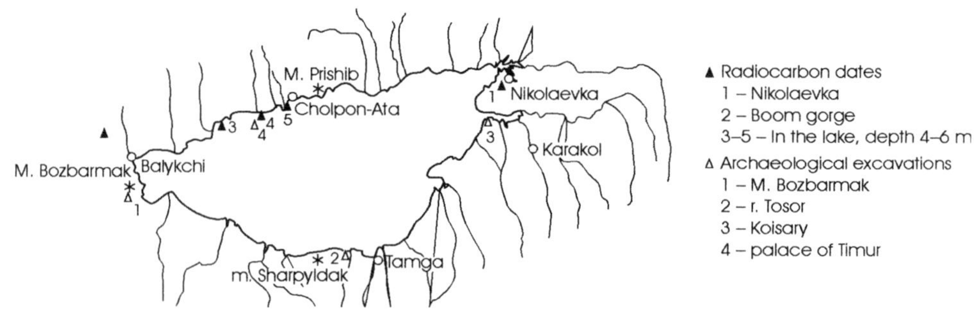

Figure 1. Map of archeological excavations and sampling sites for ^{14}C dating in the Issyk-Kul basin.

At present, only few archeological findings of Middle and Early Stone Age are known on the shores of Issyk-Kul, possibly, because of the strong level drop. If it so, the findings must be at depths 110 – 120 m, where end the submerged river valleys that mark the level of the Holocene regression.

From the 7th to the 3d century B.C., the Issyk-Kul basin was inhabited by Saks. This time is recorded in numerous mounds containing bronze and copper objects. The objects of the Sakian culture brought to the beaches by waves, as well as under-water findings, attest that at that time the lake level was lower than now. Umurzakov and Vinnik [9] studied Chinese maps and ancient chronicles and arrived at the conclusion that during the period from the 11th century B.C. to the 1st century A.D., the level of Issyk Kul was 12 – 13 m higher than now, and the lake discharged into the Boom canyon.

According to Chinese (7th century A.D.) and Persian (11th century A.D.) writings, the water in the lake was then saline and even bitter-saline. Hence, evaporation from the lake surface exceeded the riverine runoff and atmospheric precipitation, and the water level was lower than now.

The best known and best documented submerged ancient constructions of the Middle Age are the settlement in the region of Koi-Sary in the eastern part of the lake and the ruins of "Timur's palace" on the northern lake side. The Koi-Sary settlement dates back to the 10 – 12th cent. A.D. Having lived through a decay after the Mongolian invasion, the settlement kept existing till the latest 15th- earliest 16th century, when the life ceased there, as the village was flooded by the rising lake water.

According to the legend of "Timur's palace", Timur (Tamerlan), who conquered the Issyk Kul basin in 1399, had a rest after his war campaigns in a palace built on an island. Radiocarbon dates [2] on wood samples from the submerged constructions of "Timur's palace" show that the palace existed from the end of the 13th to the middle of 15th century, and then was flooded.

The Middle Age archeological site, settlement Tossor, located on the southern side of the lake, dates back to the 12th century A.D. Later the settlement was destroyed by the lake transgression, which formed a wave-cut escarpment there [10]. According to the documents of Mohammed Khaidar, in the earliest 16th century, the lake was a closed basin [11]. This evidence is confirmed by the map by Unkovsky of 1724. Later sources provide evidence of a discharge from the lake. So, a Chinese source (Daiqin-itun-zhi, 1743) mentions that Issyk-Kul in its northwestern corner "comes out of its shores and releases a flow". The map made by missionary Jesuits in 1755 and the map by Islenyev (1777) show Issyk Kul having a discharge. The high stand of the lake still upheld in the first third of the 19th century. In notes by Bubenkov, we read that "the river Chu flows from the lake", and notes by Silberstein say that "Issyk-Kul flows in a channel into the river Chuya". Merchants from Bukhara and Tashkent told Alexandre von Humboldt during his last research trip in 1829 that Issyk-Kul had a discharge. However, already in 1830, Potanin described the Chu river but wrote nothing about its link with the lake, and the map by Nifant'yev of 1849 shows the river Kutemaldy flowing into Issyk-Kul.

Therefore, the historical and archaeological evidence shows that relatively recently (in the 18th cent. A.D.) Issyk-Kul rose to an altitude 12 – 13 m above the present. Since the early 19th century, its level has been decreasing.

Important information about the dynamics of the Issyk-Kul water level comes from first special researches and eye-witness evidence of old people.

Having studied the available literature, L. Berg [12] concluded that through 1859 – 1897, all authors unanimously declare a continuous drop of the Issyk-Kul level.

G. Romanovsky, mining engineer, well-known explorer of the Turkestan region, visited Issyk-Kul in 1878 and wrote: "Indeed, some time ago the lake stand was higher than now, at least by 200 feet, and then it protruded about 50 versts into the Boom canyon". The conclusion by Romanovsky more than 120 years ago has been confirmed and specified by recent radiocarbon dating of plant detritus from the base of a lake terrace in the Boom canyon [13]. The ^{14}C age of 26100 ± 600 (sample IGAN-616) obtained by Steklenkov correlates with the age of an Upper Pleistocene lake terrace in the eastern part of the lake (dated by Aleshinskaya et al. [1]).

According to evidence by A. Golubev (1859), in the region of mountain Prishib on the northern lake side, the mountains immediately bordered the lake, and paths ran on a steep slope right along the water. In 1885, on the southern lake side, waves almost reached the scarp base of mountain Sharpyldag whose name imitates the sound of the tide. The same evidence was given by the botanist N. Sorokin who wrote that from Barskaun to the Kodji River the road runs along the Issyk-Kul shore, and mountains now step aside to the left then approached the water and faced the waves by vertical cliffy walls. At present, the lake backstepped from Sharpyldag mountain 200 m horizontally and 13 m vertically.

G. Kolpakovsky wrote in 1870: "Kirghiz people say that the lake diminishes from year to year. Manap Umbet-Ali showed me the line on cape Kasyr-Ksengyr (mt. Prishib) reached by the water earlier as he remembers; now the water is about 4 meters below this line".

Botanist V. Gern who visited the lake in 1877 noted: "Continuous lowering of the Issyk-Kul water surface and its shoaling has been noticed by people who live on the shore, while the traces of water, like the ridges across riverbeds at the mouths of former rivers, which are now gullies, make believe that the surface of the lake was much higher quite recently". Forest felling in the region of the lake was, according to Gern, one of the causes that led to lake shoaling.

The water level drop occurred at a high rate in the 19th century. Russian botanist Regel was told in 1877 by local people at Kutemaldy station that the water lowered by ~2 m in a decade. A. Selivanov wrote in 1896 that in the region of Kutemaldy station, the lake backstepped about 600 m and lowered by ~3.5 m in 23 years. Hydrologist Sheffer measured the lake level in 1898 and compared it with the points of the former stand (1878) indicated by local people. He estimated that the level had dropped by 2.88 m.

Therefore, the mean annual drop of the lake level from 1867 to 1877, according to Regel, was 21 cm/yr; this rate reduced to 15.4 cm/yr in 1872 – 1895, according to Selivanov, and to 14.4 cm/yr in 1878 – 1898, according to Sheffer.

The lake level drop caused a vivid reaction in the press. A. Vinogradsky in 1905 published a large article in the journal "Russian Turkestan" with an alarming title "Issyk-Kul – a disappearing sea-lake". He wrote: "In recent times, Issyk-Kul has been diminishing every day and every hour – soon it will become just a mentioning in legends".

According to estimates by L. Berg [12], the level of Issyk-Kul rose in the beginning of the 20th century, but dropped by 1.3 m from 1910 to 1928.

3. Instrumental measurements

Since 1926, regular monitoring has been initiated at several permanent hydrological stations. The data of this monitoring show a drop of 304 cm from 1927 to 2000. During this period of instrumental measurements, the mean annual drop was consequently 4.2 cm. The highest stand (1609.52 m) was in 1929, and the lowest (1606.17 m) in 1998 (Table 1). Over the 74 years of monitoring, there have been several years when the water level rose: 1929, 1935–1936, 1941–1942, 1956–1960, 1970–1971, 1981–1982, 1987–1989, 1992– 994. The longest period of higher stand was in 1956–1960, when a great amount of precipitation caused a 32 cm rise of water level. This rise was preceded by the longest period of lower stand from 1943 to 1955, when the water level drop was as much as 110 cm. This decrease was caused by lower precipitation and warming in the Issyk-Kul basin. The highest annual drop of 28 cm was recorded in 1948, and the highest annual rise of 26 cm occurred in 1999. The decade from 1981 to 1994 was a period of stable stand about 1606.6 m. In 1995–1998, the lake level started to decrease again but in 1999 and 2000 it rose by 11 and 10 cm respectively.

TABLE 1 : Water levels of Issyk-Kul (annual mean and level at the end of December)

Year	Annual mean	Level on 31/12	Year	Annual mean	Level on 31/12
1927	1609.48		1964	1607.86	1607.71
1928	1609.44		1965	1607.84	1607.65
1929	1609.52		1966	1607.8	1607.6
1930	1609.39		1967	1607.72	1607.49
1931	1609.23		1968	1607.6	1607.31
1932	1609.12		1969	1607.51	1607.42
1933	1609		1970	1607,56	1607,44
1934	1608.98	1608.8	1971	1607,59	1607,46
1935	1609.04	1608.84	1972	1607.5	1607.29
1936	1609.05	1608.84	1973	1607.4	1607.2
1937	1609.04	1608.77	1974	1607.31	1607.15
1938	1608.95	1608.67	1975	1607.29	1607.15
1939	1608.84	1608.63	1976	1607.19	1606.98
1940	1608.83	1608.69	1977	1607.02	1606.87
1941	1608.91	1608.71	1978	1606.9	1606.77
1942	1608.95	1608.74	1979	1606.79	1606.71
1943	1608.86	1608.54	1980	1606.73	1606.65
1944	1608.67	1608.36	1981	1606.74	1606.73
1945	1608.52	1608.32	1982	1606.75	1606.68
1946	1608.52	1608.27	1983	1606.72	1606.63
1947	1608.38	1608.15	1984	1606.64	1606.47
1948	1608.18	1607.87	1985	1606.52	1606.42

1949	1608.01	1607.87	1986	1606.49	1606.44
1950	1608.02	1607.83	1987	1606.58	1606.58
1951	1607.94	1607.79	1988	1606.68	1606.74
1952	1607.95	1607.75	1989	1606.74	1606.65
1953	1607.91	1607.74	1990	1606.66	1606.54
1954	1607.89	1607.73	1991	1606.56	1606.45
1955	1607.87	1607.75	1992	1606.49	1606.43
1956	1607.95	1607.92	1993	1606.55	1606.59
1957	1608.05	1607.94	1994	1606.65	1606.62
1958	1608.08	1607.99	1995	1606.59	1606.42
1959	1608.16	1608.08	1996	1606.4	1606.3
1960	1608.19	1608	1997	1606.27	1606.08
1961	1608.07	1607.93	1998	1606.17	1606.09
1963	1607.91	1607.7			

Annual variations of water level in the lake are controlled by interplay of the main components of water balance: water influx and evaporation from the surface. A rise in water level usually starts in April following the spring flood on rivers caused by melting of seasonal snow and continues till August being fed by glacier thawing waters. The highest seasonal stand is most often reached in August-September (once in October, in 1988). In dry years, when water withdrawal increases for irrigation, the highest stand is recorded in June and July. In the dry 1948, the spring-summer level rise was as low as 5 cm. By the end of 1948, the water level drop attained 28 cm because of depletion of groundwater resources and high evaporation. In 1956, water level was rising continuously from April to September and increased eventually by 42 cm. By the end of the year, the lake level was 17 cm higher than in the beginning of the same year.

Among other researchers Matveev [13] estimated the age of the Issyk-Kul basin. He proceeded from the content of chlorine (13 mg/l) in the inflowing river water, a river discharge of 6.31 m/yr, and the content of chlorine in the lake water (1.590 mg/l). According to his estimation, the river water should be evaporated 122 times to reach the salinity of Issyk-Kul. Matveev calculated that the complete water renewal, obtained by dividing the water volume in the lake by the annual river discharge, occurs in 274 years, hence the amount of salt in the lake (274x122) may have accumulated in 33,400 years.

Sapozhnikov and Viselkina [14] used more recent data on the surface discharge into the lake (112 m^3/s) and on the average content of chlorine in the river water (5.42 mg/l) and obtained the duration of the basinal regime as long as 135,000 years.

Later, Blagoobrazov estimated the difference of chlorine contents after Matveev [13] and Kadyrov and obtained an age of 4800 years.

Besides data on the chemistry of natural water, Chalov and Tuzova [15] approached the problem by the natural decay of U^{234} and U^{238} (effect of Cherdyntsev-Chalov). They obtained an estimate of 110 ± 40 ka, based on U^{234}/U^{238} ratios in the water of lake tributaries and in the lake itself; an estimate made with regard to the connection of the lake with the Chu river is 300 ± 40 ka. The first values agrees with the estimate of 135 ka obtained by Sapozhnikov and Viselkina [14]. However, radiocarbon dating of the Upper Pleistocene lake terrace in the Boom canyon shows that about 26,000 yr ago Issyk-Kul had a discharge through this canyon and may have been a freshwater lake. Therefore, dates older than 26 ka are poorly reliable.

Recently R. Ricketts et al. [16] have studied the Holocene history of Issyk-Kul using data on concentrations of trace elements and stable isotopes in ostracods in up to 4.5 m long cores. From 7900 to 6000 B.P., the lake was a fresh open basin; through 6000 – 4300 B.P. it evolved from an open freshwater to a close brackish basin. Since 4300 B.P., various parameters of its isotope and trace element composition attest to its close basinal regime and a salinity similar to the present-day one.

4. The problem of the lake level decline

The problem of the recent water level drop in Issyk-Kul remains a subject of discussion. Several factors are invoked for being responsible for this decline: water withdrawal from lake tributaries for irrigation (cultural component), lower humidity in the lake drainage basin (climatic component), subsidence of the lake basin (tectonic component), groundwater outflow outside the lake basin (hydrogeological component), separation of the lake from the Chu river (hydrographic component).

Water level variations in closed basins are controlled by regional humidity, which is related to river runoff determined as a difference between precipitation and evaporation from the drainage area. The rivers in the Issyk-Kul catchment are fed by snow and glaciers. The Tyup river is one of few exceptions. This second major river of the drainage basin is fed by snow; its catchment occupies the lowest and the most humid area east of the lake. Annual precipitation, according to measurements at weather station Santash, is 880 mm. The hydrological station Sarytologoi is located in the place where the river flows from the mountains into the lake plain, upstream of the withdrawal site, and the river runoff data are not distorted by cultural effects.

A comparative chronological analysis of the Tyup discharge (measured at Sarytologoi station) and the lake level variations shows that greater river discharge correlates with higher lake stand, and vice versa. At a mean discharge of 8.2 l/s, the water level rose at a discharge above 9 l/s. Proceeding from this fact, we arrived at the conclusion [17] that the lake level would approach the state of stability if the Tyup discharge increased by 0.8 l/s, which corresponds to a 50 mm increase between annual precipitation and evaporation. The relationship between the lake level at the end of each calendar year estimated relative to the level of 1940 and the sums of modulus coefficients of the Tyup discharge shows a high correlation coefficient of 0.92 – 0.99.

Therefore, a humidity deficit is observed in the drainage basin of the Tyup river, outside the zone of economic activity. The Tyup catchment is an indicator of humidity of the Issyk-Kul basin as a whole. However, this deficit is not evident in the rivers fed from snow and glaciers, as the recent warming is accompanied by an increase in glacier runoff, though the increase in river runoff at the account of more active melting of glaciers may provide only a partial compensation for the increased evaporation from the lake surface.

Thus, the drop of the water level in Issyk-Kul is related to a humidity decrease in its drainage basin. Many workers note that the notable climate warming and decrease in precipitation since the 19th century is associated with an abrupt water level drop in basinal lakes in arid regions of the globe. Lake Issyk-Kul is not an exception, and variations of its water level reflect the global climate processes.

Monitoring operated since 1956 by the Tien Shan physiographic station on the Karabatkak glacier [18] shows that the trend of total positive monthly temperatures at which thawing of glaciers occurs is negative (-0.32° C per year) from the beginning of the observations to 1972. During that period, the ice material accumulated in the years of higher stand of the lake, and the mass balance of the glacier was positive. As a rule, these were humid years with a cold summer.

In 1973, an abrupt displacement of the trend toward higher values led to an increase in the runoff from the glacier. After 1972, there were no years with a positive balance of ice mass. During that period, a considerable increase in the glacier runoff was observed on all rivers fed from snow and glaciers. Water supply into the lake increased at the account of additional melting of about 4 km^3 of glaciers, which is about 8.3% of all glaciers in the lake surroundings [17].

Therefore, prior to 1972, an increase in water level of Issyk-Kul and accumulation of water in glaciers occurred parallel in the same years at the account of water brought from outside the lake basin. After 1972, the river runoff increased at the account of degradation of glaciers on the background of a considerable climate warming, which partly compensate for more rapid water level drop by evaporation from the lake surface. In that period, a redistribution of water resources from glaciers into the lake was observed, and the water level of Issyk-Kul rose in the years of high humidity.

Deep infiltration of water from Issyk-Kul occurs into the Chu river in the region of the Boom canyon due to an elevation contrast, which is 318 m in the exit from the canyon and 946 m at the hydrological station Milyanfan. Therefore, already at the exit from the Boom canyon, the riverbed of the Chu is 40 m above the average depth of the lake (278 m), and at the Milyanfan station it is 278 m in excess of its maximum depth (668 m). Given that the Chu is the only drain, groundwater flow inevitably delivers there infiltration water from Issyk-Kul.

The possibility for deep infiltration from the lake is due not only to the elevation difference, but also to the geological structure and tectonics. The Issyk-Kul basin is bounded in the west by the Karatash discharge threshold, which is a sort of a cliffy "lock" on the way of infiltration water. The latter meets the threshold and has to rise to the surface, into the Chu , and only a minor part of water percolates through fractures in bedrock and flows toward the Chu valley.

Another site of possible discharge of water from Issyk-Kul is the zone of the Kemin fault, upstream of the Chu-Chong-Kemin confluence. The tectonic activity of this region must favour discharge of deep groundwater.

The third site of possible discharge is located in the region of mountain Tokmok where transversal uplifts pond the sub-bed flow of the Chu and squeeze it upward to the

surface producing swampy areas. River Krasnaya in this region is fed from groundwater. Between Tokmok and Chon-Kemin village, there is a so called "collapse zone" where a great amount of water is discharged by infiltration from the Chu riverbed.
Studies of deep infiltration were performed by the Institute of Limnology (Moscow) in all those regions using Cl ion as a chemical indicator of groundwater motion, taking into account that its absolute content in Issyk-Kul is almost 100 times that in the Chu. These studies show that the Issyk-Kul basin is closed, and there is no underground discharge from it into the Chu river [19].
Isotope studies carried out by the Institute of Water Problems (Moscow)[20] show that the Issyk Kul water has a low concentration of tritium (~38 T.u.) and a high concentration of ^{18}O (-0.3% δ ^{18}O). The low concentration of ^{18}O in the water of the Krasnaya and Chu rivers (-8.8 – -11.6% δ ^{18}O), which is the same as in the rivers of the Issyk-Kul basin, and high concentrations of tritium (about 200 T.u.) contradict the hypothesis that infiltration water from Issyk-Kul participates in the groundwater supply into the Chu and Krasnaya rivers.
For the Issyk-Kul lake basin, the equation of water balance is

$$x + y + y_1 - Q - z +- \Delta h = 0$$

where x is atmospheric precipitation; y and y_1 are surface and groundwater influx into the lake; z is evaporation from the lake surface ; Q is withdrawal for irrigation; Δh is water level change.
There are about 20 versions of water balance for Lake Issyk-Kul proposed by various authors [21] in which atmospheric precipitation on the lake surface varies from 230 to 350 mm, the surface river runoff varies from 184 to 567 mm, variations of groundwater influx are from 241 to 330 mm, and evaporation from the water surface is in the range of 700 to 878 mm.
The first estimates of the water balance for Issyk-Kul were obtained in 1946 by Vikulina [22]. The surface runoff was estimated from data of only 15 hydrological stations; estimates of precipitation were based on the assumption of equal precipitation for the lake and the shores; evaporation was calculated by Meyer-Tikhomirov's simple empirical relationship from data of two coastal weather stations. Nevertheless, that first water balance gives a realistic idea of the relation between its income and output.
In 1960, Tsigelnaya [23] published a water balance in which she used data from weather stations far off the lake and obtained an underestimated value of evaporation (702 mm); the residual 98 mm were assigned to the water loss for the needs of irrigation (42 mm), industry, and household (56).
Shnitnikov [2]calculated the water balance for the presence and the absence of discharge and obtained values close to those of Tsigelnaya [23], but he did not explain the details of the calculation method.
Groundwater influx was first considered as a separate component of water balance by Kaplinskii and Timchenko [24] who showed that in the water balance averaged over many years the groundwater component makes about 30% of income.

Ratkovich [25] considered water balance using calculations by S. Presnetsova, which were based on extensive observation data, including measurements from aboard a ship on the lake. However, the value of water loss for irrigation, which is 2/3 of withdrawal, appears strongly overestimated.
The most reliable water balance was suggested by Krivoshei and Gronskaya [26][19] for the period from 1935 to 1978. In this work, based on many years of field observations, the authors analyze details of the formation of surface runoff and its transformation in the lake plain, consider separately the underground component of water income, analyze the water loss related to withdrawal for irrigation, and estimate evaporation by four independent methods. The income component of the water balance is made by precipitation (274 mm), surface runoff (295 mm), groundwater runoff (299 mm); the output component consists of evaporation (836 mm) and withdrawal loss for irrigation (78 mm); the increment is 50 mm, and the residual is 4 mm.
Evaporation from the lake surface was estimated using a floating raft where were installed two evaporators, a rain meter, and instruments for measuring water temperature, air humidity, and wind strength. This raft, set up about 100 m offshore, was used for many years for evaporation monitoring at the lake station Cholpon-Ata. As a rule, floating evaporation measuring systems do not run at high waves, as water overflows the evaporation tanks. The gaps in data are compensated in this case by calculations, using empirical relationships by Meyer-Tikhomirov, Ivanov, Reizvikh and Popova, Molchanov-Laptev, Rymshi-Donchenko, etc. Romanovsky [27] obtained a regional formula of evaporation, which looks as:

$$E = 0.19n\,\Delta e_{200}\,(1 + 0.40\,U_{200}),$$

where Δe_{200} is the difference between partial pressure of water vapor 200 cm above the water surface and the pressure of saturated vapor at the temperature of the water surface; U_{200} is wind strength at 200 cm above the water surface, in m/s; n is time in days.
Evaporation strongly varies over the lake. To a first approximation it may be divided into two main zones, the littoral and the pelagial zones. In a narrow (1-2 km) littoral zone evaporation is strongly controlled by the influence of land: air is weakly saturated with water, wind transforms at the shore-water transition. In the open lake air is relatively saturated with lake water, the wind profile is uniform, as it is not influenced by underlying surfaces. Evaporation in the littoral zone strongly exceeds that in the open lake. In summer, at still weather, evaporation at 1 km offshore is 20% lower than near the shore. At night, when breeze is seaward, the width of the evaporation zone in the littoral is about 2 km, and in the daytime, when the lake breeze changes its direction, it narrows down to 100 – 200 m.
Estimates of annual evaporation for 1986 [27] show that it is the lowest (620 – 650 mm) in the center of the lake and the highest (930 – 960 mm) in the west of the lake in the Rybachiy Gulf. In the eastern part of the lake, annual evaporation is 780 – 840 mm. Water temperature is an important factor that controls evaporation from the lake surface. Through most of the year (April – October), the temperature of the water

surface is higher in the shallow littoral zone. Another important controlling factor is the eastward increase in humidity.

Variations of water balance in various parts of Issyk-Kul are controlled by space heterogeneity of evaporation, river runoff, and precipitation [28]. Let us illustrate this by an example of 1979. In the lake, conventionally divided into three zones, the annual water balance for 1979 is as follows:

- In the western zone, with an area of 571 km^2, that covers the depths <100 m, influx is 1 l/s or 56 mm/yr, precipitation is 71 mm, evaporation is 939 mm, water level increment is negative, -812 mm or –0.46 km^3/yr.
- In the middle zone, area 4725 km^2, influx was 24 l/s or 160 mm/yr, precipitation 208 mm, evaporation 730 mm, water level increment negative –362 mm or –1.71 km^3/yr.
- In the eastern zone, area 940 km^2, that covers depths <100 m, influx is 47 l/s or 1576 mm, precipitation 337 mm, evaporation 682 mm, water level increment positive +1231 mm, or 1.16 km^3yr.

The area of irrigated land in the Issyk-Kul basin is over 150,000 ha, which is about the area of its water surface. According to data of the Hydrological Institute (St. Petersburg) [26], the loss for irrigation is in average 12% of the influx from the zone of runoff formation, or 45% of withdrawal. The loss for irrigation is so high because it is of gravity type on about 90% of land, and a great portion of water is evaporated and transported outside the lake basin. Moreover, at gravity irrigation fertilizers and pesticides are washed out into the lake. In these conditions, the problem of improvement of irrigation in the lake basin should be solved without additional withdrawal of water from lake tributaries but should be based on water-saving machine irrigation technologies.

The water-balance equation by Krivoshei and Gronskaya [19] in which the loss for irrigation is 0.49 km^3/yr or 78 mm/yr was used in the calculation of the "gravity level" of Issyk-Kul. By "gravity level", we mean the position of the water surface at which the loss for evaporation is compensated by water income with precipitation and river runoff. With the present tendency for lake level drop, the evaporation loss will decrease at the account of contraction of the evaporating surface. Calculations with the use of the morphometric parameters that characterize the lake basin show that with the present-day water withdrawal for irrigation, the area of Issyk-Kul will be 512 km^2 smaller and the level will be 10 m lower. Without withdrawal for irrigation, the surface area may increase by 362 km^2, and the level will be at 1615 m, or 9.1 m above its present position.

5. References

1. Aleshinskaya, Z.V., Bondarev, L.G., Voskresenskaya, T.N., and Leflat, O.N. (1971) *Recent Fill of the Lake Issyk Kul Basin* [in Russian], Moscow University Press, Moscow, 164 p.

2. Shnitnikov, A.V. (1979) *Lake Issyk Kul: Nature, Protection and Economic Prospects* [in Russian], Ilim, Frunze, 196 p.
3. Berdovskaya, N. and Egorov, A.N. (1986) Reconstruction of environment at the Pleistocene-Holocene boundary [in Russian], in Sevastianov, D.V. and Smirnova, N.P. (eds.), *Lake Issyk Kul and its Evolution*, Nauka, Leningrad, 165-179.
4. Sevast'yanov, D.V. (1991) Historic and archaeological evidence of changes in the shoreline and water level in Issyk Kul [in Russian], in Treshnikov, A.F. (ed.), *The History of Lakes Sevan, Issyk Kul, Balkhash, Zaisan, and Aral*, Nauka, Leningrad, 90-96 p.
5. Trofimov, A.K. and Trunilin, S.I. (1991) Quaternary tectonics of the Issyk Kul basin [in Russian], in Chedia, O.K. (ed.), *Tien Shan at the Recent Stage of its Geological History*, Ilim, Bishkek, 58-66.
6. Melnikova, A.P. (1987) *The Holocene History of Vegetation in the Northern and Central Tien Shan (from Spore and Pollen Data)* [in Russian], Abstract, Cand. Dissertation, Moscow, 23 p.
7. Ranov, V.A. (1968) Glaciation and Paleolithic man in Central Asia [in Russian], in Zabirov, R.D. (ed.), *Advances of Glaciology in the USSR*, Ilim, Frunze, 421-429.
8. Yunusaliev, M.B. (1967) Paleolithic time in Kirghizia [in Russian], *Izv. AN Kirg. SSR*, N 4, 91-96.
9. Umurzakov, S.U. and Vinnik (1975) Historic and archaeological evidence of water level fluctuations in Lake Issyk Kul [in Russian], in Umurzakov, S.U. (ed.), *Problems of Geography of Kirghizia* (Proc. II Congr. of the Kighiz Geographical Society), Ilim, Frunze, 208-211 .
10. Bondarev, L.G. (1978) Water level fluctuations, from archaeological evidence and historic accounts [in Russian], in Koralev, V.G. (ed.), *Lake Issyk Kul*, Ilim, Frunze, 112-117.
11. Zabirov, R.D. (1978) Water-level changes since the first quarter of the 19th century [in Russian], in Koralev, V.G. (ed.), *Lake Issyk Kul*, Ilim, Frunze, 98-102.
12. Berg, L.S. (1930) Hydrological studies in Lake Issyk Kul in 1928 [in Russian], *Bull. Mining Institute* Vol. 28, 14-37.
13. Matveev, V.P. (1935) Hydrological studies on Lake Issyk Kul in 1932 [in Russian], in Berg, L.S. (ed.), *Lake Issyk Kul: Hydrology, Ichthyology, and Fish Production*, Proc. Kirg. Special Expedition of 1932-1933, AN SSSR Press, vol. 111, issue 2, Moscow-Leningrad, 7-56.
14. Sapozhnikov, D.G. and Viselkina, M.A. (1960) *Recent Bottom Fill in Lake Issyk Kul and its Gulfs* [in Russian], AN SSSR Press, Moscow, 159 p.
15. Chalov, P.I. and Tuzova, T.P. (1968), Dating of natural sites from uranium series disequilibrium, *Bull. Commission for Absolute Age Determination of Rocks*, issue 8, AN SSSR Press, Moscow, 177-181.
16. Ricketts, R.D, Johnson, T.C., Brown, E.T., Rasmussen, K.A., and Romanovsky, V.V., Holocene paleolimnology of Lake Issyk Kul (Kirghizstan): contents of trace elements and stable isotopes in ostracodes (in press).
17. Mamatkanov, D.M., Dikikh, A.N., Romanovsky, V.V., Suyunbaev, M.N., and Shambetov, Z (1997) Current dynamics of air temperature and the response of

different water resources of the Issyk Kul basin [in Russian], *Ekho Nauki* N 2-3, 96-101.

18. Dikikh, A.N. (1986) Dynamics of glaciers in the basin of Lake Issyk Kul through the 20th century, in Sevastianov, D.V. and Smirnova, N.P. (ed.), *Lake Issyk Kul and its Evolution*, Nauka, Leningrad, 59-68.
19. Krivoshei, M.I. and Gronskaya, T.P. (1986) Water balance of Lake Issyk Kul [in Russian], in Shnitnikov, A.V. (ed.), *Studies of Large Lakes in the USSR*, Nauka, Leningrad, 276-280.
20. Romanov, V.V., Vlasova, L.S., Meskheteli, A.V., and Salnova, L.V. (1989) Formation of waters in the Issyk Kul basin: results of tritium studies [in Russian], *Vodnye Resursy* N, 82-95.
21. Smirnova, N.P. (1986) Water balance of Issyk Kul: state of the art [in Russian], in Sevastianov, D.V. and Smirnova, N.P. (ed.), *Lake Issyk Kul and its Evolution*, Nauka, Leningrad, 69-78.
22. Vikulina, Z.A. (1946) Water balance [in Russian], in Molchanov, I.V., *Lake Issyk Kul*, Gidrometeoizdat, Leningrad, 130-145.
23. Tsigelnaya, I.D. (1960) Water balance of Lake Issyk Kul [in Russian], in Zabirov, R.D. (ed.), *Glaciological Studies in Tien Shan*, Ilim, Frunze, iss. 2, 91-96
24. Kaplinsky, M.I. and Timchenko, L.S. (1977) Water balance of Lake Issyk Kul over many years [in Russian], *Bull. SARNIGMI* iss. 50 (131), 87-101.
25. Ratkovich, D.Ya. (1977) Water balance and lake level fluctuations in Issyk Kul [in Russian], *Vodnye Resursy* Vol. 5, 20-33.
26. Gronskaya, T.P. (1983) *Water balance and anticipated water levels in Lake Issyk Kul* [in Russian], Abstract, Cand. Dissertation, Leningrad, 16 p.
27. Novak, V.A. and Romanovsky, V.V. (1988), Evaporation from the surface of Lake Issyk Kul in 1932 [in Russian], in Korataev, V.N. (ed.), *Physiographic Studies of Lake Issyk Kul and its Surroundings*, Ilim, Frunze, 52-73.
28. Cherkasov, L.G. (1975) Short-period water balance of Lake Issyk Kul [in Russian], in Umurzakov, S.U. (ed.), *Geography of Kirghizia,* (Proc. XI Congr. of the Kighiz Geographical Society). Ilim, Frunze, 88-90.

HYDROMINERAL RESOURCES OF THE ISSYK-KUL REGION

V.E. MATYCHENKOV
Institute of Water Problems and Hydropower
533, Frunze str., Bishkek, Kyrgyzstan

1. Hydromineral resources of the Issyk-Kul region: general background

The hydromineral resources of the Issyk-Kul region include the lake itself, small satellite lakes: Kara-Kul (or Karakel), Akterek and Kadji Sai, a dozen deposits of balneal mud, and over twenty occurrences of mineral water (Fig. 1), most often thermal water (thermomineral water, TMW), which is the main focus of this paper.

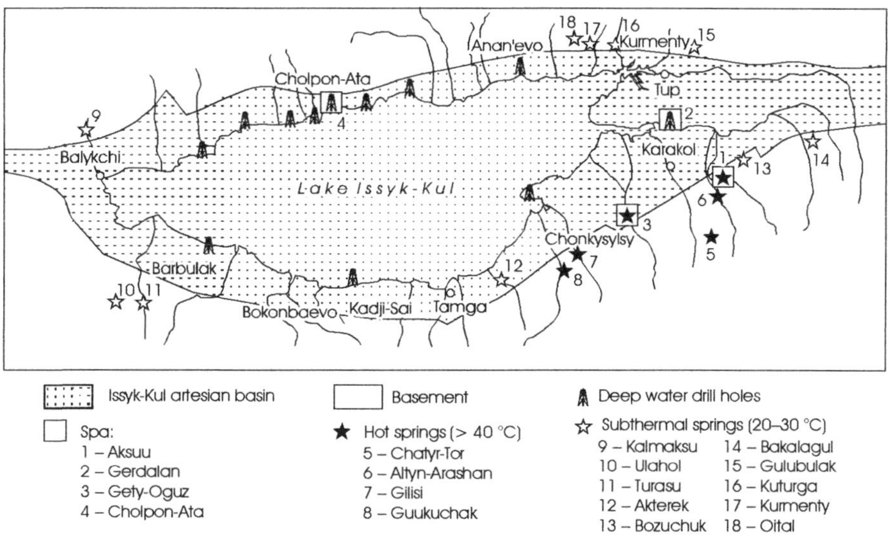

Figure 1. Hydromineral resources of the Issyk-Kul basin.

The Issyk-Kul artesian basin which is the principal hydrogeological unit of the Issyk-Kul region, is structured in three units. The upper unit consists of 450 m thick Quaternary and Pliocene-Quaternary water-bearing sediments composed of boulders and pebbles as well as alluvial pebble beds of piedmont plains and stream valleys. It shows active water exchange, predominantly of fresh water, which constitutes the

J. Klerkx and B. Imanackunov (eds.), Lake Issyk-Kul: Its Natural Environment, 59–70.

greatest portion of the economic water resources of the region (including drinking and irrigation groundwater). The fresh water resources are the main component of groundwater drainage of Lake Issyk-Kul with a discharge of 60 m^3/s.

The 4500 m thick middle unit is made up of Mesozoic to Upper Pliocene stratified terrigenous deposits (mostly fine-grained clay, silt, sand, less often conglomerates and salt-bearing deposits) containing stratal water aquifers. Drilling strips thermal mineral water ($T > 40°$ C, mineralization 0.35 – 45 g/l) from the depths of 800 – 1600 m. The TMW chemistry displays a lateral zonation and anomalies related to the fault block structure of the Meso-Cenozoic sedimentary fill of the artesian basin. Salt and water saturation of various blocks depends on their lithology and geological evolution. The elevated temperature of water is related to enhanced geothermal gradients of 2.0 to 3.5 °C/100 m in drill holes to depths below 2000 m, from 2.3 to 3.5 °C/100 m in holes attaining a depth of 4900 m, and somewhat higher in blocks on the northern side of Lake Issyk-Kul. Hydrothermal anomalies are also associated with low-mineralized nitric water that originates at great depths and discharges in springs along faults.

The lower unit within the artesian basin is composed of Proterozoic to Upper Paleozoic bedrocks, which are faulted metasedimentary and igneous rocks whose equivalents are exposed in the ridges around the lake. Fresh infiltration water is most often associated with weathered or faulted rocks, though cold leaching brines may form in recently exposed blocks as NaCl water, as for instance water with a salinity of 100 g/l discharging in the eastern part of the region near the Djergalan coal mine.

Associated with the same rocks is also vein-fracture water, locally thermal, whose discharge sites often mark greater deposits of thermal mineral water. The total daily discharge of thermal fracture water (>40° C) in the Issyk-Kul region is 60 l/s [1].

2. Typical TMW deposits in the Issyk-Kul region

The numerous hydromineral sites of the Issyk-Kul region are of a few balneal water types: water free from specific balneal components, cold water of very limited use at the Issyk-Kul shore spas, nitric, and radon thermal waters. The Issyk-Kul region apparently lacks acidulous water, which is of broad occurrence elsewhere in Kyrgyzstan. Moreover, the Issyk-Kul basin lacks oil and gas, and thus sulfide water that is typical for the Ferghana region in western Kyrgyzstan, is absent.

The *Djety-Oguz deposit* is located at an altitude about 2000 m a.s.l on the northern slope of the Terskei-Alataoo ridge 30 km southwest of the Karakol mountain and 25 km away from the Issyk-Kul shore (Fig. 2). The deposit is unique by its hydrogeological and balneal properties: it shows extremely strong variations in temperature, radioactivity, and mineralization (about an order of magnitude or more over less than 1 km^2 in area and less than 100 m in depth), having no match elsewhere in the world. The deposit is attributed to a double-plunging fold in Meso-Cenozoic sediments, in the core of which are exposed intensely faulted Paleozoic and Proterozoic rocks. The evidence from high-resolution aerial photographs suggests that the Djety-Oguz hydrotherm is located on the NW contour of a crater-type circular structure about 10 km in diameter.

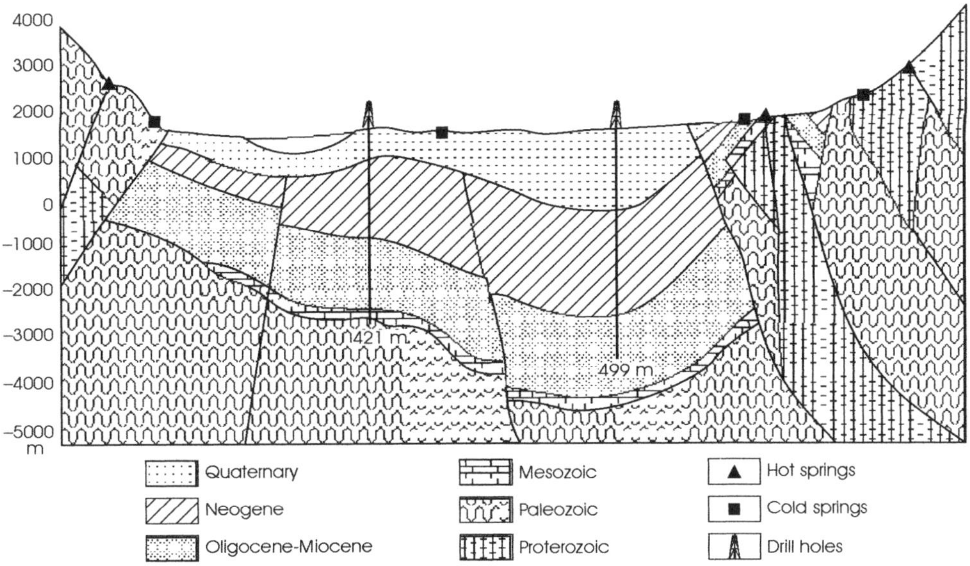

Figure 2. Hydrogeological cross section of the Issyk-Kul basin.

The hottest mineral and highly radioactive water was encountered by drilling at a depth of 66.5 m in Carboniferous limestones and in the overlying alluvium, at the intersection of a small W-E striking fault with the major fault followed by the Djety-Oguz valley.
The thermal waters are of four main balneal types, distinguished on the basis of their composition, temperature, radioactivity, and mineralization [2]:

(1) low-thermal (20 – 25° C), low-mineralized (less than 1 g/l) high-Rn (about 50 nKu/l) water (well 15), with a chemistry of:

$$M_{0,8} \quad \frac{Cl_{62}\ SO^{4}{}_{27}}{(Na{+}K)_{65}\ Ca_{30}}$$

(2) low-thermal (about 25° C), medium-mineralized (under 3 – 4 g/l), medium-to-high radioactive (from 35 to 70 nKu/l) (water (well 5):

$$M_{4,1} \quad \frac{Cl_{85}\ SO^{4}{}_{15}}{(Na{+}K)_{50}\ Ca_{45}}$$

(3) thermal (35 – 37° C), mineralized (9 – 11 g/l), low-radioactive (about 10 nKu/l) water (well 6):

$$M_{4,1} \frac{Cl_{90}\ SO^4{}_7}{(Na+K)_{55}\ Ca_{44}}$$

(4) thermal (35 – 37° C), mineralized (9 – 11 g/l), low-radioactive (about 10 nKu/l) water (well 1-k):

$$M_{2,6} \frac{Cl_{95}}{(Na+K)_{55}\ Ca_{44}}$$

The mineralization and Cl-ion content of all these waters fit the mixing curve, i.e., the Ca-Na chloride water with a mineralization 10 – 15 g/l typical of the Issyk-Kul region, is diluted to various degrees with the ultrafresh groundwater typical of the Djety-Oguz valley. The original high chloride enrichment and temperature of the water provides intense Ra leaching from acid intrusions that occur within the zone of its circulation, and the disturbance of carbonate equilibrium by mixing of thermal and fresh cold water leads to leaching of calcium and radium carbonates. It is radium that provides the high emanation properties of carbonate-rich rocks and Quaternary alluvium at relatively shallow depths, where the carbonate equilibrium is more disturbed.

The dissolved gas composition of TMW, recalculated into air-free TMW, is shown in Table 1.

The Ar/N_2 ratio of 0.83 and 0.45 (2.6 if corrected for solubility difference) indicates that the low-mineralized water contains atmospheric gases in concentrations twice as high as the higher mineralized less diluted water.

TABLE 1. Dissolved gas composition of groundwater of Djety-Oguz deposit

Sampling site	Gas concentration cm^3/l	Gas content, vol. ‰					
		N_2	CO_2	CH_4	Ar	He	Total
Well 5	26.0	92.31	2.15	1.77	2.00	1.76	99.99
Well 6	21.5	86.22	6.05	3.52	1.85	2.35	99.99

The Djety-Oguz deposit is used for a spa of the same name in which the balneal effect of its unique radon water is fortified by a pictoresque landscape and climate of medium-elevated mountains.

The *Cholpon-Ata deposit* is located on the northern shore of Lake Issyk-Kul over an area from village Sugetty in the east to village Karaoi in the west. The first well (N 725) that reliably reached stratal TMW, at a depth of 1375 m, was drilled for the "Avrora" spa that was then under construction in the eastern segment of the deposit, 2 km southeast of a granitic bedrock exposure at the Prishib mountain and 0.4 km away from the lake shore. Further deep drilling in the territory of the spas "Goluboi Issyk-Kul",

"Ala-too", and "Kazakhstan", and then throughout the northern shore from village Toru-Aigyr in the west to village Chon-Uryukty in the east, revealed ubiquitous nitric TMW of various rates of mineralization.

The technical and hydrogeological characteristics of the deposit can be illustrated by data from well 773 drilled on a low terrace in Cholpon-Ata town within the territory of the "Goluboi Issyk-Kul" spa.

The drilled section is as follows:

0 – 200 m: Quaternary boulder-pebble beds with sand interbeds;

200 – 400 m: Pliocene-Lower Quaternary boulder-pebble and pebble beds weakly cemented with clay-carbonate;

400 – 590 m: thinly intermittent Pliocene-Lower Quaternary sandstones, gravelstones, and conglomerates with thin fine-grained interbeds;

590 – 1506 m: intermittent Pliocene sandstones and argillaceous clays with scarce gravelstone interbeds (Upper Issyk-Kul Formation).

A water flow of 6.0 l/s, 48° C, and a chemical composition of

$$M_{20,3} \frac{Cl_{92}\, SO^{4}{}_{7}}{(Na+K)_{64}\, Ca_{35}}$$

was obtained from a filtration column 108 mm in diameter at perforations every 3 to 13 m in the depth interval 1135.9 – 1380.4 m, at a static level of 8.9 m and a lowering of 46 m.

The trace element composition of the Cholpon-Ata TMW is as follows (in mg/l): 0.6 – 2.0 F, 1.0 – 1.2 I, 8.0 – 14.4 Br, 0.002 – 0.004 Cu, <0.002 As, <0.005 Pb, 0.001 – 0.015 U.

The contents of radioactive elements in the water from well 773 are: $(18.5 \pm 3.7)\cdot 10^{-3}$ Bq/l ^{226}Ra, 32.18 ± 14.8 Bq/l ^{222}Rn, and 1.30 ± 0.03 $^{234}U/^{238}U$, according to estimates of the Radiometric Laboratory of the Institute of Physics and Mathematics NAC KR.

Exploration work, including test drilling of more than ten deep wells, has reliably confirmed the possibility for extraction within the Cholpon-Ata deposit, at a daily discharge of 1736 m^3 of 40 – 55° C water, with a composition described by

$$M_{[15-40]} \frac{Cl_{[80-90]}\, SO^{4}{}_{[5-20]}}{(Na+K)_{[60-80]}\, Ca_{[20-35]}}$$

A similar section is crossed by deep drilling for water elsewhere on the Issyk-Kul shore, but the drilled TMW shows a broad mineralization range, from saline to fresh water. Low-mineralized water is poor in Br and rich in F, As, and some major elements, which may indicate possible influx of nitric water from the basement.

Note that about twenty exploration and test wells for oil and gas, to a depth of 4930 m, have been drilled within the territory. One well in the southeastern part of the lake plain reached the crystalline basement at a depth of 4210 m. However, the hydrogeological sampling was of low quality and is thus poorly informative in almost all wells, except for a test well near the Djergalan river mouth. In 1960, a TMW flow was obtained from this well, from the depth interval of 863 – 1521 m, for the first time in the Issyk-Kul region, and was used for the Djergalan spa. The initial discharge of >43° C water was 37.5 l/s and then reduced to 20 l/s. There were also cases of increased initial discharge of a well that strips groundwater in bedrock or in faulted sediments. For instance, well 1111 that collected TMW with a mineralization of 18.4 g/l on the Issyk-Kul shore in the vicinity of village Kadji Sai, from a depth interval of 975 – 1306 m. In 1991, 36° C water discharged at a rate of 1.1 l/s and a pressure of 0.3 atm above the surface and in October 2000 the dsicharge increased to 6 l/s at a pressure of 2 atm, and water temperature was 56.7° C.

The chemistry of the *Djergalan* water, which has been monitored for 40 years, is described by the general formula

$$M_{0,9} \frac{Cl_{[62-72]}\,SO^{4}_{[18\ 30]}}{(Na+K)_{[85-89]}\,Ca_{[8-10]}}$$

This water is remarkable for its high F concentration of 6 – 7 mg/l, which makes it somewhat similar to vein-fracture low-mineralized nitric TMW typical of the mountainous part of the Issyk-Kul basin. The concentration of heavy hydrogen in this water is 0.98 Tashkent tap water standard [4]. On the basis of this parameter, the Djergalan water is classified as light water of infiltration origin. The dissolved gas composition of stratal TMW of the Issyk-Kul region is shown in Table 2.

TABLE 2. Dissolved gas composition in stratal TMW in Issyk-Kul artesian basin (in vol. ‰)

Well and location	725 Sugetty	773 Cholpon-Ata	846 Cholpon-Ata	999 Spa "Kazakhstan"	1-A Djergalan
Salinity, g/l	15.7	20.3	15.1	36.0	0.92
Temperature, ^{0}C	42.8	48.3	45.0	53.0	53.4
Gas concentration, cm^3/l	18.3	24.4	15.0	14.2	13.0
Carbon dioxide	8.25	3.35	4.4	2.31	20.86
Hydrogen sulfide	-	-	-	-	-
Methane	0.77	0.13	0.25	0.21	0.05
Hydrogen	0.167	<0.001	0.014	<0.001	0.0023
Nitrogen	82.96	86.38	82.88	84.99	75.87
Argon	1.62	2.20	1.50	1.52	1.80

Helium	1.403	0.1575	0.3389	0.7028	0.005
Oxygen	4.82	7.78	10.61	10.26	1.40
Total	99.99	99.99	99.99	99.99	99.99

Groundwater discharge of five thermal (>36° C) springs and about 10 occurrences of low – thermal (20 – 36° C) water exist in the mountains around the lake. At the Ak-Suu deposit, there is a children spa, which successfully treats children after poliomyelitis; at the Bozuchuk, Djuukuchak, and Djilisu springs there are medical and recreational balnearies.

The *Altyn-Arashan site* (Kirghiz for "Gold balneal groundwater") shows the best possibilities for future development. It includes three groups of nitric thermal springs on the right side of the Arashan river valley 20 km southeast of Karakol mountain and is situated in a pictoresque forest landscape at an altitude of 2350 – 2435 m (Fig. 3).

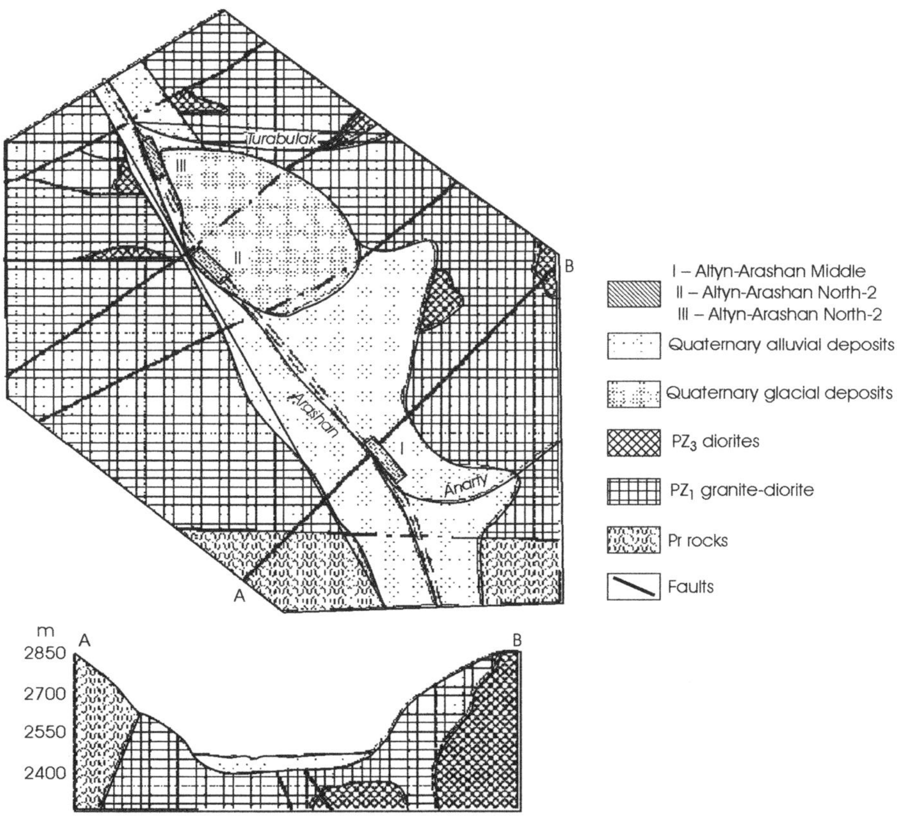

Figure 3. Generalized hydrogeology of the Altyn-Arashan thermal water deposit.

The three sites line up in NW direction along a zone of Upper Ordovician faulted granitoids that steeply dip northeast. The position of the hydrotherms is controlled, to a certain degree, by a transversal system of nearly parallel faults associated with Upper Paleozoic (?) diorite dikes and small stocks.

No drilling has been as yet performed in the site. The composition of the least diluted springs with temperatures of 44 to 51° C is described by the formula

$$M_{(0,28\text{-}0,35)} \frac{Cl_{(40\text{-}45)}\ HCO^{3}{}_{(38\text{-}43)}}{(Na+K)_{(85\text{-}90)}\ Ca_{(9\text{-}14)}}$$

The Altyn-Arashan water is remarkable for a high H_2S concentration (8.5 mg/l), elevated content of silicic acid (up to 60 mg/l), extremely high concentrations of fluorine (12 – 16 mg/l) and rare alkalis (up to 0.3 mg/l Li, and up to 0.1 mg/l Rb and Cs). The springs that occur within the site show a general relationship between temperature and concentration of total H_2S, Cl and Na ions, F, Li, Sm, Zn, W, and total mineralization, which may reflect the composition of original nitric water flows that discharge there through a zone of groundwater.

The hottest spring is accompanied by gas release with a gas factor 0.007. Its composition is shown in Table 3.

TABLE 3. Gas composition in Altyn-Arashan springs, in vol.‰ (Analyzed at Gas Laboratory, Geological Surveys, Moscow)

Sampling date	16/09/1974	04/09/1983	04/09/1983
Gas type	Spontaneous	Dissolved	Spontaneous
Gas concentration, cm^3/l	7.0	11.8	7.0
Carbon dioxide	1.24	2.39	<0.01
Hydrogen sulfide	Not found	Not found	Not found
Methane	<0.02	0.73	<0.16
Hydrogen	<0.001	0.041	<0.001
Nitrogen	93.44	90.87	96.80
Argon	1.57	2.75	1.86
Helium	0.5375	0.2671	0.8437
Oxygen	3.21	2.94	0.33
Total	99.99	99.99	99.99

The Altyn-Arashan water, as well as other similar nitric springs in the Issyk-Kul region, have elevated concentrations of He, which reflects the great depths of water circulation. The daily discharge of thermal water at the Altyn-Arashan site is at least 300 m^3. In the area of the springs there is a local balneary with two water pools. Being located in favorable physiographic conditions near a regional center and accessible due to a drive

way off the main road, the Altyn-Arashan springs, with their resources and quality of water, show good prospects for various uses, including local house heating.
The *low-thermal springs* of the Issyk-Kul region discharge mostly groundwater-diluted nitric water of deep origin. At the same time, there are three sites (Aksai, Turasu, and Ulakhol) accompanied with carbonate tuff deposits (travertine), which indicates the presence of acidulous water in the region.
The *Turasu site*, considered here as an example of a site associated with travertine, is located on the northern slope of the Terskei-Alatoo ridge 28 km south of Lake Issyk-Kul, 2 – 2.5 km upstream the Turasu valley, in medium-to-high mountains at an altitude of 2810 – 2890 m, and includes about twenty thermal springs divided into an Upper and a Lower groups. The springs discharge at intersections of a major ancient N-S striking fault with recent NW faults within a faulted terrain of the Proterozoic Taldysui Formation composed of limestones, micaschists and quartzite lenses and interbeds, overlain by Quaternary glacial deposits. At a mean annual air temperature about 2° C, the temperature of discharged groundwater attains 21° C. The total discharge of the two hydrotherm groups spaced at 0.5 km is 30 l/s, with a discharge of individual flows attaining 8 l/s. The total heat release by these springs is 460 ccal/s, which is equivalent to a daily amount of 4.7 tons of conventional fuel. At two hottest springs there are baths occasionally used by shepherds and tourists.
The chemistry of the springs of the Upper group is described by the formula

$$M_{0,32} \frac{HCO^{3}{}_{86}}{Ca_{62}\ Mg_{22}} \qquad T\ 20{,}7^{0}\ C$$

That for the Lower group springs is:

$$M_{0,32} \frac{HCO^{3}{}_{73}\ SO^{4}{}_{19}}{Ca_{60}\ Mg_{26}} \qquad T\ 20{,}7^{0}\ C$$

The Turasu springs are the highest radioactive of all low-thermal springs in the Issyk-Kul region, as they contain 95 – 145 Eman/l Rn (measured at different periods of time by different radiologists). We observed an interesting fact when radon concentration in a spring located in the Ulakhol site, 5 km southwest of Turasu, with a similar hydrogeological structure and water properties, first increased from 8 to 15 eman/l before the M=4 Kochkor earthquake and then reduced to the background level. A similar pre-seismic behavior can be hypothesized for the Turasu springs but no appropriate observations were carried out. The composition of the Turasu water is as follows (in mg/l): <0.005 Co, <0.005 Zn, <0.002 Cu, <0.001 As, <0.005 Pb, 0.05 B, <0.005 Cr^{3+}, Cr^{6+}, <0.001 Ge, 0.3 Li_2O, 0.5 – 2.4 F, 0.5 I, <0.1 Br, 0.0027 U. The concentrations of Sr, Mo, Sn, Cb, and Zn, elevated against the background values, were determined by spectral analysis of dry residual.

The Turasu springs contain elevated concentrations of carbon dioxide, which is responsible for intense travertine formation at the discharge, associated with an abrupt disturbance of the carbonate equilibrium and precipitation of excess $CaCO_3$.
Being located in an inhabited terrain and having an ordinary water composition, the Turasu springs show poor prospects for balneal uses. At the same time, high radioactivity of the water, active and long-existing travertine formation, possible correlation of water composition and regional seismicity, as well as relatively easy accessibility makes this site, along with the neighboring Ulakhol site, an attractive scientific target, largely visited by geologists since the beginning of the 20th century.

3. Balneal mud

Balneal mud is of especially wide occurrence on the shores of Issyk-Kul, though some deposits occur outside it (e.g., the Konurulen deposit 7 km south of the lake). Mud deposits immediately related to the modern or higher stand of the lake have been found in the regions of Pokrovskoye, Djergalan (two sites), Kamensk, Kurmenty, Ananyevo, Cholpon-Ata, and Dolinkino backwater and gulfs. They consist of ordinary terrigenous silty H_2S-rich saline mud suitable for various balneal uses with a subsequent regeneration. Their organic content rarely attains 7%, which determines the character of their biological activity.
Geologically explored sites occur on the surface or at depths within 10 m below it; the thickness of productive beds is from 0.3 to 4 – 6 m. The resources of small sites are no more than 25,000 – 30,000 m^3, though major occurrences (Djergalan group, Pokrovskoye) amount to 1,000,000 m^3 and more. In total, 3,000,000 m^3 has been explored. Balneal mud is used at almost all spas of the Issyk-Kul region.

4. Satellite lakes

There are several satellite lakes near Issyk-Kul. The best known are Karakel Akter and Karakul Kajisai. The former, the most saline lake of Kirghizstan, is located 0.25 km south of Issyk-Kul, 25 km southwest of village Bokombaevskoye, and 3 km east of the Akterek mouth. Its surface area varies yearly from 6 to 3 ha while its southern boundary departs shifts for more than 150 m between the high-stand level and the lowest-stand in the cold season. The lake is chiefly fed by groundwater, most intensely in its northeastern part (3 – 5 l/s), and from ephemeral streams, especially in spring or during heavy rainfall.
Given that the salinity of the Issyk-Kul water is about 6 g/l (see Karmanchuk, this volume), that the salinity of groundwater varies from 3.2 g/l (measured in July 1985) to 4.2 g/l (measured in October 1999), and that there is a complete coincidence of water chemistry in Issyk-Kul and Karakel, we may infer that the water in Karakel is that from Issyk-Kul 1.5 – 2 times diluted with fresh water.

The water in Karakel is variable in composition and mineralization during the year (see Table 4, sample 92 from the main lake and sample 91 from its separated southern shallow basin in which mirabilite and/or tenardite crystals precipitate in amounts about 0.4 kg/m^2).

TABLE 4. Water chemistry of Lake Karakel Akterek as a function of time

Composition	Sampling date 01.07.1985		06.09.1989		18.10.1999 sample 92		18.10.1999 sample 91	
Results	mg/l	‰-equ	mg/l	‰-equ	mg/l	‰-equ	mg/l	‰-equ
Na+K	14690	52	36600	72	198577	57	46096	76
Mg	6962	46	6850	26	7381	40	7430	23
Ca	58	2	880	2	720	3	701	1
F	N.d.	-	N.d.		6.3	-	22.8	-
Cl	39998	91	41400	53	44375	83	46150	49
SO 4	4629	8	49000	47	11974	16	63988	50
HCO 3	976	1	907	-	488	1	1049	1
Total ions	67401		135240		84712		16559 8	

Increase in salinity leads to changes in mineralisation type from Mg-Na chloride to Na chloride-sulfate. The modern chemistry of Karakel is apparently a natural model of what Issyk-Kul may be in the geological future if it keeps its closed basinal regime, and if 10 to 25 % of its volume of water is evaporating.
The total content of salts in Karakel is about 20,000 t. It is a unique natural site attractive for tourists and promising as a source of bishofite solution.

5. References

1. Matychenkov, V.E. and Imankulov, B.I. (1987) *Mineral Waters in Kirghizia* [in Russian], Ilim, Frunze.
2. Kataeva, N.I., Melnikova, Z.I., Barsutskaya, R.D., Fomin N.G. (1969) *Mineral Groundwaters in Kirghiz SSR* [in Russian], Kyrgyzstan, Frunze.
3. Vasilieva, V.N. (1960) Formation of Dzhety-Oguz radioactive thermal springs [in Russian], in V.V.Ivano (ed), *Formation and Distribution of Mineral Waters in the USSR Territory*, Gosgeoltehizdat, Moscow, pp. 47-50.
4. Sultankhodzhaev, A.N., Latipov, S.U., and Ibragimov, D.S. (1972) Disequilibrium U in groundwaters of some artesian basins in Central Asia [in Russian], in G.A.Mavlianov (ed), *Isotope Composition of Groundwater*, GIDROINGEO, Tashkent, pp. 53-133.

5. Volkenshtein, S.K. (1901) Ak-su hot springs [in Russian], В.И. Андрусов *Mineral Waters of the Semirechensk Region*, Kazan', pp. 3-19.
6. Prochukhan, D.P. (1935) Ak-su and Dzhety-Ogyz hot springs (Kirghizia) [in Russian], in M.M,Gatalskiy. (ed), *Geology and Geochemistry of Tien Shan*, Proc. Kirg. Special Expedition of 1932-1933, vol. II, part V, Izd. AN SSSR, Moscow, pp. 155-184.
7. Kopotilov, Yu.P. (1971) Mineral and thermal waters of the Issyk Kul basin and prospects for discovering new deposits [in Russian], in E.Pressman (ed), *Geology and Engineering Geology in Kirghiz SSSR*, Kyrgyzstan, Frunze, pp. 44-48.
8. Belev, A.V., Romankulov, B.A., and Tarasov, S.A. (1978) Thermal mineral water on the shore of Lake Issyk Kul [in Russian], *Razvedka i Okhrana Nedr* **9**, 48-50.
9. Matychenkov, V.E., Matychenkova, E.A., and Aganov, Kh.T. (1976) Comparative activity of natural hydrothermal vents in the Issyk Kul region [in Russian], in . A.N.Sultanhodgarv (ed), *Regional Hydrogeology and Engineering Geology in Central Asia*, Publishing House "Fan", Tashkent, pp. 81-87.

GROUNDWATER IN THE ISSYK KUL BASIN

A.N. MANDYCHEV
Institute of Water Problems and Hydropower,
533, Frunze str., Bishkek, Kyrgyzstan

The groundwater system of the Issyk Kul basin is limited by the divides of Terskei Ridge in the south, Kungei Ridge in the north, Santash pass in the east, and Karatash sill in the west. A plane interface within the Terskei and Kungei Ridges is marked by a groundwater-filled highly permeable fractured zone in Paleozoic bedrock. The deepest faults of this zone may channel crustal and mantle fluid flows into the basin.

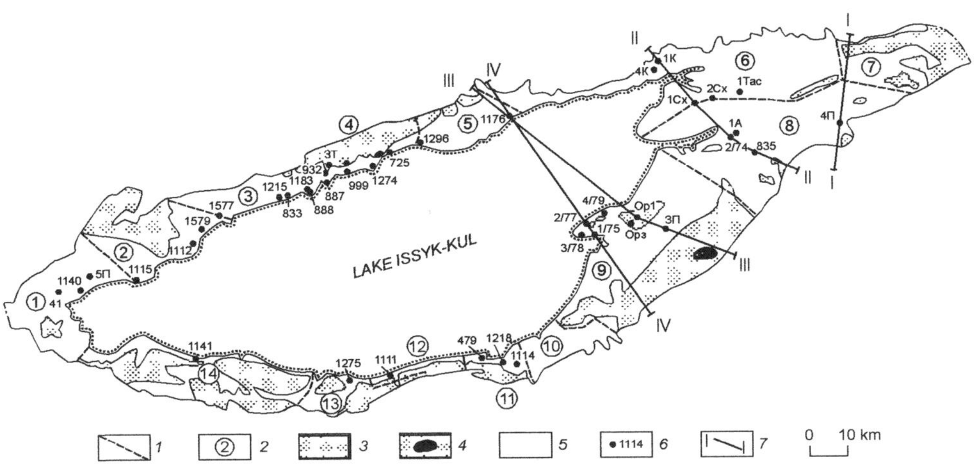

Figure 1. Issyk Kul Mesozoic-Cenozoic groundwater system

1. Boundaries of groundwater districts;
2. Number of groundwater district ;
3. Exposures of Mesozoic – Cenozoic groundwater aquifers in sediments;
4. Bedrock exposures;
5. Quaternary deposits;
6. Test well;
7. Hydrogeological profile

J. Klerkx and B. Imanackunov (eds.), Lake Issyk-Kul: Its Natural Environment, 71–76.

The bedrock fractured zone is buried under a Mesocenozoic sedimentary cover composed of pebbles, sands, clays, and cemented sand, clay, and siltstone conglomerates (Fig. 1). Groundwater circulates within the sediments through the most permeable coarse clastics with a porosity reaching 10-20 %, far above that of the basement. The sedimentary cover, attaining a thickness of 5 km in the east of the basin (Fig. 2), is impregnated with groundwaters of various salinity and residence ages, mostly of meteoric origin. The youngest groundwater occurs in the ~300 m thick Quaternary section of the least consolidated coarsest clastics that are thus the most permeable (infiltration rate 0.04 to 4 m/hr). The Quaternary aquifer is mostly fed with fluvial water infiltrated in the upper pediment parts of alluvial fans. A small portion of water (up to 15 % of total natural renewable groundwater resources of the Quaternary aquifer, which is 62.8-74.4 m^3/s) comes through fractures in the surrounding ridges and in the basement. All groundwater of the Quaternary aquifer discharges into Lake Issyk Kul through littoral and submarine vents on the submerged slopes and lake bottom. This groundwater has a salinity about 1 g/l and mostly bicarbonate-calcium chemistry. The total natural storage of fresh water in the Quaternary aquifer, or the volume of interstitial water in the Quaternary sediments, is 58 km^3 [1].

Therefore, it is the Quaternary groundwater aquifer of the Issyk Kul basin that places the hydrodynamic and hydrostatic constraints on the lake regime. The distribution of groundwater runoff along the lake periphery is structurally controlled by the low pediment sides of the basin composed of Mesozoic and Paleogene-Neogene argillic, often salt-bearing rocks, as well as by the river network [3,4].

Yet, the exchange between the Quaternary groundwater aquifer and Lake Issyk Kul does not account for the full water-exchange cycle which also includes discharge from deeper aquifers (Pliocene-Quaternary Sharpyldak, Paleogene-Neogene Issyk Kul and Kirghiz, and Mesozoic aquifers). The ~ 500 m thick Sharpyldak aquifer with cold fresh bicarbonate-calcium water is less permeable (a few cubic meters per second) and hence poorer in natural renewable resources than the Quaternary aquifer but has a comparable storage capacity. The 1500 – 2000 m thick Issyk Kul aquifer with a permeability of the order of 4-5 m^3/s [3] has however about the same storage capacity as the Quaternary aquifer, as its lower porosity is compensated by a greater thickness. Groundwater in this aquifer is thermal, due to the natural geothermal gradient of 3° C/100 m in average, and has either a fresh bicarbonate-calcium chemistry, at an active (open) water exchange, or is chloride-sodic and calcium with a salinity over 60 g/l, at a passive (close) regime (Fig. 3). The active or passive hydrodynamic regime of the aquifer is controlled by tectonic conditions of sedimentary blocks within the low pediment zone. Passive water exchange is illustrated by anomalously high confining pressure and chloride-calcium brines with a salinity up to 475 g/l revealed in the Kirghiz aquifer by test oil drilling in the B. Orgochor uplift (Fig. 3). This groundwater will eventually discharge into the lake as soon as the isolated system is seismically broken. Residual thermal anomalies detected by geophysical soundings in the Toruaigyr and Karabulun regions might also be associated with thermal water vents on the lake bottom.

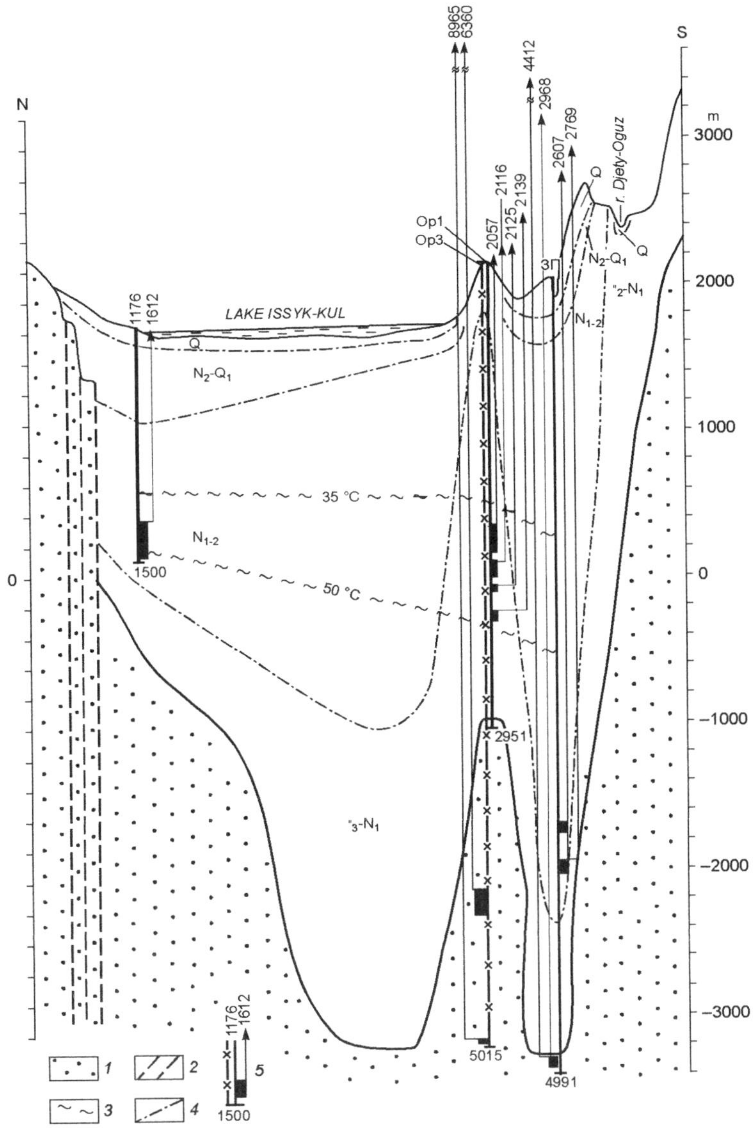

Figure 2. Exposures of basement groundwater aquifers

1. Paleozoic basement
2. Faults.
3. Isotherms
4. Boundaries of geologic units
5. Drill holes indicated by number at the top and depth at the base. Black rectangle is sampling interval.

The Kirghiz and Mesozoic aquifers are mostly passive, i.e., do not receive meteoric water and lack natural renewable water resources. These aquifers enclose old waters with residence ages of hundreds or thousands years. Their storage capacity is an order of magnitude lower than in the Issyk Kul aquifer because of consolidated and cemented wall rocks.

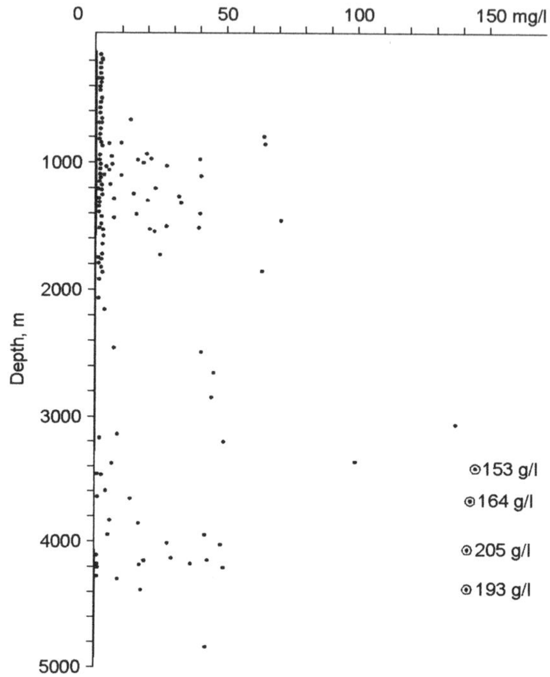

Figure 3. Mineralisation of groundwater

Thus the complete groundwater formation cycle in the Issyk Kul basin starts with atmospheric precipitation accumulated as snow on mountain slopes and percolated as rain through fractured rocks in the mountains. Then it is either channeled by faults to as deep as 3000 m or is vented onto the surface as thermal mineral water (e.g., Djety-Oguz, Chon-Kyzylsu, Teploklyuch deposits, with a total discharge of 0.049 m^3/s [1]). The major portion of percolated groundwater again becomes part of the surface runoff within the ridges, and 15 % circulates within the basin fill.

At the following stage, groundwater in the basin fill is fed from river flow outside the mountains, mostly in the head of alluvial fans and along the streams. There the main portion of the surface runoff is infiltrated at a rate about 50 m^3/s from riverbeds and about 18 m^3/s from irrigation channels. Atmospheric precipitation provides only a minor source of groundwater being restricted to vadose zones (where it infiltrates at a rate about 1.5 m^3/s) and groundwater reservoirs. Therefore, the groundwater within the

Issyk Kul basin is mostly fed from long-existing concentrated surface runoff. Note that ephemeral streams are of very low importance in this respect. Evaporation immediately from the groundwater surface in shallow aquifers (< 3 m) is about a few cubic meters per second [2].
The Paleogene-Neogene aquifers are also mostly fed from rivers within the low pediment where they rise to the surface. Water exchange in these aquifers is strongly controlled by tectonics [3]: active exchange (renewable resources and lower salinity of water) is associated with weakly deformed monoclines that dip toward the basin center, whereas passive exchange (no renewable resources and high salinity) of groundwater is associated with strongly dissected monoclines often dipping towards groundwater reservoirs.
At deep levels of the Issyk Kul aquifer, active water exchange occurs in groundwater districts 1, 2, 3, 5, 6, 7, 8, 9, 10, 11, 14 (Fig. 1), and districts 4, 12, 13 have passive exchange. The groundwater involved in active exchange has a salinity of 1-2 g/l and a pressure corresponding to that in feeding areas; that in the passive regime has a salinity exceeding 20 g/l and a lower pressure.
The water budget of the Issyk Kul basin includes a static and a dynamic components. The static component exists as glaciers and snow fields, lake water, and groundwater reservoirs. The dynamic component includes surface and underground runoff, and the lake is the final basis of accumulation of all water resources. The output portion of the budget consists of evaporation and meteoric water transport outside the water basin. The Issyk Kul basin was established to be an isolated hydraulic system, as groundwater outflow occurs neither in the eastern nor in the western terminations of the basin. At the western termination, underground discharge would be theoretically probable from the lake and from the Quaternary groundwater aquifer into the Boom canyon along the hypsometrically lower Chu basin but for a buried sill. The sill (Kutemalda) is build up by groundwater flow infiltrated from the NS striking segment of the Chu before it turns into the canyon. West of the Kutemalda sill, the Karatash sill dams the Paleogene-Neogene groundwater aquifer. The Paleogene-Neogene sedimentary cover pinches out within the Karatash sill, and the Paleozoic basement is exposed. Fault zones in the bedrock walls of the Boom canyon allow very limited groundwater infiltration over a distance about 40 km [5].
The available estimates of the hydrologic budget of the Issyk Kul basin are restricted to the lake itself and the groundwater of the Quaternary aquifer, whereas the total budget have never been estimated.
The input of 86.3 - 112 m^3/s in the water balance calculated by Vikulina in 1946, Tsigel'naya in 1960, and Bol'shakov in 1969 [1,2] and the 62.8-74.4 m^3/s groundwater budget of the Quaternary aquifer estimated by projects of the Kirghiz Research Institute of Mineral Resources (1971, 1984) and HYDROINGEO (1979), and by the Kyrghiz hydrogeological expedition (1978) are in principle consistent and do not require much revision. Further studies of the hydrologic budget of the Issyk Kul basin are expected to address its poorly investigated specific components to provide a basis for more precise estimates of the total balance.

Conclusions

Investigation of the groundwater system of the Issyk Kul basin allowed revision as to the groundwater component of the water balance associated with deep Paleogene-Neogene and Mesozoic aquifers. The greatest part of the groundwater input into the basin (over 70 %) consists of fresh cold water from the Quaternary aquifer and thermal mineral waters. Therefore, groundwater causes a strong influence on water chemistry of Lake Issyk Kul and is at the same time the main channelway of cultural chemical contamination. The Quaternary aquifer places hydrostatic and hydrodynamic constraints on the lake regime.

The geological structure of the eastern and western terminations of the Issyk Kul basin shows no evidence for groundwater outflow from the lake and its aquifers into the Boom canyon and on to the Chu catchment basin.

These specific features should be taken into account in studies of chemistry and water exchange in Lake Issyk Kul. The updated estimates of hydrologic budget are of particular importance in reasonable planning the household and drinking use of cold fresh and thermal mineral groundwater in order to avoid excess withdrawal and contamination, especially in view of aggravated cultural impact in the region of Balykchi and Karakol towns. Moreover, it appears necessary to control withdrawal of thermal mineral groundwater to prevent exhaustion of its resources. The groundwater stored in the rock reservoirs beneath the Issyk Kul basin may become an exceptionally important source of fresh water in the region in future in the conditions of climate drying.

References

1. Sidorenko, A.V. and Tyutyukin, V.S. (eds.) (1971), *Hydrogeology of the USSR. Kirghiz SSR* [in Russian], vol. 40, Nedra, Moscow, 487 pp.
2. Kadyrov, V.K. (ed.) (1978), *Lake Issyk Kul. Essays on physiography* [in Russian], Ilim, Frunze, 210 pp.
3. Mandychev, A.N. (1996) Groundwater resources of deep aquifers in the East Chu and Issyk Kul artesian basins, in *Mountain studies: changes and prospects for the 21st century* [in Russian], Ilim, Bishkek, pp. 136-137.
4. Mandychev, A.N. and Mandycheva, V.N. (1986), Typization of hydrogeolgical conditions of the upper part of the Issyk Kul artesian basin [in Russian], *Proc. 8th Conf. Kirghiz*, Ilim, Frunze, pp. 129-131.
5. Mandychev, A.N. and Suyunbaev, M.N. (1981), Possible discharge of groundwater in the western part of the Issyk Kul artesian basin toward the Boom canyon [in Russian], *Proc. 1st Conf. Young Scientists of Kirghizia*, Ilim, Frunze, pp. 44-46.

CURRENTS AND VERTICAL WATER EXCHANGE IN LAKE ISSYK KUL

V.V. ROMANOVSKY
Institute of Water Problems and Hydropower
533, Frunze str., Bishkek, Kirghyzstan

and

G.D. SHABUNIN
Issyk-Kul Lake Observatory
Cholpon Ata, Kyrgyzstan

1. Historical background

The first evidence of counter-clockwise currents in Issyk-Kul was reported by Matveev [1] based on few measurements of currents and ship drift. In 1970-1985, Rotatova and Filatova [2], Bukin [3], Stavissky and Ovcharenko [4], Romanovsky [5], and Shabunin [6] (1984) studied currents in Issyk Kul using the dynamic method and drift cards ("bottle mail").

To calculate drift by the dynamic method, data of mean monthly water temperatures were used, averaged over many years of measurements on standard profiles as well as the results of more detailed (up to 56 profiles) measurements in the upper 100 m of water (Fig. 1 a). The depth of the dynamic surface was then assumed to be 100 m, which is not always reasonable, as temperature contrasts at this depth in autumn and in early winter may attain 2 – 3° C. So, observations on buoy stations [5] showed that the surface with zero resulting water transport in Issyk Kul was at the depth of 100 m in July 1979 but moved down to 150 m in September.

Another assumption of the uniform vertical and lateral distribution of salinity, was based on chemical analyses of more than 500 water samples from various depths and various regions of the open lake [7]. According to earlier estimates by Kadyrov and Danilina [8] on 122 samples, the specific weight of the Issyk Kul water at 20° C is 1.0052 g/ml in the open water and less in gulfs and bays (1.0023 – 1.0040 g/ml). Romanovsky [5] compiled a table of conventional specific volumes of the Issyk Kul water, which was used by many researchers in calculations of geostrophic currents by the dynamic method. The results confirmed the presence of cyclonic circular currents in the lake. Figure 1 shows the scheme of surface currents based on temperature measurements and drift card experiments (500 cards), and directions of currents estimated by the dynamic method. The two data sets are in good agreement.

J. Klerkx and B. Imanackunov (eds.), Lake Issyk-Kul: Its Natural Environment, 77–87.

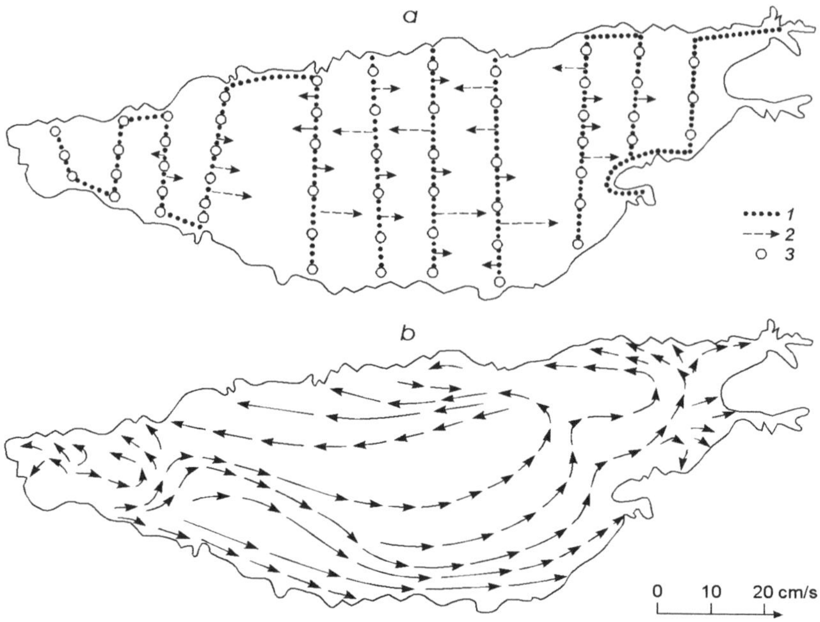

Figure 1. (a) Pattern of currents in Issyk-Kul estimated by dynamic method, measurements taken on 27 – 29 May 1980. Lengths of vectors are proportional to velocity values.
(b) Pattern of currents based on drift card data.
1 – drift card profiles; 2 – current velocity vectors; 3 – observation sites.

2. Currents in Lake Issyk-Kul

Observations from aboard a ship in 1982 – 1985 included monthly instrumental measurements of the velocity and directions of currents on 8 horizontal profiles evenly distributed along the lake perimeter (Fig. 2) plus four or five vertical profiles within each horizontal line for measurements at depths 4, 6, 10, 20, 40, 60, and 100 m. Below 100 m no measurements were taken because of anchoring difficulties. Therefore, those measurements represent the littoral water transport rather than the whole-lake pattern. However, it is in the littoral zone that currents have the most intricate pattern produced by the shore and bottom topography and man-induced effects. The measurements generally confirmed the presence of counter-clockwise currents in all seasons, but in each specific situation more complex lateral and vertical patterns have been evidenced (Fig. 2). On the background of predominant cyclonic water transport in the open lake, there are local anticyclonic currents in the littoral zone or the currents may change with depth to the opposite direction. Maximum measured velocities of currents varied from 60 m/s in summer to 20 m/s in winter.

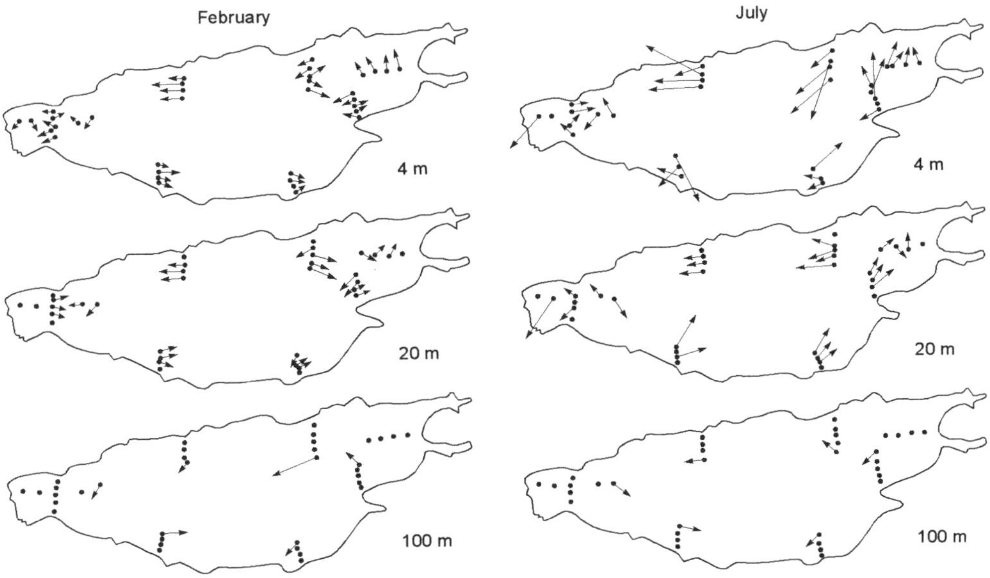

Figure 2. Vectors of currents, measured in February and July 1983 at depths of 4, 20, and 100 m.

Currents in Issyk Kul are either of drift or density nature. The drift currents are controlled by winds and the density currents by horizontal temperature contrasts between the colder central zone and the warmer periphery of the lake. Note that shore winds in Issyk Kul are directed seaward, especially the "hydrodynamically significant" winds that influence temperatures and currents. These winds induce drift currents that depart to the right off the direction of the casual air flow, i.e., cyclonic currents. The density currents are also cyclonic as a result of the Earth's rotation and the temperature field. Superposition of currents of the two types (and of the same sense) results in higher velocities of the general cyclonic water turnover. The theoretical velocities of density currents estimated from temperatures are 20 – 25 cm/s, while the measured values amount to 65 cm/s [5]. A rather high velocity of currents, especially in summer, is a particular feature of the Issyk Kul thermodynamics. In seas and oceans and in large lakes, the peripheryward (in the Northern hemisphere) flow in cyclonic counter-clockwise circulation driven by the Coriolis force is compensated by upwelling in the center [9]. As a result, the discontinuity layer is shallower in the center of the cyclone and deeper on its periphery because of warmer surface water. Cyclonic circulation in Issyk Kul likewise causes upwelling of cold deep water in the central part of the lake and downwelling of warm water in the littoral zone. Therefore, the temperature structure of the Issyk Kul water is dome-shaped, as cyclonic density circulation develops around a cold-water dome in the center.

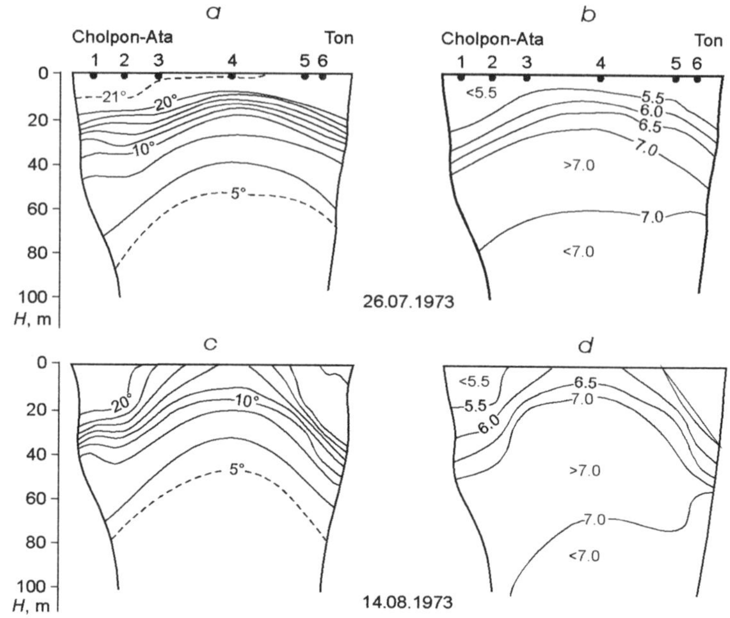

Figure 3. Distribution of water temperature (a, c) and dissolved oxygen in mg/l (b, d) along the profile Cholpon-Ata – Ton bay.

As already stated, the specific wind regime of the Issyk Kul basin induces drift circulation of the same sense. Storm winds that affect the water surface may strongly accelerate (twice or three times) currents in the whole-lake cyclonic turnover, as greater rotation velocity of water causes a more intense peripheryward movement. In some years, thermocline water rising to the surface abruptly cools down the water in the lake center (Fig. 3, a, c). Figure 3 shows water temperature variations along a cross section through the central part of the lake before and after a strong invasion of air into the basin that lasted 13 days with short breaks. The surface water temperature in the center (upwelling zone) dropped from 21.1° to 13.2° C. A similar change occurred in the distribution of dissolved oxygen (Fig. 3 b, d).

The average rate of upwelling, estimated from the vertical shift of the 13° C isotherm in the upwelling center (measured from 26 July to 14 August 1973), is $1.3 \cdot 10^{-3}$ cm/s, compared with 10^{-3} in Lake Baikal [10] and $3.8 \cdot 10^{-3}$ cm/s in the open Atlantic [11].

The temperature drop in the zone of upwelling reaches a depth of 150 m, decreasing from 0.6° at 60 m, 0.16° at 100 m, to 0.05° at 150 m. In the late August of 1973, the period of warming was followed by cooling. The horizontal and vertical temperature gradients gradually lowered but the thermal pattern with characteristically pinched-out isotherms in the central part of the lake persisted till the beginning of the next warming period in March 1974.

These thermodynamic regularities were confirmed by studies of Revyakin [12] and Arkhipov and Revyakin [13] [14] [15]. They applied mathematical modeling to investigate the effect of storms on water circulation and the influence of temperature stratification at variable wind conditions on the evolution of currents and temperature fields. The modeling results show that the wind conditions in Issyk-Kul produce cyclonic drift circulation in the lake, whereby a divergence zone forms in the center and a convergence zone on the periphery of vortices. A cold-water dome forms within the divergence zone, and thermal fronts with horizontal temperature gradients up to 4 – 5 °C/km arise between the divergence and convergence zones. After the wind effect ceases, the dynamic zones experience the reversal of divergence under baroclinic pressure gradients; the dome-shaped temperature pattern smoothens out, and the thermal fronts become gradually washed out and disappear till the following storm. Winds redistribute the density field in such a way that the density cyclonic circulation strengthens the drift circulation and persists long after the storm has stopped. The results of numerical experiments showed a good agreement with the observed temperature dynamics.

Thus, the main conclusions are as follows:

1. Cyclonic horizontal circulation of water exists in the pelagial zone of Issyk-Kul all the year round. Wind-induced anticyclonic vortices arise in the littoral zone (especially, in the western part of the lake).
2. The wind regime over the lake is the principal controlling factor of cyclonic circulation and its dynamics. It produces drift circulation, while the density circulation is of secondary importance.
3. Vertical water circulation is controlled by zones of convergence and divergence of currents in summer, and in winter also by convection.

3. Thermal regime

By its thermal regime, Issyk-Kul is a subtropic warm water basin with a vertical exchange occurring once a year in winter [5]. Water temperatures in the zone of maximum depths (600 – 650 m) vary from 3.7° to 4.2° C, according to measurements taken in different years. Measurements along vertical profiles in the central part of the lake from 1972 to 1991, never showed temperatures below 4° C. Temperature measurements in the zone of greatest depths were carried out by a team of the Academy of Sciences of the USSR headed by Matveev in 1928, 1930, and 1932, and then by the Kirghiz Hydrometeorological Agency from aboard the research vessel Multor once a season, or four times a year, from 1981 to 1991. During those experiments, the temperature of 3.7° C was measured in August 1930, September 1932 [1], and in February, May, August, and November in 1981.

According to data of water temperature monitoring published in an Annual Bulletin, measurements at all eight stations of the profile Cholpon-Ata – Kol'tsovka bay, including the five stations at depths >200 m, yielded temperatures below 4.0° C on 9 March 1985. At depths from 5 to 200 m, water temperature was 3.6° C. The winter of

1985 was anomalously cold. The thermobar that forms at the maximum density temperature of 2.75° C [3] appeared in the east of the lake (station Koisary) in the second half of December 1984 and persisted till the first half of March; near the northeastern shore (station Grigorievka). It existed through January and February and degraded in mid-March; near the western shore (station Rybachye), the thermobar formed in mid-December 1984 and disappeared in mid-March. Note for comparison that in 1981, when the winter was anomalously mild, no thermobar was observed at inshore stations on the eastern and northeastern shores, and on the western shore it existed for a short time in the middle of March.

In spite of the anomalously cold winter of 1985, no low temperatures were recorded in the deepest water. Measurements at 650 m were taken on 18 February 1985 and yielded a temperature of 4.1° C, possibly, because it was before the minimum temperature stratification had set in. The following measurements at this depth on 15 August 1985 showed a value of 3.9° C. It could be thus hypothesized that the temperature of 3.7° C was not reached because of insufficient vertical water exchange. The reason, however, was different.

Measurements at station 4 taken on 25 April 1985 showed anomalously high temperatures at the depth of 652 m; at depths below 200 m the temperature was 5.1° C, while other stations yielded 4.0° and 4.2° C at the same depths. Variations of the 5° isotherm show that in the end of 1985 a strong anticyclonic circulation existed in the lake center in which littoral warm water moved centerward and then down into the funnel of the anticyclonic vortex.

At the end of April, temperature stratification was poorly pronounced, and currents in the density cyclonic circulation were slow. A strong (25 m/s) western wind on April 23 and 24, 1985 destroyed the density cyclonic circulation and formed an anticyclonic circulation in the central part of the lake, whereby the water temperature in the zone of maximum depths was increased. That is why, the measurements in August 1985 did not record low temperatures caused by anomalous cooling of the upper 200 m of water in winter.

The effect of vertical water exchange in the first half of March 1985 was evident in the presence of highly transparent pelagial water on the lake surface, when the transparency of the Issyk-Kul water was the highest (53 m) and exceeded that of Lake Baikal (40 m). Therefore, full vertical water exchange occurred at least twice in 1985: in early March and in late April.

Proceeding from the vertical distribution of water temperature in the center of the lake, (monitoring data from station 5), the high transparency of water in 1986 and 1987 was associated with low temperatures (4.1 – 4.2° C) on the lake surface and was due to convective and wind mixing (Table 1; the tabulated data do not include the surface water temperatures as they are highly sensitive to weather changes).

The lower the temperature in the upper layer, the lower the density stability of water and the higher the bottom water temperature. In the winter of 1981, at a water temperature of 4.8° C at 5 m below the surface, mixing involved 100 m of water, whereas in 1983 – 1987, water temperature at 5 m was 4.1 – 4.4° C, and mixing occurred throughout the entire water column. Water transparency of >40 m was observed in the lake center in

TABLE 1 : Winter vertical distribution of water temperature in the central part of Lake Issyk-Kul at station 5, H = 658 m.

Depth, m	1981	1982	1983	1985	1986	1987	1988
0	5.0	-	4.6	4.8	4.2	4.6	4.6
5	4.8	4.5	4.4	4.3	4.1	4.2	4.4
10	4.8	4.5	4.4	4.3	4.1	4.2	4.3
15	4.8	4.5	4.4	4.3	4.1	4.2	4.3
20	4.8	4.5	4.3	4.2	4.1	4.2	4.3
30	4.8	4.4	4.3	4.2	4.1	4.2	4.3
50	4.8	4.3	4.3	4.2	4.1	4.2	4.3
100	4.7	4.3	4.2	4.3	4.1	4.2	4.3
200	4.1	4.2	4.1	4.3	4.1	4.0	4.2
300	4.1	4.0	4.1	4.2	4.1	4.0	4.1
400	3.9	4.0	4.1	4.1	4.0	4.1	4.0
500	3.9	3.9	4.0	4.1	4.0	4.0	4.0
600	3.9	3.9	3.9	4.2	3.9	3.9	4.0
643	3.7	3.9	3.9	4.1	3.9	3.9	4.0
$\Delta t°_{5\text{-}643}$	1.1	0.6	0.5	0.2	0.2	0.3	0.4
Trans-parency	15.5	23.8	33.5	41.0	42.5	45.0	26.5

1985 – 1987 at temperatures 4.1 – 4.3° C at the depth of 5 m, which attests to upwelling of pelagial water to the surface.

In the winter of 1981, vertical distribution of temperature was homogeneous in the upper 100 m of water; in the layer between 100 and 200 m, a weak thermocline was observed with a temperature decrease of 0.6° C; from depths below 200 m to the bottom, the temperature gradually decreased to 3.75° C. The high temperature gradient in the 100 – 200 m layer and low temperatures at the bottom indicate that the vertical exchange in 1981 involved only the upper 100 m of water, which is also confirmed by uniformly distributed low transparency (15.5 – 16.5 m). The low bottom water temperatures in the region of maximum depths, which were observed through 1981, may be due to strong cooling of the surface water in mid-March 1979, when the water temperature measured at station Tamga was 4.1° C (average over 10 days). In 1979 no measurements were taken for logistic reasons, and the temperature of 3.7° C at 650 m was then not recorded.

Thus, vertical water exchange through 1980 and 1981 was incomplete, and the temperature of 3.7° C measured at 650 m in 1981 was related to low water temperatures and convective mixing in the entire water column in March 1979.

TABLE 2 : Water temperature in Lake Issyk-Kul measured in different years at station Tamga in January, February, and March

Years	January	February	March
1945	4.8	4.9	4.1
1946	6.0	5.2	5.9
1947	5.4	5.1	5.9
1948	5.8	5.2	5.4
1949	4.6	3.9	4.9
1950	6.1	5.4	5.9
1951	5.0	4.0	4.6
1952	6.3	5.7	5.8
1953	5.4	5.1	5.3
1954	5.4	4.7	4.8
1955	5.1	4.6	5.1
1956	5.8	4.7	5.3
1957	5.8	5.0	4.9
1958	5.9		
1959	5.1	5.1	4.8
1960	5.7	5.0	5.2
1961	5.4	4.6	4.8
1962	5.8	4.8	5.6
1963	5.4	5.1	5.4
1964	6.4	5.1	5.1
1965	5.3	4.9	4.7
1966	6.2	5.4	5.2
1967	5.0	4.2	4.8
1968	6.4	4.8	5.3
1969	4.7	4.2	4.6
1970	6.7	5.1	5.3
1971	5.9	5.0	5.2
1972	6.6	4.4	4.7
1973	5.5	4.3	4.4
1974	6.1	4.6	4.7
1975	4.9	3.9	4.8
1976	5.6	5.0	5.3
1977	4.8	4.0	4.6
1978	5.6	4.1	4.6
1979	5.8	5.0	4.5
1980	6.6	5.2	5.3
1981	7.3	5.7	6.1
1982	5.2	5.1	5.5

1983	5.0	4.6	5.2
1984	5.6	4.3	4.6
1985	4.4	4.0	4.3
1986	5.9	5.1	5.1
1987	5.2	4.7	5.7
1988	5.7	4.8	5.5
1989	6.6	5.2	5.5
1990	6.4	5.4	5.8
1991	5.9	5.0	5.3
1992	6.5	5.3	5.3
1993	5.8	4.9	5.3
1994	6.1	5.1	5.5
1995	6.1	5.0	4.9
1996	5.6	4.7	5.2
1997	5.9	5.1	5.9
1998	5.7	5.4	5.6
1999	6.2	6.2	6.3

The evidence of vertical water exchange in Issyk-Kul comes from data on surface water temperatures in the zone of maximum depths, measured at station Tamga. Temperature distribution in the period of weakest thermal stratification (January-March), averaged over many years of observations (Table 2), suggests the following inferences:

1. Full vertical water exchange by convective mixing occurred in the years when mean monthly temperature of surface water was between 3.9 and 4.2° C, i.e., about the same value as at 650 m.
2. At surface water temperatures between 4.3 and 5.1° C, full vertical water exchange occurred by wind mixing, and was incomplete in the years when wind intensity in January-March was low.
3. Full vertical water exchange did not occur in the years when mean monthly temperature of surface water in January-March exceeded 5.1° C, as it was inhibited by a density gradient, at any wind activity.

On the basis of these conclusions from observations at Tamga through 1945 – 1999, we distinguished years with full and partial vertical water exchange. Full exchange during three months (January – March) occurred in 1945, 1949, 1951, 1967, 1969, 1975, 1977, and 1985, and partial exchange occurred in 1946, 1948, 1950, 1952, 1953, 1966, 1980, 1981, 1989, 1990, 1992, 1998, and 1999. Therefore, the vertical water exchange in Issyk-Kul is partial approximately once in four years (observed in 12 years of 54).

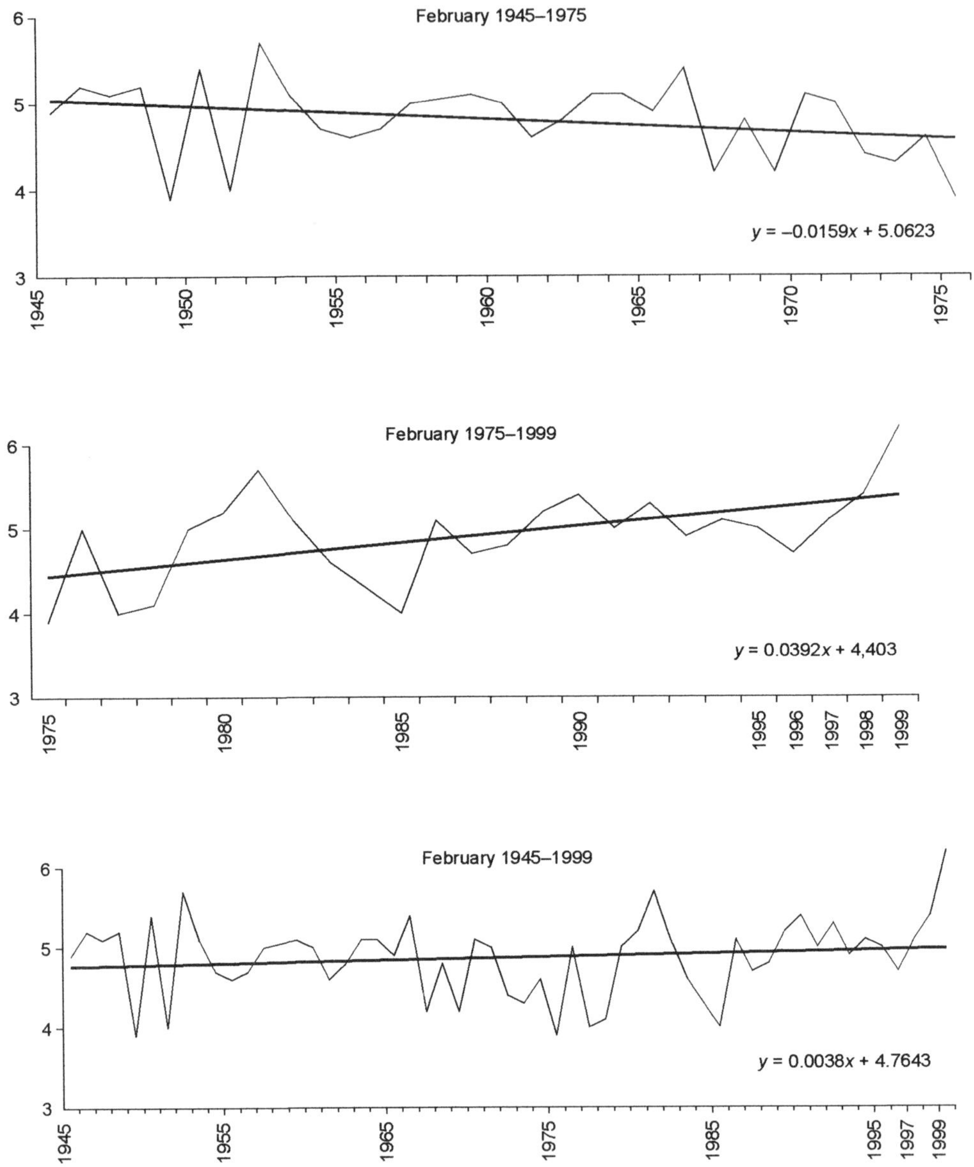

Figure 4. Water temperature variations in Issyk-Kul (measured at station Tamga).

Note the tendency of winter water temperature increase and worsening of mixing conditions in recent years: from 1989 to 1999 water exchange was partial five times, or in 45% of cases. Temperature variations in February measured at station Tamga (Fig. 4) show a distinct trend of growth from 1975 to 1999, obviously related to global warming evident in Tien Shan since 1973 [16].

4. References

1. Matveev, V.P. (1935) Hydrological studies on Lake Issyk Kul in 1932 [in Russian], in L.S. Berg (ed), *Lake Issyk Kul: Hydrology, Ichthyology, and Fish Production*, Proc. Kirg. Special Expedition of 1932-1933, vol. 111, issue 2, 7-56.
2. Rotatova, T.V. and Filatova, T.N. (1973) Dynamic method and its application in studies of currents in continental lakes and seas, *Bull. GGI* **203**, 80-124.
3. Bukin, V.M. (1974) Special post cards for gathering information on surface currents in lakes [in Russian], *Meteorologia i Hidrologia* **2**, 82-85.
4. Stavissky, Ya.S. and Ovcharenko, T.P. (1978) Geostrophic circulation of water in Lake Issyk Kul [in Russian], *Bull. SARNIGMI* **58(140)**, 55-59.
5. Romanovsky, V.V. (1990) *Lake Issyk Kul as a Natural System* [in Russian], Ilim, Frunze.
6. Shabunin, G.T. (1984) *Thermal Regime and Dynamics of Water in Lake Issyk Kul* [in Russian], Candidate Dissertation, Frunze.
7. Kadyrov, V.K. (1986) *Water chemistry in Lake Issyk Kul and its Catchment Basin* [in Russian], Ilim, Frunze.
8. Kadyrov, V.K. and Danilina, N.G. (1965) Specific weight of water in Lake Issyk Kul [in Russian], *Voprosy Vodnogo Khozyaistva* **3**, 30-34.
9. Glagoleva, M.G. and Skriptunova, L.I. (1979) *Prediction of Water Temperature in Oceans* [in Russian], Gidrometeoizdat, Leningrad.
10. Verbolov, V.I. (1977) Vertical near-slope circulation and vertcal water exchange [in Russian], in G.I. Galazii (ed), *Currents in Lake Baikal*, Nauka, Novosibirsk,45-66.
11. Fedorov, K.N. (1979) Slow relaxation of the thermal train of storms in oceans, *Dokl. AN SSSR* **245 (4)**, 960-963.
12. Revyakin, V.I. (1987), *Annual Variations in Current Fields and Water Temperature in Lake Issyk Kul* [in Russian], Candidate Dissertation, Frunze.
13. Arkhipov, B.V. and Revyakin, V.I. (1986) Three-dimensional barotropic model of currents in Lake Issyk Kul [in Russian], in A.S. Chklyaev (ed), *Current Problems of Geography*, Izd. PSU, Perm', 108-109.
14. Arkhipov, B.V. and Revyakin, V.I. (1986) Numerical modeling of wind currents in Lake Issyk Kul in winter and spring [in Russian], in V.A. Znamenskii (ed), *Hydrology of 2000*, Nauka, Moscow, 167-168.
15. Arkhipov, B.V. and Revyakin, V.I. (1987) Numerical modeling of currents in Lake Issyk Kul [in Russian], *Bull. Moscow University, Ser. Geogr.*, **1**, 43-48.
16. Mamatkanov, D.M., Dikikh, A.N., Romanovsky, V.V., Suyunbaev, M.N., and Shambetov (1997) Current dynamics of air temperature and the response of different water resources of the Issyk Kul basin [in Russian], *Ekho Nauki* **2-3**, 96-101.

USING TRANSIENT CONSERVATIVE AND ENVIRONMENTAL TRACERS TO STUDY WATER EXCHANGE IN LAKE ISSYK-KUL

R. KIPFER and F. PEETERS
Environmental Isotopes Group, Dept.of Water Resources and Drinking Water (W&T), Swiss Federal Institute of Environmental Science and Technology (EAWAG)
CH-8600 Dübendorf, Switzerland

1. Introduction

Mixing and transport processes are of major importance for the ecology of lakes [1]. In particular, deep-water renewal determines to a large extent the distribution of nutrients and dissolved oxygen in deep lakes such as Lake Issyk-Kul (Kyrgyzstan). The characterization and quantification of deep-water exchange is therefore essential for the understanding of the functioning of such lake ecosystems.

Rates of vertical water exchange can be estimated from concentrations of transient tracers, whereas the processes responsible for transport can be identified by environmental tracers such as temperature, salinity, dissolved oxygen and light transmissivity. In the following we first describe the use of transient and other environmental tracers for water age determination and for the characterization of water exchange processes. Then we discuss the application of these methods to Lake Issyk-Kul and present first tentative results on the mechanism of deep-water exchange based on data from two expeditions, one conducted in September 1996 and one in March 2001.

1.1. TRANSIENT TRACERS

The analysis of conservative trace gases, such as the noble gases ^{3}He [2][3], ^{222}Rn [4], ^{85}Kr [5][6] and/or purely anthropogenic, persistent substances such as chlorofluorocarbons (e.g. CFC-11, CFC-12 [7]), sulphur hexafluoride (SF_6 [8]) and tritium (^{3}H [9]) provides a very powerful tool for tracing physical processes in the aquatic environment, because these transient tracers do not undergo biological or chemical transformation and their distribution in the environment is therefore determined only by physical processes. Tracer concentrations and the isotopic composition of noble gases in water have been successfully employed to reconstruct soil temperature during groundwater recharge in the past [10][11], to date young and old groundwaters [2][3][12] and to analyse deep-water exchange in large lakes [13][14][15] and oceans

J. Klerkx and B. Imanackunov (eds.), Lake Issyk-Kul: Its Natural Environment, 89–100.

[16]. All these experimental techniques, which were successfully developed by oceanographers over the last 25 years, today represent tools that are routinely used to determine the dynamics of aquatic systems.

In principle, two different dating methods can be distinguished. The first utilises the fact that certain atmospheric trace components, such as CFCs, ^{85}Kr, and ^{3}H, have a time-dependent input function to the aquatic system. The atmospheric concentrations of these tracers have varied substantially over the last 50 years. Industrial production released most of the CFCs, especially CFC-11 and CFC-12, to the atmosphere, whereas ^{85}Kr stems from nuclear fuel reprocessing plants. In the early 1960s, nuclear bomb tests in the atmosphere increased the global tritium inventory by 2 to 3 orders of magnitude. The tritium concentration in precipitation has steadily decreased since the bomb-testing, but today tritium activity in meteoric water is usually still one order of magnitude larger than the expected background level due to tritium production by cosmic rays. If a given water parcel is in gas exchange with the atmosphere the observed concentration of a gaseous tracer reflects the atmospheric partial gas pressure at this time (Henry's Law). If the atmospheric partial pressure of the conservative gaseous tracer changes over time, the tracer concentration in the water parcel carries information about the time elapsed between the last gas exchange and the sampling date.

The second dating method is based on the phenomenon of radioactive decay. ^{222}Rn is produced by the decay of ^{226}Ra, mainly bound in the sediments, from which Radon escapes to the water column. The short half-life of ^{222}Rn (3.86 days) leads to strong vertical concentration gradients in the water column which can be employed to determine mixing rates near the sediment/water interface [4].

The determination of the so-called '^{3}H-^{3}He water age' utilises the radioactive decay of ^{3}H to ^{3}He. The half-life of tritium is 12.4 years. In a closed system, in which all concentration changes are solely the result of radioactive decay, the ratio of the concentration of the stable daughter isotope (^{3}He) to that of the radioactive mother isotope (^{3}H) yields a direct measure of the time elapsed since the water was last in contact with the atmosphere [2]. Note that ^{3}H-^{3}He dating does not depend on a knowledge of the atmospheric tritium input function.

In addition to dating aspects, the combined use of several transient tracers allows the problem of mixing processes in groundwater and open waters to be addressed (Fig. 1). Since the above-mentioned tracers are chemically and biologically inert, they trace only the physical transport in the water column of lakes. Thus chemical and biological transformations in the water column, e.g. oxygen depletion and gas exchange, can be separated out from the physical processes controlling the overall (deep) water exchange (advection and diffusion).

The various dating methods have very different links to the water body. CFCs and rare gas isotopes are dissolved as gases in the water, whereas tritium is bound in water molecules as HTO and is therefore part of the exchanged water itself. Since the influence of processes at the air/water interface on water molecules, on CFCs and on rare gases is different, the combination of tracer techniques allows the study of mixing processes within the water body and of the exchange at the air/water interface.

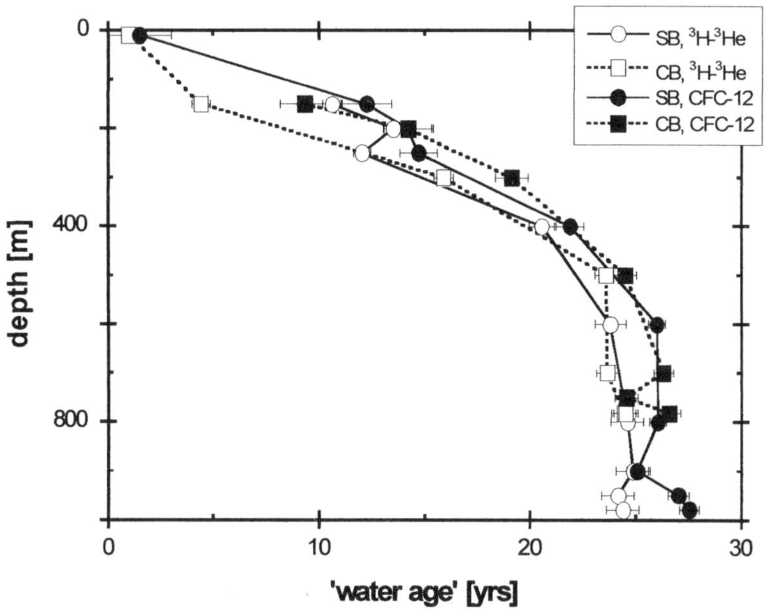

Figure 1: Water ages in the southern (SB) and central (CB) basins of the Caspian Sea derived from CFC-12 concentrations and the ^3H-^3He method. In both basins the water age of the deep water is the same and remains constant over time, suggesting similar mixing rates in the two basins. The difference in the water ages between surface and deep water has been used as a measure of the time required to exchange the water below 200 m. This time of internal exchange is about 15 years and hence is significantly shorter than the time required by runoff and precipitation to (re)fill the Caspian Sea (Peeters et al. 2000).

1.2. ENVIRONMENTAL TRACERS

Temperature and salinity, which can be measured with high spatial resolution using modern CTD probes (conductivity, temperature, depth (pressure)), are commonly employed in oceanography and limnology to characterize water masses and to trace their transport trajectories. In contrast to most other substances that are used as tracers, temperature and salinity are non-passive, i.e. they have a major influence on water density and in consequence affect transport and mixing processes.

In addition to temperature and salinity, adequately equipped CTD probes can simultaneously record dissolved oxygen concentrations and light transmissivity. Oxygen is introduced into the water body via gas exchange at the air/water interface and by primary production in the euphotic zone in the uppermost part of the water column. Oxygen depletion occurs in the deeper part of the water column and in particular at the sediment/water interface as a result of the mineralisation of organic matter. Consequently, oxygen concentrations commonly decrease with increasing depth, resulting in gentle concentration gradients. Deviations from rather smooth oxygen gradients toward larger values indicate the presence of oxygen-rich surface water and hence also indicate active deep-water exchange.

Light extinction occurs in waters mainly due to the presence of suspended particles from various sources. Hence light transmission decreases as the load of suspended particles in the water column increases. Often a decrease in light transmission can be interpreted to indicate either riverine inflow containing a high particle load or the remobilisation of particles from the sediments. Note that suspended particles also affect the water density and therefore cannot be considered passive.

2. The Use of Tracers to Study Deep-Water Mixing in Lake Issyk-Kul

According to historical data, deep-water temperatures at 600 m depth and below always range between 3.7 - 4.2 °C [17] and hence are significantly higher than the temperature at which Issyk-Kul water has its maximum density (T_{md}). Since the mean salinity in the open water of Issyk-Kul is close to 6.0 per mille (calculated from the chemical composition given by Tsigelnaya [18]) T_{md} is approximately 2.6 °C at surface pressure. The value T_{md} = 2.75 °C given in Romanowski [17] might be based on a slightly smaller salinity. In any case, T_{md} at the lake surface is lower than the deep-water temperatures. For surface temperatures down to 2.6 °C the thermal expansion coefficient $\alpha = -\rho^{-1}\partial\rho/\partial T > 0$ in the entire water column. Thus seasonal cooling at the lake surface down to T_{md} leads to an increase in the density of the surface water and can generate large-scale vertical convection down to the deepest parts of the lake, probably resulting in fast renewal of the water below 400 m. This hypothesis is supported by the existence of very high concentrations of dissolved oxygen – close to atmospheric equilibrium – throughout the water column ([19] and Fig. 2) and by the brownish colour of the topmost 15 cm of sediment cores indicating oxygenated sediments (Santiago Giralt, personal communication). In addition, in winter 1976 tritium concentrations were found to be homogeneous throughout the water column with the exception of the bottom 50 m [20], indicating that mixing had already smoothed out the tritium bomb-peak of 1963. This suggests that the residence time in the deep water can be expected to be less than 13 years.

Because atmospheric CFC concentrations have not changed significantly since 1996, measurements of CFCs in 2001 or later cannot provide reliable information on residence times shorter than 5 years. Thus, CFCs are possibly only of limited value for the study of mixing processes in Lake Issyk-Kul. However, the tracer SF_6 might provide detailed information on residence times because its atmospheric concentration is still increasing strongly with time. Alternatively, the ^3H-^3He method might be employed, because it is capable of resolving water ages ranging from one month to several decades. This suggests that SF_6 and ^3He and ^3H will be the most promising tracer combination to allow a reliable quantification of deep-water residence times and deep-water exchange in Lake Issyk-Kul.

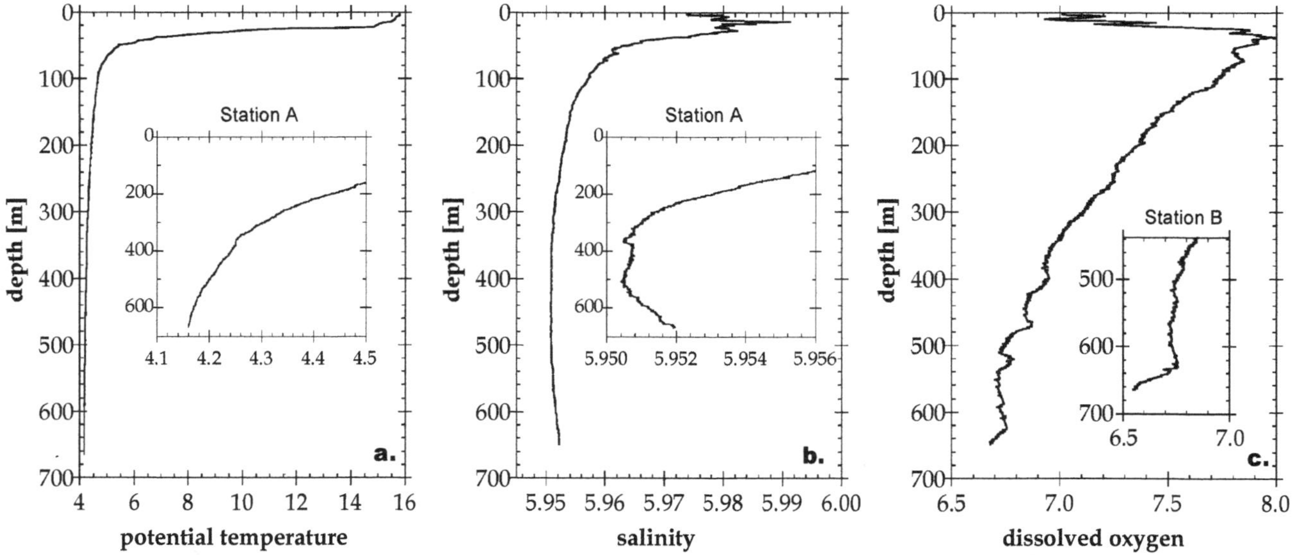

Figure 2: Potential temperature, salinity and dissolved oxygen measured in the deep central part of Lake Issyk-Kul in September 1996 (Station A). The insets in Fig.2a and b show the profiles in more detail, whereas in Fig. 2c the inset shows the deep section of dissolved oxygen measured at station B. Note that the O_2 data stem from a sensor for dissolved oxygen and have not been calibrated by chemical measurements.

3. Preliminary Results from CTD Measurements

A first attempt to characterize the processes responsible for deep-water exchange based on high-resolution CTD measurements was undertaken in September 1996 (Fig. 2). With increasing depth, potential temperature decreases steadily down to 4.16 °C. Because the temperature profile shows a minimum at the largest depth, advective transport processes must play a major role in venting even the deepest region of Lake Issyk-Kul.
Salinity was calculated from conductivity using the UNESCO equations for ocean waters [21]. This procedure underestimates the absolute values of salinity and overestimates the effect of pressure on conductivity because the ion composition of Issyk-Kul water [18] differs from that of the ocean on which the UNESCO equations are based. We therefore applied an additional calibration factor and a pressure correction which will be described elsewhere.
Except for the uppermost 30 m, salinity is rather constant throughout the water column. Whether the slight increase in salinity with depth below 400 m (Fig.2b) is real or is an artefact of the applied pressure correction cannot be determined. Temperature and salinity profiles are fairly smooth and give no indication of lateral intrusions except possibly at 350 m depth. The profile of dissolved oxygen, however, has distinct structures at depths 400 m, 470 m and 520 m (Fig. 2c) that suggest the lateral inflow of oxygen-rich waters. In all 6 profiles measured at deep stations in September 1996, dissolved oxygen concentrations decrease with depth in the near-bottom waters immediately above the sediments. In most profiles the decrease with depth is more pronounced than at station A and similar to that measured at station B (Fig 2c, inset). This specific feature in the vertical O_2 distribution is most likely related to mineralisation processes at the sediment-water interface, which become the dominant factor of oxygen consumption near the lake bottom due to the large sediment area to water volume ratio. In Fig. 2 the O_2 concentrations are not calibrated by chemical measurements and thus might be shifted relative to the corresponding true absolute values. However, the relative changes within the O_2 profile indicate that dissolved oxygen might be the tracer of choice to study advective transport processes and intrusions in Lake Issyk-Kul.
Temperatures measured in September 1996 along a transect from north to south through the centre of the lake at about 77.06°E indicate warm water in the shallow near-shore regions (Fig. 3). At the same depths the water in the centre of the lake is significantly colder than the near-shore water. This general temperature structure has already been described in the literature [22] as a doming effect and has been interpreted to indicate the large-scale upwelling of cold water in the open lake. According to Romanowski and Shabunin [22] the prevailing cyclonic wind field leads to the downwelling of warm water in the coastal regions and the upwelling of cold water in the open lake. Note that this process appears to be limited to the top 100-200 m (Fig. 3), probably because density stratification hinders a deeper penetration of the warm downwelling water. In addition to the dynamic process outlined above, differential warming of the near-shore regions during the warm season might enhance the temperature difference between open and coastal regions.

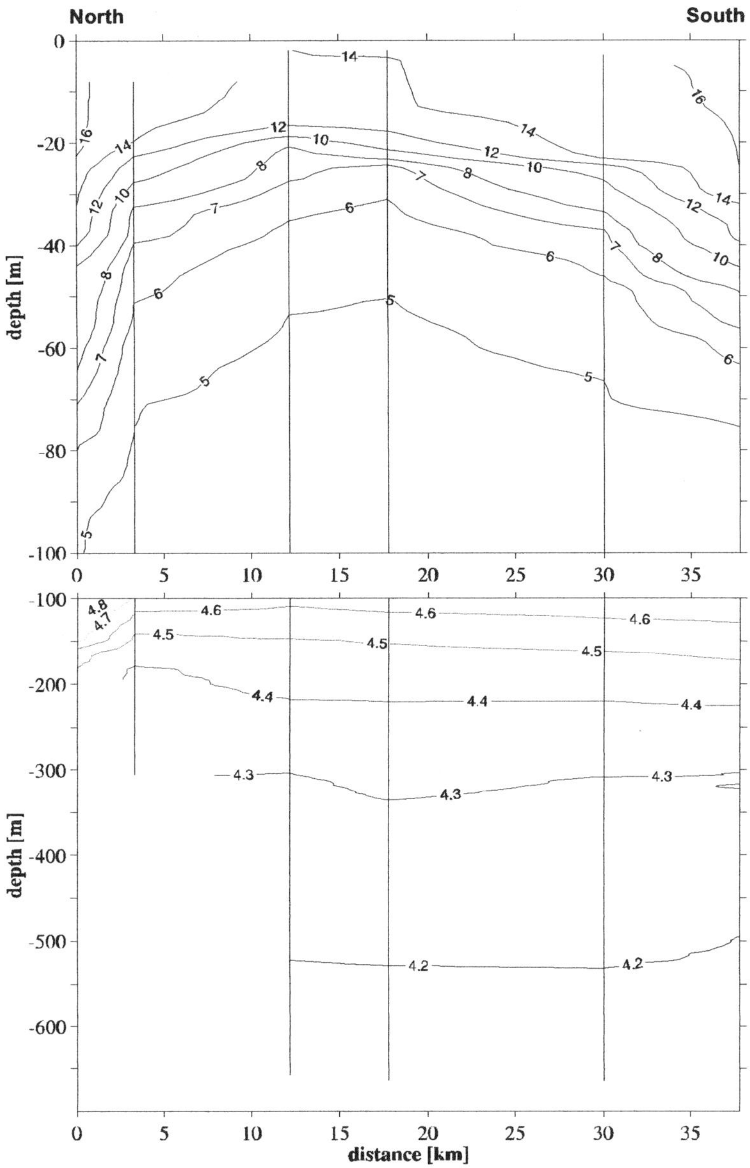

Figure 3: Distribution of potential temperature along a North-South transect about 77.06°E through the deepest part of Lake Issyk-Kul. The upper panel shows the temperature distribution in the 0-100 m depth range and the lower panel the temperature distribution below 100 m. Note that the intervals between the temperature isolines, given in [°C], are different in the two panels.

Because during the warm season temperatures at shallow depths are significantly higher than those of the deep water, advective processes transporting oxygen-rich surface water to very large depths can only be driven by salt gradients or gradients in suspended particle load. The former might be generated by evaporation in shallow bays, whereas the latter could be caused by inflowing meltwater from glaciers. However, it is unlikely that significant deep-water convection occurs during the warm season because of the strong temperature gradients, which imply strong density gradients.

Deep-water exchange should be expected to occur in the cold season between late autumn and early spring, when surface temperatures are low and vertical temperature gradients and the corresponding density gradients are small. In winter 1976, Romanov et al. [20] observed a strong increase in tritium activity in the lowermost 50 m of the open water column, indicating the lateral input of tritium-rich water which must originate from upper regions of the water body ('mixing around the edge'). Similar observations are known from Lake Baikal, where near-bottom anomalies in tracer concentrations are caused by massive advective deep-water exchange [23][24][14].

Further evidence for strong advective mixing in spring was found in March 2001 (Fig 4). Profiles from two CTD casts are shown (dotted and solid lines) which were taken within 75 minutes of one another at a distance of less than 2.5 km apart. The distinct differences between the profiles indicate that the conditions in Lake Issyk-Kul in spring are spatially and/or temporally very inhomogeneous. Dissolved oxygen and light transmissivity profiles at station 8 (solid lines) show distinct features which are correlated and apparently characterize water masses which are transported from shallow depths to the deepest regions of the lake, as indicated by the high oxygen concentrations. Fig. 4 demonstrates that, in addition to dissolved oxygen, light transmissivity proves to be an excellent tracer for the identification and characterization of convective processes in Lake Issyk-Kul.

Comparison of the salinity profiles from stations 7 and 8 measured in 2001 (Fig. 4) indicates several intrusions. At around 550 m, for instance, the intruding water at station 8 has a higher salinity than that at station 7. Associated with the salinity peak are values of temperature, O_2 concentration and suspended particle load (inversely related to light transmissivity) that are higher than background values. This suggests that the water at 550 m depth at station 8 originates from shallower depths where O_2 concentrations and suspended particle loads are generally higher.

At both stations salinity appears to be highest at intermediate depths. Possibly this intermediate salinity maximum is related to intruding higher-salinity water from shallower regions, where evaporation during the warm season has resulted in increased salt concentrations. However, as mentioned before, the salinity maximum at intermediate depth might not be real, but instead be the result of the imperfect empirical relations employed to calculate salinity from conductivity.

The change in the absolute values of salinity from 1996 to 2001 is most likely due to the use of different conductivity sensors in the two expeditions. The salinity difference between 1996 and 2001 suggested by the CTD measurements are smaller than measurable by chemical methods and can therefore not be tested independently.

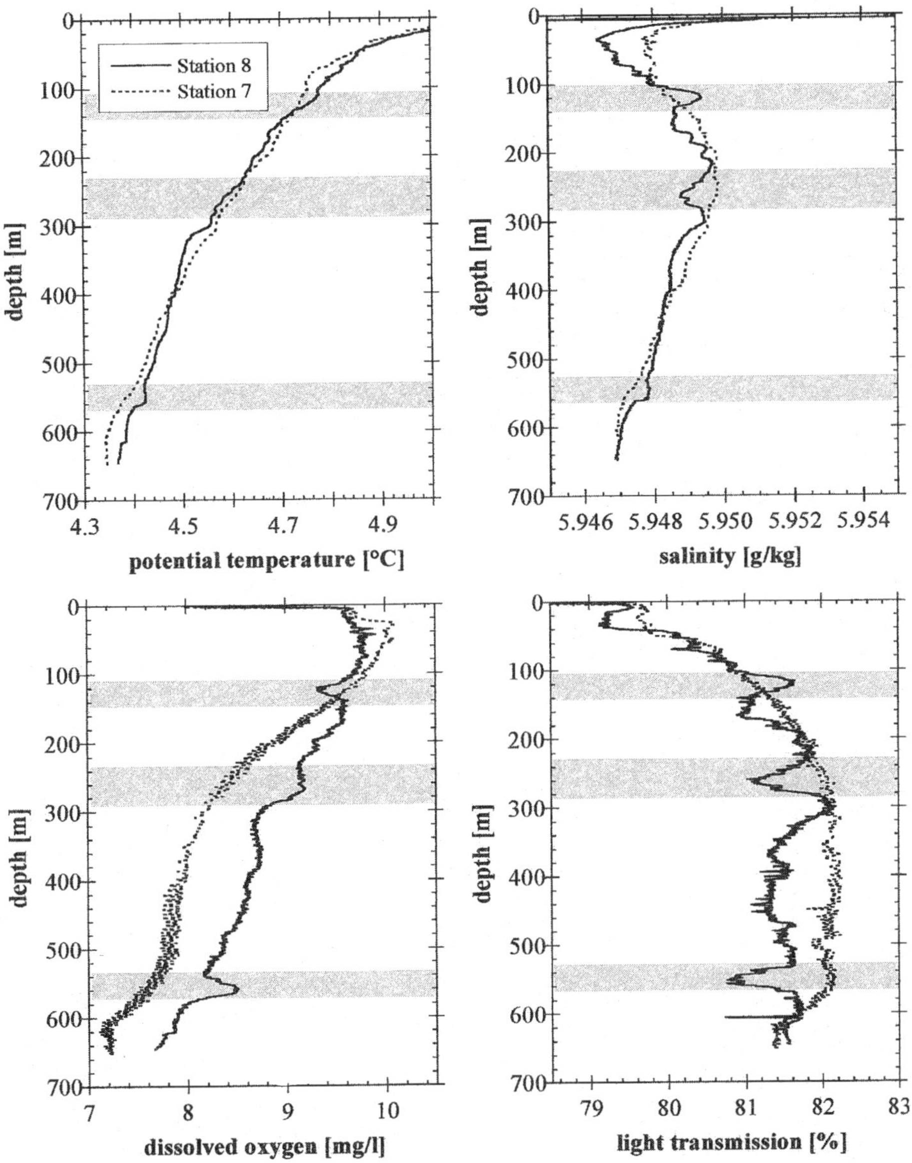

Figure 4: Potential temperature, salinity, dissolved oxygen and light transmissivity measured in March 2001 in Lake Issyk-Kul. The data stem from two CTD casts (station 8, solid line; station 7, dotted line) which were taken within 75 minutes of one another at a distance of less than 2.5 km. The O_2-sensor measurements agree well with the concentration of dissolved O_2 determined by Winkler titration

4. Conclusions

In addition to the classical tracers temperature and salinity, dissolved oxygen and light transmissivity are key variables in the study of processes leading to deep-water renewal in Lake Issyk-Kul. Deep-water exchange most likely occurs predominantly in winter and spring, when temperature gradients are small. Clear evidence for rapid deep-water mixing in spring is provided by CTD profiles measured in March 2001. During the summer season, vertical advection might be driven by salinity gradients generated by evaporation in shallow bays. Vertical mixing at shallower depths is probably dominated by wind-driven processes leading to downwelling in coastal areas and upwelling in the open water [22]. Homogeneous tritium concentrations in 1976 [20] and high values of dissolved oxygen close to atmospheric equilibrium concentration throughout the water column suggest that deep-water residence times must be rather short. Therefore, the combination of the tracer SF_6 with the ^3H-^3He method would appear to be the most suitable approach to quantify reliably deep-water exchange in Lake Issyk-Kul.

5. Acknowledgements

Thanks are due to Prof. Jan Klerkx and Santiago Giralt for making available the CTD data from the 1996 expedition to Lake Issyk-Kul. This research was made possible by funding from the Swiss Federal Office of Education and Science (BBW; Contract No. 99.0720) within the framework of the European INCO/COPERNICUS Project APELIK ('Assessment and Prognosis of Environmental Changes in Lake Issyk-Kul (Kyrgyzstan)'; Contract No. ICA2-CT-2000-10003

6. References

1. Imboden,D.M. (1990) Mixing and transport in lakes: Mechanisms and ecological relevance, in M. Tilzer and C. Serruya (eds.), *Large Lakes: Ecological Structure and Function.*, Springer, Berlin, pp. 47-80.
2. Tolstikhin, I. N., Kamenskiy, I. L. (1969) Determination of ground-water ages by the T-^{3}He Method. *Geochemistry International* **6**, 810-811.
3. Schlosser, P. (1992) Tritium/3He dating of waters in natural systems. Isotopes of noble gases as tracers in environmental studies, Vienna, IAEA.
4. Imboden, D.M., Emerson, S. (1978) Natural radon and phosphorus as limnologic tracers: Horizontal and vertical eddy diffusion in Greifensee. *Limnol. Oceanogr.* **23**, 77-90.
5. Loosli, H.H., Oeschger, H. (1979) Argon-39, carbon-14 and krypton-85 measurements in groundwater samples, IAEA, Ed., Isotope Hydrology 1978, Neuherberg IAEA.
6. Loosli, H.H., Lehmann, B.E., Balderer, W. (1998) Argon-39, argon-37 and krypton-85 isotopes in Stripa groundwaters. *Geochim. Cosmochim. Acta* **53**, 1825-1829.

7. Dunkle, S. A., Plummer, L. N., Busenberg, E., Phillips, P. J., Denver, J. M., Hamilton, P. A., Michel, R. L., Coplen, T. B. (1993) Chlorofluorocarbons (CCl_3F and CCl_2F_2) as dating tools and hydrologic tracers in shallow groundwater of the Delmarva Peninsula, Atlantic Coastal Plain, United States. *Water Resour. Res.* **29**, 3837-3860.
8. Busenberg, E. and Plummer, L. N. (2000) Dating young groundwater with sulfur hexafluoride: natural and anthropogenic sources of sulfur hexafluoride, *Water Resour. Res.* **36**, 3011-3030.
9. Münnich, K.O., Roether,W., Thilo, W. (1967) Dating of groundwater with tritium and ^{14}C. IAEA, Ed., Isotope Hydrology, Vienna, **19**, 305-320.
10. Stute, M., Clark, J. F., Schlosser, P., Broecker, W. S. (1995) A 30,000 yr continental paleotemperature record derived from noble gases dissolved in groundwater from the San Juan Basin, New Mexico. *Quatern. Res.* **43**, 209-220.
11. Beyerle, U., Purtschert, R., Aeschbach-Hertig, W., Imboden, D. M., Loosli, H. H., Wieler, R., Kipfer, R. (1998) Climate and groundwater recharge during the last glaciation in an ice-covered region, *Science* **282**, 731-734.
12. Beyerle, U., Aeschbach-Hertig, W., Hofer, M., Imboden, D.M., Baur, H., Kipfer, R. (1999) Infiltration of river water to a shallow aquifer investigated with $^3H/^3He$, noble gases and CFCs. *J. Hydrol.* **220**, 169-185.
13. Aeschbach-Hertig, W., Kipfer, R., Hofer, M., Imboden, D. M., Baur, H. (1996) Density-driven exchange between the basins of Lake Lucerne (Switzerland) traced with the ^3H-^3He method. *Limnol. Oceanogr.* **41**, 707-721.
14. Hohmann, R., Hofer, M., Kipfer, R., Peeters, F., Imboden, D.M. (1998) Distribution of helium and tritium in Lake Baikal. *J. Geophys. Res.* **103**, 12823-12838.
15. Peeters, F., Kipfer, R., Achermann, D., Hofer, M., Aeschbach-Hertig, W., Beyerle, U., Imboden, D. M., Rozanski, K., Fröhlich, K. (2000) Analysis of deep-water exchange in the Caspian Sea based on environmental tracers. *Deep-Sea Res.* **I 47**, 621-654.
16. Jenkins, W.J. (1976) Tritium-helium dating in the Sargasso Sea: A measurement of oxygen utilization rates. *Science*, **196**, 291-292.
17. Romanowski, V. V. (1991) Lake Issyk-Kul as natural system. Frunze (in Russian).
18. Tsigelnaya, I. D. (1995) Issyk-Kul Lake, in A.F. Mandych (ed.), *Enclosed seas and large lakes of eastern Europe and middle Asia.* SPB Academic Publishing.
19. Kadyrov, V. K. (1986) Hydrochemistry of the Lake Issyk-Kul and its basin. Ilim Publishing House, Frunze, p. 212. (in Russian)
20. Romanov, V., Vlasova, L. S., Meskheteli, A. V., Salnova, L. V (1989) A Tritium study of the formation of natural waters of Issykul basin, *Water Resources*, **4**, 82 – 94 (in Russian).
21. UNESCO, ICES, SCOR, and IAPSO, (1981) Background papers and supporting data on the practical salinity scale 1978. *Unesco technical papers in marine science*, **Nr. 37**, UNESCO.

22. Romanowski, V. V., Shabunin, G. (1981) Upwelling in Lake Issyk-Kul. In: Dynamics of modern coastal processes in Lake Issyk-Kul, Ilim Publishing House, Frunze, 160-169 (in Russian).
23. Weiss, R. F., Carmack, E. C., Koropalov, V. M. (1991) Deep-water renewal and biological production in Lake Baikal, *Nature*, ***349***, 665-669.
24. Hohmann, R., Kipfer, R., Peeters, F., Piepke, G., Imboden, D.M., Shimaraev, M.N. (1997) Processes of deep-water renewal in Lake Baikal. *Limnol. Oceanogr.* **42**, 841-855.
25. Imboden, D.M., Lemmin, U., Joller, T., Schurter, M. (1983) Mixing processes in lakes: Mechanisms and ecological relevance. *Schweiz. Z. Hydrol.* **45**, 11-44 .

BATHYMETRY AND SEDIMENTARY ENVIRONMENTS OF LAKE ISSYK-KUL, KYRGYZ REPUBLIC (CENTRAL ASIA): A LARGE, HIGH-ALTITUDE, TECTONIC LAKE

M. DE BATIST, Y. IMBO, P. VERMEESCH
Renard Centre of Marine Geology, University of Gent
Krijgslaan 281 s.8, B-9000 Gent, Belgium

J. KLERKX, S. GIRALT
International Bureau for Environmental Studies (IBES)
Rue Audrey Hepburn 9/13, B-1090 Bruxelles, Belgium

D. DELVAUX
Royal Museum of Central Africa
Steenweg op Leuven 13, B-3080 Tervuren, Belgium

V. LIGNIER, C. BECK
Laboratoire de Géodynamique des Chaînes Alpines, Université de Savoie
F-76766 Le Bourget du Lac, France

I. KALUGIN
Institute of Geology, Siberian Branch of the Russian Academy of Sciences
630090 Novosibirsk, Russia

K.E. ABDRAKHMATOV
Institute of Seismology, Kyrgyz National Academy of Sciences
720060 Bishkek, Kyrgyz Republic

1. Introduction

Lake Issyk-Kul, located in the northern Tien Shan of the Kyrgyz Republic, in the heart of Central Asia, is one of the deepest and largest lakes in the world. Although Kyrgyz and Russian scientists have studied the lake quite intensively since the 1850's, not much of the scientific literature has found its way to the international community. It is only recently that Lake Issyk-Kul has also started to attract international attention, and this is essentially thanks to the fact that the area occupies a potentially interesting location for paleoclimate research as well as for geodynamic studies.

The sediments that have accumulated in the lake are believed to have functioned as

J. Klerkx and B. Imanackunov (eds.), Lake Issyk-Kul: Its Natural Environment, 101–123.

accurate "recorders" of the processes that were active during or after their deposition. As such, they potentially hold a record of tectonic events that affected the area and/or of past changes in regional climate. However, sediments in large lakes like Lake Issyk-Kul are not all generated and deposited by the same processes and are far from uniform, and their "recording capacity" is thus highly variable. A thorough knowledge of the sedimentary processes and environments in the lake is therefore indispensable for the selection of sites that are suitable for the collection of sediment cores for such investigations.

Such knowledge is also vital for the assessment of the potential impacts on the lake system of the influx and dispersion of solid contaminants. This is a particularly acute problem for the Lake Issyk-Kul region, which is currently attempting to deal at the same time with plans for a further, sustainable growth of the recreational and tourist industry around the lake shores and with the environmental effects of the intensive gold and uranium exploitation that took or is still taking place in the basin and the nearby mountain ranges and of the increasing agricultural use of the lake shores.

In this paper we will define and characterise the main sedimentary environments in Lake Issyk-Kul. We will do this on basis of the interpretation of existing bathymetry data, of a comprehensive first reflection seismic study carried out on the lake in 1997, and of a series of reconnaissance gravity cores collected from the lake in 1998. All the data presented here were acquired as part of an ongoing Belgian-Kyrgyz-Russian-French research initiative.

2. General characteristics of Lake Issyk-Kul

Lake Issyk-Kul (Fig. 1) is a closed mountain lake, located at about 77° E and 42°30' N, in the northern part of the Tien Shan mountain belt in the Kyrgyz Republic (Central Asia). It is situated at an altitude of 1,607 m above sea level and surrounded by high mountain ranges: the Kunghei Alatau Range in the north with the highest peaks reaching 4,770 m, and the Terskei Alatau Range in the south with peaks exceeding 5,200 m.

The lake has the shape of a trapezium with its base oriented to the north (Fig. 1). It is 178 km long in the E-W direction, its width is about 60 km and it covers an area of 6,236 km^2, making it the second largest high-altitude lake of the world. It is also the world's fifth deepest lake, with a maximum depth of 668 m. The average water depth is 278 m, the length of the coastline is 688 km, and the total water volume is 1,736 km^3 [1] [2] [3].

About 118 rivers enter the lake. They drain an area of 22,080 km^2 and are predominantly fed by melt-water from snow and glaciers, which occupy about 509 km^2 of this drainage basin, at altitudes of 3,000 m and above [4]. The main inflowing rivers are the Djyrgalan River and the Tioup River, entering the lake axially at its eastern extremity.

At present the lake has no outlet. Previously, the Chu River was flowing into and out of the lake at its western extremity. Now, the river changes its course a few kilometers before reaching the lake, and flows into the Boom Canyon towards west. This change in the course of the Chu River and its disconnection from Lake Issyk-Kul is inferred to have occurred at the end of the Late Pleistocene, and most likely resulted from tectonic movements [5]. Since the disconnection, the lake has still been intermittently overflowing via the paleo-outlet of the Chu River during periods of higher lake level; the last of such periods ended about 150 years ago.

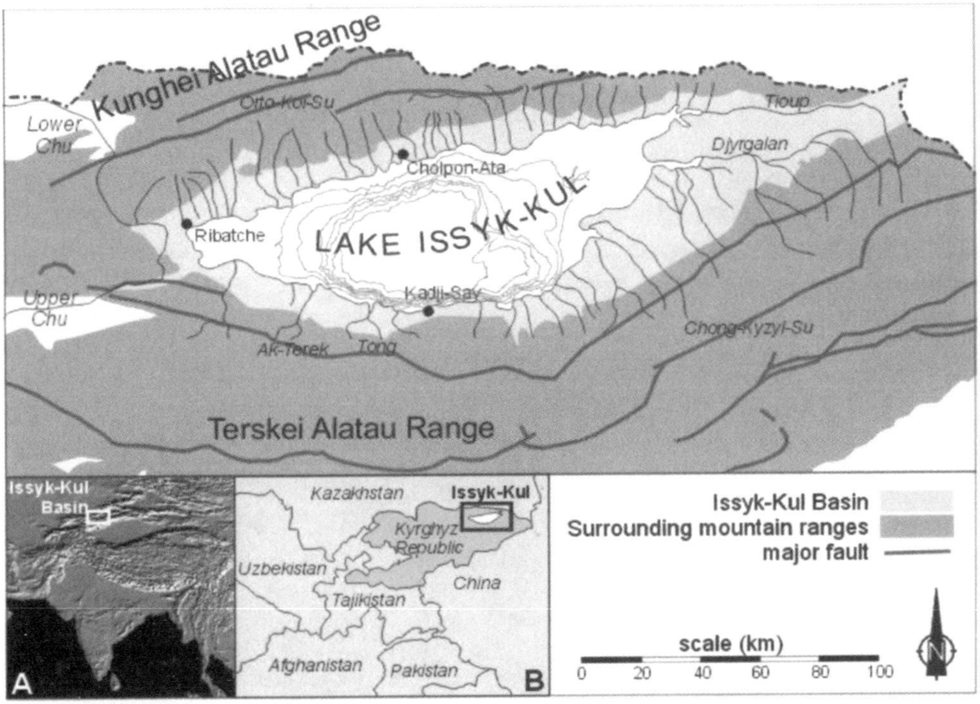

Figure 1. General localisation map of Lake Issyk-Kul, indicating the surrounding mountain ranges, the depth of the lake (with some selected contours), the main towns, rivers and other localities mentioned in the text. Inset A shows the location of Issyk-Kul Basin in the regional context of the Tien-Shan. Inset B indicates the political-geographical location.

The lake surface temperature in January does not drop below 2-3°C, and in July it reaches up to 19-20 °C [6]. At depths of more than 100 m, the water temperature remains constant all the year at 3.5-4.5 °C. Water circulation is counter-clockwise and surface water currents of up to 0.3 m/s occur in the northern part of the lake [7].

Lake Issyk-Kul is monomictic, slightly saline (6 g/l), oligotrophic to ultra-oligotrophic and it has high values of dissolved oxygen even at the lake bottom [6]. The water chemistry is dominated by Na^+ and K^+ among the cations, and by Cl^- among the anions [8]. The slightly saline waters are oversaturated in calcite, monohydrocalcite and vaterite, minerals which are found in the sediments [9] [10].

The total annual water input by rivers amounts up to 437 mm/yr (about 3,720,000 m^3), and the average annual rainfall is 251 mm/yr. The western part of the lake is slightly drier than the east. With evaporation at about 702 mm/yr, the lake has a negative hydrological budget [11] and the lake level has the tendency to gradually become lower since ± 150 years [7]. This effect appears to be even amplified by the increasing use of river water for human activities and agricultural purposes.

3. Geological setting

The northwestern Tien Shan is characterized by a succession of E-W-trending mountain ranges with elevations of more than 4 km and intervening intermontane basins containing a Cenozoic infill of up to 6 km in thickness. Ranges are generally overthrusted onto adjacent basins [12] [13] [14], and major active fault zones mark the boundary between them.

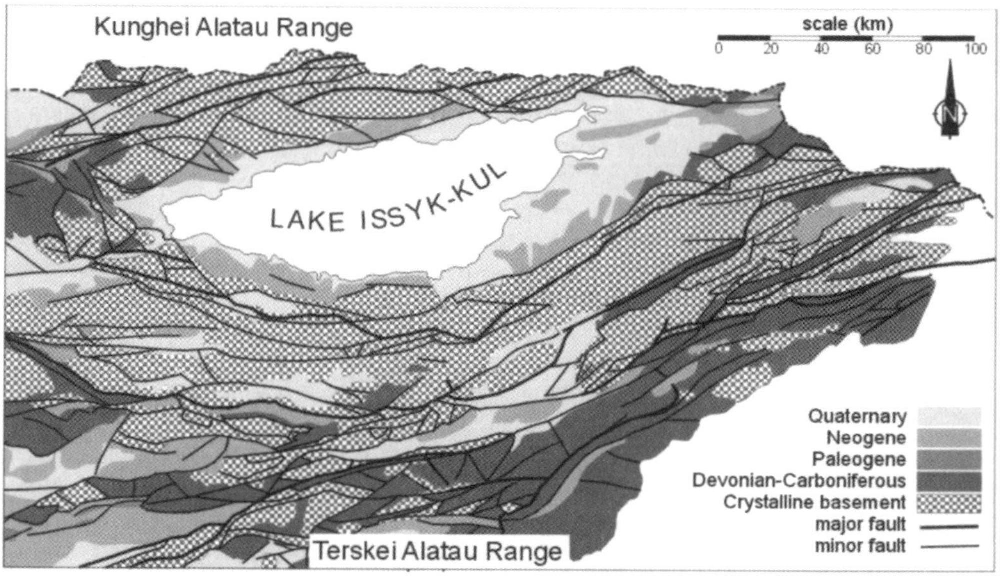

Figure 2. Simplified geological map of Issyk-Kul Basin and surrounding mountains (after [7]).

The Issyk-Kul Basin is one of the largest of these intermontane basins in the Kyrgyz Tien Shan. It has an oval shape, a length of 250 km in the E-W direction and a width of about 145 km in the N-S direction. Lake Issyk-Kul occupies the central part of the basin [15] [16] [17].

The mountain ranges surrounding the Issyk-Kul Basin are the Kunghei Alatau Range in the north and the Terskei Alatau Range in the south. They are composed of crystalline basement rocks of Archean to Middle Paleozoic age, covered by volcano-sedimentary and sedimentary strata of Devonian-Carboniferous age (Fig. 2). Their uplift and exhumation is believed to have started in Oligocene-Miocene times [18], although recent data from the neighbouring Chu Basin point to a Late Miocene onset of mountain building [19].

The mountain ranges are separated from the Issyk-Kul Basin by broad strike-slip zones that have accommodated most of the Late Cenozoic strain and have protected the basin interior from any significant deformation: the Chon-Kemin-Chilik zone in the north and the Pred-Terskey zone in the south [17] (Fig. 2).

The oldest deposits in the basin are thought to date back to Oligocene-Miocene times [12] [18] [15]. Since that period up to 3,500 m of sediments have accumulated in the depression [20]. These comprise a series of continental and lacustrine formations of Miocene to Quaternary age, that witness the initial expansion of the lake and the subsequent shrinking to its present-day size (Fig. 2). Uplifted lacustrine deposits of the Pliocene Djuka Formation are currently exposed in the southern part of the basin, along the southern lakeshore, where they are cut by horizontal Quaternary lacustrine terraces. The northern shoreline of the lake is dominated by glacial and fluvioglacial deposits, which cover older lacustrine deposits and reflect the expansion of the Kunghei Alatau valley glaciers to the foot of the mountain range during Pleistocene glaciations.

The area is tectonically still highly active: some of the largest earthquakes of the last century occurred in the Kyrgyz Tien Shan [14] [21]. Most of the present-day tectonic activity is focused at the margins of the intermontane basins. For example, the Chon-Kemin earthquake of 1911 (Ms = 8.2) has activated a fault within the Kemin fault zone, north of the Issyk-Kul Basin, over more than 200 km [22]. A considerable component of strain is currently also being accommodated by younger fault systems that have propagated into the basin centres during the late Quaternary [19].

4. Methods and data

The bathymetry, seismic and gravity coring information discussed in this study comprise both existing data sets and recently acquired new data.

The 1:100,000 bathymetry map of Lake Issyk-Kul has been digitised at the Royal Museum of Central Africa, and the data were merged with those of digitised 1:200,000 topography maps of Kyrgyzstan and Kazakhstan, in order to create a digital terrain model (DTM) of the entire Issyk-Kul Basin and surrounding mountain ranges [23].

In 1982, a very first reflection seismic survey was conducted on the lake by Moscow University. A total of 31 seismic profiles were collected across the lake, using a single-tip sparker source and a single-channel receiver. Only a few interpreted line drawings of these profiles have ever been published [24].

A second reflection seismic survey was carried out in 1997, by The Renard Centre of Marine Geology in cooperation with the Royal Museum of Central Africa and the Seismological Institute. About 990 km (62 profiles) of seismic profiles were acquired, using a 500 J multi-electrode sparker, a single-channel streamer, GPS positioning and digital recording on a Delph2 acquisition system (Fig. 3). Post-acquisition processing involved frequency band-pass and spiking deconvolution filtering. The data have a theoretical (Rayleigh criterion) vertical resolution of < 1 m; penetration is highly variable, but in places it exceeds 350 ms TWT.

In 1998, the Royal Museum of Central Africa, the Institute of Geology and the Université de Savoie collected 28 short gravity cores (up to 1.76 m long) from different parts of the lake, along a longitudinal and a latitudinal transect (Fig. 3). The core locations had been selected on basis of the results of the 1997 seismic survey. These cores have been sub-sampled and characterized using gray-scale analysis, magnetic susceptibility measurements, grain-size analysis, x-ray diffraction measurements, thin sections studies and pollen analysis. A chronological framework has been established on basis of ^{14}C AMS and ^{210}Pb dates [25] [10] [26].

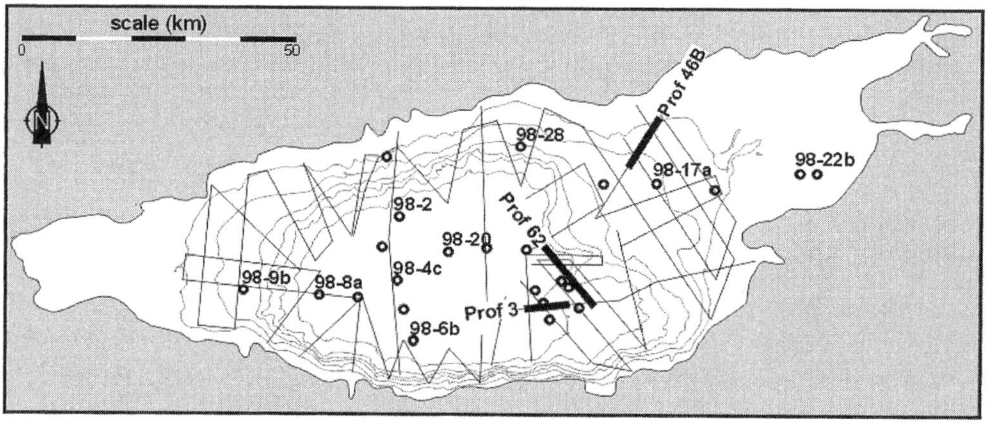

Figure 3. Location map of the seismic profiles (shot in 1997) and sediment cores (collected in 1998) used for this study.

5. Morphology of the lake floor

Lake Issyk-Kul is characterized by a highly unusual lake-floor morphology (Fig. 4). Different distinct morphological "provinces" can be discerned. The most striking ones are two shallow platforms, and a flat-floored central deep, which are separated by very steep slopes.

5.1 THE WESTERN AND EASTERN MARGINS

The western and eastern parts of the lake consist of gently dipping shelf-like areas or platforms that extend over distances of 40 km (west) to 60 km (east) from the shorelines in water depths between 0 and about 300 m (Fig. 4). These platforms represent about half of the total lake-floor surface. They both comprise two distinctly different parts: a shallow part between 0 and 110 m, and a deeper part between 110 and 300 m (Fig. 4).

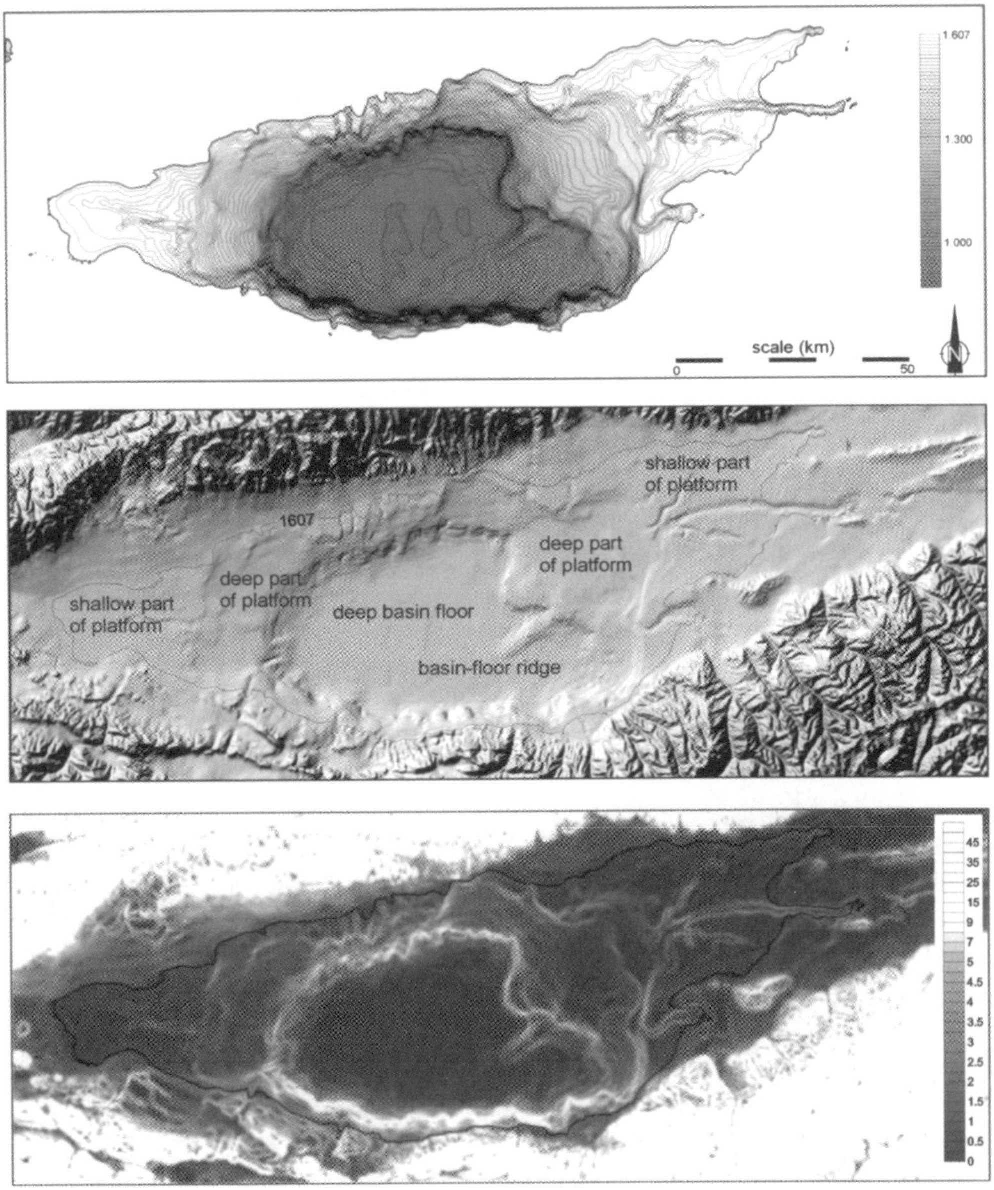

Figure 4. Lake-floor morphology maps of Lake Issyk-Kul, based on DTM [23].
Top: Grey-shaded bathymetric contour map (with isobaths every 100 m, in m above sea-level).
Middle: Shaded-relief map with indication of the main morphological "provinces".
Bottom: Grey-coded slope gradient map.

The shallow part of the platforms, of which most is shallower than 50 m, has an average dip angle of 0.5° and is incised by a basinward converging system of channels (Fig. 4). These channels are about 2-3 km wide, up to 50 m deep, and abruptly disappear at 110 m of water depth. The main channel in the eastern platform extends all the way to the lakeshore, to the mouth of the inflowing Djyrgalan River; further to the southwest a solitary channel links up to the mouth of the Chong-Kyzyl-Su River. The main channels in the western part become very subdued in water depths of less than 25 m but appear to extend to the former mouths of the inflowing and outflowing Chu River.

The mid-platform break at about 110 m (corresponding to 1,495 m above sea level) is a very prominent feature in the platform morphology. It is particularly well expressed on the eastern platform, but also clearly visible on the western platform. It occurs at exactly the same water depth around the lake.

The deeper part of the platforms is not incised and has an average dip angle of about 1° (Fig. 4). It is characterized by a series of low-relief terraces (with heights of up to 20-30 m) that exhibit a slight basinward convex shape in map view; these terraces are best developed on the eastern platform.

The platforms transit abruptly into steep slopes (Fig. 4) that extend to central basin floor, at water depths of over 600 m. The slope of the western margin has an average dip angle of 3-6° and its southern part is incised by a number of small canyons. Along the eastern margin, the transition from the platform edge to the slope is more gradual than in the west, and the slope is not incised by canyons. The northern part of the eastern slope has a two-step appearance, with a gradually dipping (2-3°) upper part between 300 and 400 m, and a steeply dipping (4-6°) lower part between 400 and 600 m. The southern part has an average dip angle of 1-5°.

5.2 THE NORTHERN MARGIN

The northern margin of the lake is characterized by a small, shallow-water shelf (Fig. 4). It occupies the central part of the northern shoreline over a distance of about 30 km, and extends for about 3-5 km from the shore in water depths of 25-50 m. It has an average dip angle of 1°. It is incised by a series of short trunk channels that connect to the shoreline, to the mouths of the small rivers draining the nearby Kunghei Alatau Range (e.g. the Orto-Koi-Su River, west of Cholpon-Ata).

In basinward direction, the shelf transits into a two-step slope (Fig. 4). The upper slope extends from 25-50 to 300-350 m water depth, with an average dip angle of 2-4°; the lower slope from 300-350 to 600 m with an average dip angle of 4-9°. The upper slope has an irregular, bumpy morphology, and some of the shelf channels appear to extend down to these depths. Where the shelf is only very small, e.g. off Korumdy (just west of Cholpon-Ata), the upper slope starts virtually at the shoreline. The lower slope is very steep and strongly incised by several canyons.

5.3 THE SOUTHERN MARGIN

The southern margin of the lake consists of a single steep slope, extending from the shoreline to the central basin floor (Fig. 4). It has an average dip angle of 8-10°, and is highly irregular with small terraces and a number of canyons that seem to originate from some small rivers draining the southern part of the Issyk-Kul basin (e.g. the Tong and the Ak-Terek Rivers).

5.4 THE DEEP BASIN FLOOR

The deep basin floor is essentially flat, apart from some minor undulations. Water depths range from 600 m at the foot of the slopes to the maximum depth of 668 m in the central part of the basin (Fig. 4).
The eastern part of the basin floor is characterized by the presence of a ridge-like structure (Fig. 4). It has a SW-NE orientation, a maximum width of 4 km and a length of 20 km. It reaches a maximum height of 150 m with respect to the surrounding basin floor. It is connected to the eastern slope and plunges to the southwest where it disappears as a morphological feature.

6. Seismic profiles of the lake structure and infill

The 1997 seismic data cover most of the above-described morphological "provinces" (Fig. 3). Because of the highly variable penetration of the seismic profiles (due to e.g. the presence of shallow gas or very coarse-grained deposits), the often very steep slopes and the strong lateral variations in seismic facies, it has not been possible to establish a whole-lake seismic stratigraphic framework. Below, the seismic data will thus be discussed per morphological province.

6.1 THE WESTERN AND EASTERN MARGINS

Seismic profiles from the western and eastern platforms usually have a penetration of 100-200 m, except where they are characterized by acoustic blanking, probably due to the presence of shallow gas. Deeper areas show a complex stratigraphy. Sedimentary units of variable thickness (typically up to 40-50 m, but laterally changing in thickness quickly) are separated by distinct unconformities marked by onlap, downlap, toplap, regional erosional truncation or localized erosional incisions. Unconformities typically amalgamate in basinward as well as in landward direction.
Three main seismic facies types can be discerned in the platform strata (Fig. 5). The most common type consists of relatively parallel, continuous reflectors of medium to high amplitude. Packages of these reflectors alternate with units with a much more subdued facies that can be almost reflection-free or contain some weak parallel reflectors. The third facies type, which occurs interbedded between the others, consists of lens- or wedge-shaped units (typically up to 50 m thick and 10-15 km across) characterized by a typical offlap pattern in longitudinal (i.e. east-west directed) cross-section and by bi-directional downlap in transverse cross-section. Topsets are often associated with high-amplitude returns and gas blanking. The wedges have the typical

characteristics of prograding delta lobes, similar to those observed in other large tectonic lakes [27] [28]. Interlobe areas are sometimes highlighted by channel incisions with high-amplitude basal reflectors and limited penetration, suggesting a coarse-grained basal channel fill. Several of such lobes occur at different stratigraphic levels and at different distances from the shores in a complicated stacking pattern. Some of the lobes are virtually at the lake floor, some are completely buried and others are only partially buried and still have a lake-floor expression in the form of the convex terraces in the lower part of the platforms (Fig. 5).

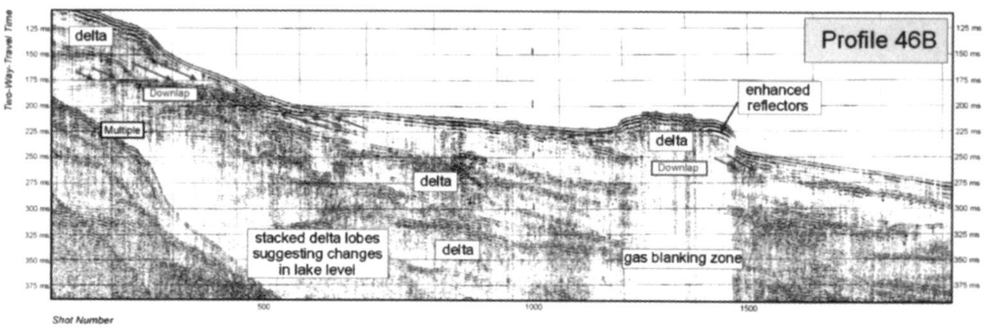

Figure 5. Seismic profile across the eastern platform, showing present-day and buried proximal delta deposits and the various seismic facies associated with it.

Towards the distal parts of the lower platforms, the stratigraphic variability becomes more moderate and the units become more homogeneously stratified, especially in the northern part of the eastern platform. Elsewhere, the units tend to thicken towards the platform edges, where they form large progradational lobes on the upper part of the slope. The steep slopes themselves are difficult to image on seismic records, but they appear to be characterized by chaotic units, suggesting that they are gravitationally unstable.

In the southern part of the eastern platform, strata are locally cut and displaced by a fault that continues up to the lake floor, suggesting it is currently active.

6.2 THE NORTHERN MARGIN

Seismic profiles from the northern margin generally have a very limited penetration. This is partly due to the steep slopes, but also profiles across the rather flat small shelf and the less steep upper slope show strongly reduced penetration. A number of irregular wedges with a variable, chaotic to roughly basinward prograding facies can be discerned, stacked on top of each other. Their seismic facies and the limited penetration suggest they are composed of very coarse-grained deposits. They downlap onto the upper slope, or prograde across the slope edge.

The presence of irregularities in the slope morphology and of mounded chaotic units further downslope indicate that intensive slope failure is taking place.

The structuration of the margin into a shelf and an upper and lower slope appears to be structurally controlled.

6.3 THE SOUTHERN MARGIN

The southern margin is characterized by very limited seismic penetration due to the high slope angles. Locally, there are some small mid-slope terraces, which are most probably fault controlled and have accumulated small patches of sediment.
The accumulation of chaotic units near the foot of the slope indicates that also the southern margin is subject to intensive slope failure.

6.4 THE DEEP BASIN FLOOR

Seismic profiles from the central deep basin floor typically have a penetration of 200-300 m. They generally show a uniform seismic facies, consisting of parallel, laterally very continuous reflectors, without apparent unconformities. These reflectors occur in 15-30 m thick packages with alternating higher and lower amplitudes. At the foot of the slopes, individual reflectors or groups of reflectors show a slight thickening in the direction of the margin.

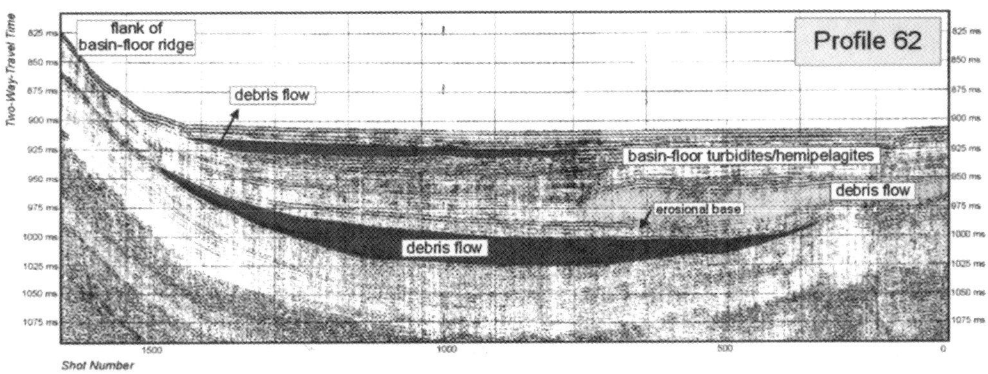

Figure 6. Seismic profile across the deep basin floor, showing a series of superimposed debris-flow deposits north of the basin-floor ridge.

In the eastern part of the basin floor, several large lenses with an acoustically transparent seismic facies occur in between the conformable, stratified units (Fig. 6). These lenses may reach up to 30 m in thickness, and are up to 3-5 km across. They usually have a basin-filling, ponding appearance, with smooth upper and lower boundaries, although sometimes they may also display irregular, erosive basal unconformities or have a mounded appearance with a convex-upward upper boundary. The lenses have the typical characteristics of debris-flow or other mass-flow deposits that have been observed in other lakes [29] [30]. Their occurrence appears to be restricted to the area just downslope from the major progradational lobes on the eastern platform edge.

Just south of the area with the debris flow lenses, seismic profiles reveal the internal structure of the basin-floor ridge, which consists of anticlinally folded basin-floor strata, transected by a gently dipping fault parallel with the fold axis. The ridge thus appears to be a structural feature, rather than a depositional or erosional one.

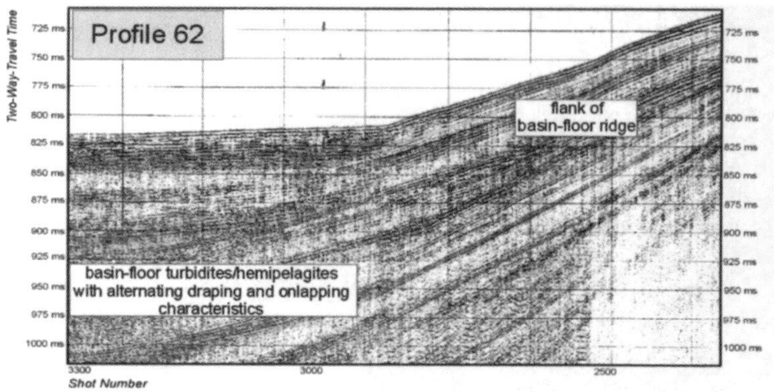

Figure 7. Seismic profile across the deep basin floor, showing the onlapping turbidites against the flanks of the basin-floor ridge.

In the vicinity of the ridge, the basin-floor strata have a slightly different appearance: they consist of an alternation of higher-amplitude reflector packages (typically about 15 m thick) that onlap and thin towards the ridge, and of lower-amplitude reflector packages (about 20-30 m thick) that maintain their stratigraphic thickness across the ridge (Fig. 7).

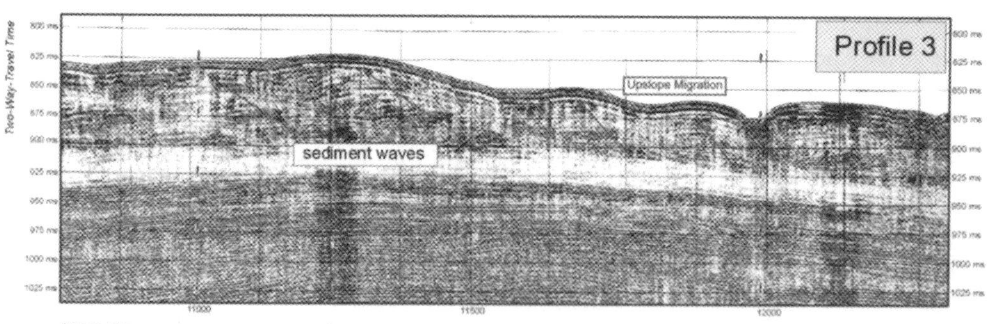

Figure 8. Seismic profile across the deep basin floor, showing the sediment waves occurring at the nose of the basin-floor ridge.

At the nose of the ridge, the uppermost 50 m of the basin-floor strata are characterized by the presence of large wavy structures, overlying horizontally-bedded strata. The waves are typically about 10 m high, they have a wavelength of about 800-900 m and appear to be migrating up-slope, towards the crest of the ridge (Fig. 8). They resemble sediment waves, typical for current-controlled deposits. Similar features have been observed in other large lakes and are attributed to the effects of bottom currents [31] [32] [33].

7. Interpretation of depositional environments

The sediments in Lake Issyk-Kul are either generated in situ by chemical precipitation of several carbonate phases, or they are terrigenous in origin and brought into the lake by inflowing rivers [9]. The main source of terrigenous sediment is presently the Djyrgalan River, which drains the basin axially and flows into the lake at its easternmost end. Much shorter rivers drain the northern and southern margins and carry a much less significant sediment load. The upper Chu river is larger than the Djyrgalan River and when it was flowing into the lake, it should have been a major source of sediments in the lake, at its westernmost end.

Depositional environments within the lake are thus controlled by the interplay between authigenic formation and terrigenic input, and by dispersion and redistribution processes under the influence of currents and along steep slopes.

Based on the interpretation of our seismic profiles and bathymetry data, and supported by sedimentological evidence from shallow gravity cores (Fig. 9), we distinguish up to 6 main depositional environments: proximal delta deposits, distal delta deposits, glacial outwash deposits, mass-flow deposits, basin-floor turbidites/hemipelagites and basin-floor current-controlled deposits (Fig. 10).

7.1 PROXIMAL DELTA DEPOSITS

We interpret the strata making up most of the western and eastern platforms as proximal delta deposits, consisting primarily of delta plain deposits and of prograding delta lobes exhibiting the characteristic topset-foreset-bottomset configuration. Angles of foreset slopes are around 1.5°. The delta-lobe strata are deposited out of the bedload and the coarser-grained component of the suspended load from rivers entering the lake. The main tributaries are those entering the lake at the eastern side and draining the basin axially, i.e. the Djyrgalan River and the Tioup River. Delta lobes in the western platform have probably been deposited by the Chu River, which was the main tributary at the western part of the lake before it was deflected to the west. Delta lobes are typically about 5 km^3 in volume, and each represents a significant period of sedimentation under relatively stable conditions.

The position of the individual lobes within this succession has switched laterally, due to autocyclic processes, and longitudinally, due to allocyclic processes that are primarily driven by changes in lake level or significant changes in sediment load of the tributaries. During periods of high lake level, delta lobes developed high up the platforms, while during periods of lake-level lowering these deposits got cannibalised and re-deposited on the lower-lying parts of platforms.

Figure 9. Some of the sediment cores discussed in the text. See Fig. 3 for location.

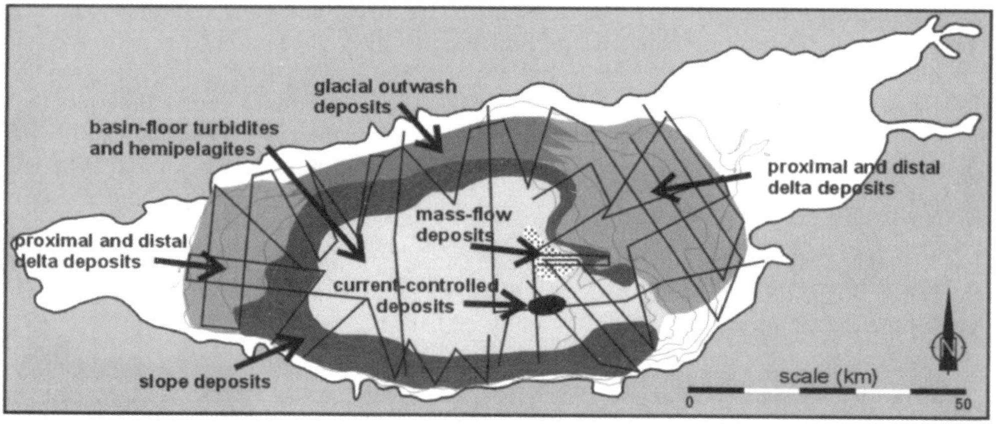

Figure 10. Morphosedimentary map of Lake Issyk-Kul showing the main depositional environments as interpretated from the seismic profiles, bathymetry data and shallow sediment cores.

Several sub-environments can be distinguished in the proximal delta deposits. Inter-lobe channels funnel parts of the coarser bedload/suspended load further across the platform in basinward direction, where it is deposited on the lower platform. There is thus an important amount of sediment bypass in the delta area. Core 98-17a from the lower platform area, about 100 m deeper than the mouth of the main channel consists of a layer of fine sands at the bottom, which progressively changes upwards to a centimetric to millimetric alternation of light and dark layers of clays and silty-clays. Delta plain deposits are characterised on seismic records by toplap, high-amplitude reflection returns and gas blanking. This is probably due to the presence of shallow gas, originating from the biogenic disintegration of organic material in the coarse proximal delta deposits, i.e. most likely soils that developed in the delta plain area as this prograded and gradually became sub-aerially exposed. Core 98-22b shows the presence of very compact clayey, sandy to gravelly deposits [26].

These proximal delta deposits thus essentially consist of coarse-grained bedload sediments, intermixed with slightly finer-grained suspended load sediments deposited out of interflows. These deposits may have been sub-aerially exposed one or several times after their formation, and they may have been often reworked as a result of lake-level fluctuations and of autocyclic or allocyclic switching of channel courses.

7.2 DISTAL DELTA DEPOSITS

On the outer parts of the western and eastern platforms, more homogeneously stratified deposits locally occur, intermixed with the proximal delta deposits. We interpret these deposits as distal delta deposits, consisting primarily of the fine-grained component of the suspended load from rivers settled out on the delta fronts. This fine-grained material has been transported across the platforms as interflows, at the level of the thermocline during summer stratification. The suspended particles get carried along with the counter-clockwise-moving currents and deflected to the right by the Coriolis force and finally settle out in the distal parts of the platforms, at the right-hand side from the downstream direction of the inflowing rivers. The most important accumulation occurs in the northern part of the distal eastern platform (originating from the main inflowing Djyrgalan River). Core 98-28, which consists of a rhythmic alternation of light and black silty-clay sediments [10], confirms that these distal delta deposits are essentially fine-grained suspended load sediments that have settled from the water column (Fig. 9). They have accumulated continuously through time controlled by the current regime, the thermal stratification characteristics of the water column and the suspended sediment load of the inflowing rivers, and are virtually undisturbed.

7.3 GLACIAL OUTWASH DEPOSITS

By comparison of the seismic facies with onshore outcrops, we interpret the acoustically chaotic, generally prograding sediment wedges on the small shelf and upper slope of the northern margin as fluvioglacial outwash deposits. Onshore, the whole north-central part of the shoreline is dominated by sandur deposits, consisting of very coarse-grained, poorly sorted sediments, containing blocks of several dm in diameter. Such deposits are believed to have been generated by the breaching of the enormous terminal moraines at the foot of the Kunghei Alatau Range (i.e. well developed above Cholpon-Ata, at 2,500 m altitude) and subsequent drainage of formerly dammed proglacial melt-water lakes. These melt-water outbursts probably extended into Lake Issyk-Kul, where the sediment load was quickly deposited on the small shelf due to rapid deceleration of the flow.
Presently, the rivers draining the Kunghei Alatau Range carry very little sediment and only a thin layer of suspended load fall-out covers the fluvioglacial outwash deltas in the lake.

7.4 MASS-FLOW DEPOSITS

The steep fault-controlled walls of the deep basin are evidently prone to slope failure. This is particularly true for the slopes that receive larger amounts of sediment, such as those of the western and especially the eastern platforms. The fault-controlled slopes of the northern and southern margin are extremely steep and have not allowed large quantities of sediment to accumulate, but also here are indications for slope failure. Mass-flow deposits are evident from the seismically chaotic mounded heaps (slumps) at the foot of the northern and southern slopes, and from the seismically transparent debris-flow deposits along the platform-edge delta fronts of the eastern margin. The debris-flow deposits appear to not have had long run-out distances as they occur close to the foot of the slopes. Nevertheless, it is not excluded that they may have partially developed into turbidites, which may have extended much further in the basin. Sediment cores 98-09b, located at the upper part of the delta front, and 98-08a (Fig. 9), located at the foot of it, contain inversely graded gravels and sands, together with slumped and folded laminated sediments overlying an erosive base [26] [10]. They confirm that these

foot-of-slope mass-flow deposits may contain quite coarse sediments, which thus most likely find their origin in the failure of proximal prograding delta deposits or basinward funnelled channel deposits.
Slope failure may have occurred simply as a consequence of rapid sediment accumulation due to delta progradation onto the slopes, or it may have been triggered by external factors. Earthquake triggering seems a likely cause for generating slope instabilities in a region of active mountain building like the Tien Shan. Our records indicate the presence of 5 major debris flow units incorporated in the upper 200 m of basin-floor strata along the eastern margin [34]. Assuming an average sedimentation rate of 0.3 mm/yr, as deduced from sediment cores from the basin floor [26], this means 5 large-scale slope failures in about 670 kyr.

7.5 BASIN-FLOOR TURBIDITES/HEMIPELAGITES

The seismic facies on the basin floor strata, with the distinct alternation of intervals with higher and lower reflection amplitudes, is typical of deep-water depositional environments, e.g. a mixture of biogenic or fine terrigenous hemipelagic deposits and terrigenous sediments transported to the basin floor by turbidity currents. This is confirmed by our sediment cores from the deep basin, which invariably consist of gravels and sands in fining-upward sequences (near the steep slopes, such as in cores 98-02 and 98-06b) representing a proximal turbidite facies, or of homogenous dark muddy sediments with sub-millimetre-scale sand lenses (at the centre of the lake, such as in cores 98-4c and 98-20; Fig. 9) representing a distal muddy turbidite facies or a homogenite facies [35]. These central basin-floor deposits may thus represent run-out turbidites resulting from gravitational slope failures (seismically triggered or not) along the steep slopes or they may represent underflow deposits resulting from sediment charged melt-water input (mainly from the northern margin, as can be deduced from the magnetic susceptibility records) in a thermally not stratified lake during spring. In any case, the cores indicate that all these sediments are heavily re-worked.
The configuration around the basin-floor ridge in the eastern part of the deep basin, offers the possibility to estimate the relative proportions of turbidites and hemipelagites in the basin-floor sedimentation. The anticlinal ridge has functioned as a topographic feature during sedimentation. Around the ridge, the basin-floor deposits are characterised by an alternation of higher-amplitude reflector packages that onlap and thin towards the ridge, and of lower-amplitude reflector packages that maintain their stratigraphic thickness across the ridge. The latter most probably represent essentially the draping hemipelagite component of the basin floor deposits, while the former are more likely to represent essentially the ponding and onlapping turbidite component. The seismic data thus indicate that there have been alternations of periods with increased turbidite input and of periods with predominantly hemipelagic sedimentation. These alternations may reflect climatically driven changes in underflow intensity (precipitation changes or periods of enhanced glacier melting), and/or tectonically driven pulses of increased slope failures (higher seismicity). Assuming an average sedimentation rate of 0.3 mm/yr for the basin-floor sediments [26] and for the sake of simplicity disregarding

the evident differences in sedimentation rates between turbidites and hemipelagites, the observed facies alternations suggest that the periods of enhanced turbiditic activity had a duration of about 50 kyr and a recurrence rate of about 117-150 kyr.

7.6 BASIN-FLOOR CURRENT-CONTROLLED DEPOSITS

We interpret the large sediment waves that affect the basin-floor deposits in the vicinity of the ridge as induced by bottom-current activity. In agreement with existing views on deep-ocean current-controlled or "drift" sedimentation, we believe that deep-water currents in the central basin depression are responsible for the deflection and entrainment of hemipelagic particles causing them to accumulate in the down-current direction. Similar drift deposits are increasingly being identified in large lakes: i.e. the East African Rift lakes [31] [32], Lake Baikal [33]. Their formation, and more particularly the nature of the currents causing them, remains however still poorly understood.

8. Lake-level changes

The channel incisions, terraces and buried delta lobes that characterize the western and eastern platforms clearly demonstrate that the lake has witnessed several important fluctuations in lake level. Morphological steps, such as those associated with terraces and delta lobes, may thus be abrasive or depositional in nature, or a combination of both, but do point to a lower-than-present lake level.

The shape and pattern of the incised channels in the upper part of the platforms, the fact that they visibly connect to the mouths of inflowing rivers and their sudden disappearance at the edge of the upper part of the platforms at about 110 m of water depth (1,495 m above sea level) indicate that they were generated by fluvial incision during a time when the lake level was about 110 m lower than present. The age of this lake-level lowstand has been dated variably, ranging from Late Pleistocene [11] [5] to Middle Holocene [36].

Older terraces, at about 18 and 40 m above the present lake level, are also well preserved along the south-western part of the lake, and especially between Kadji-Say and Ribatche. The lowest one corresponds to the altitude of the overflow level of the lake into the lower Chu River (1,622 m above sea leavel). This level was reached by the lake in the first half of the 19th century. A major terrace is also visible at an altitude of 1,645-1,650 m above sea level. It corresponds in the field to the limit between low smoothed slopes and higher rugged badlands. This terrace can be interpreted as an abrasive/depositional scarp of a stable lake-level highstand, the age of which is also debated and ranges from Late Pliocene [11] to Middle Pleistocene [37].

Lake-level changes can be due to two main causes: climate (changes in precipitation/evaporation conditions, changes in temperature leading to increases or decreases in melt-water production) and tectonics (diversion of inflowing rivers due to surface deformation, closure of outflowing rivers due to margin uplift, deepening or shallowing of the basin). The 110 m lowstand of Lake Issyk-Kul is generally attributed to tectonic processes. Some authors postulate that these tectonic processes are related to and coeval with those that caused the deflection of the Chu River and its disconnection from Lake Issyk-Kul at about 13-18 kyr BP [5]. Trofimov [36] attributes the lowstand to a major tectonic event causing a.o. the assumed collapse of the central part of the lake (see below).

The presence of other terraces and delta lobes in the lower part of the underwater platforms, at water depths below 110 m, suggests that there have been periods of even lower lake level earlier in the Pleistocene: i.e. at 150 m, 250-270 m and 380-400 m.

The terrace at 40 m above lake level, attributed to a stable lake-level highstand, is also interpreted as resulting from compressional tectonic activity along the Chon-Kemin fault zone that crosses the Lower Chu River, which was previously outflowing from the lake.

9. The myth of the central basin collapse

Based on morphology and on information from seismic profiles [24], Trofimov [36] interpreted the central flat-floored basin deep with its steep and evidently fault-controlled margins as resulting from a sudden tectonic collapse, over a height of about 200 m, during the middle Holocene. He invokes this collapse as the cause of the 110 m lake-level lowstand.

In retrospect, this collapse scenario seems, however, a rather unlikely hypothesis. A 200 m tectonic subsidence of the central basin floor since the middle Holocene (i.e. 5 kyr) would imply tectonic subsidence rates of at least 4 cm/yr, assuming the process has continued up to present, and even much higher if the collapse is considered as a shorter-lived event as proposed [36]. These are extremely high deformation rates for extension in an intracontinental environment and difficult to reconcile with the present-day, geodetically determined shortening rates in the area, which are in the order of 1 cm/yr [14].

The new seismic and core data also bring no evidence in support of the collapse hypothesis. Although the seismic profiles across the steep northern and southern basin slopes confirm their fault-controlled origin, they do not provide clues as to the timing of the main fault activity. However, most of the slope-controlling faults are covered by significant amounts of relatively undeformed deposits, suggesting they have not been very active in recent times. Also, a major tectonic event as a basin-floor collapse would have a basin-wide impact on the sedimentary infill, visible on seismic records and in sediment cores: slumping, sliding, in-situ deformation, faulting, seiche development... Our data show no evidence of such basin-wide synchronous effects. Given the penetration depth of our seismic records on the central basin floor of about 350 m and average sedimentation rates of 0.3 mm/yr [26], we postulate that the subsidence of the central part of the basin – if catastrophic at all – must be older than about 1.2 Myr, and thus much older than previously believed. Multi-channel, deep seismic profiling would be required to solve this problem...

10. Tectonic deformation of the basin interior

Current models for the structural evolution of the northern Tien Shan [14] [21] [19] indicate that most of the Cenozoic strain has been accommodated by the broad deformation zones separating the intermontane basins from the surrounding mountain ranges, and that the basin interiors have been largely protected from tectonic deformation. However, field data indicate that since the late Quaternary younger fault systems have started to propagate into the basin centers where they currently absorb a significant amount of the deformation.

Our reflection seismic data bring clear evidence for recent compressive deformation in the centre of the Issyk-kul Basin [34]. The ridge affecting the eastern part of the deep basin floor consists of anticlinally folded basin-floor strata, transected by a gently dipping fault parallel with the fold axis. The fact that the anticline still persists in the lake-floor morphology suggests that it is a relatively young feature. If not, it would already have been buried as a result of the ponding and onlapping behaviour of the turbidite component of the basin-floor deposits. In fact, detailed analysis of the lap-out patterns of this turbidite facies against the flanks of the ridge indicates that the onlap limit does not gradually approach the crest of the anticline as would be expected with a gradual burial (Figure 15). This suggests that the anticline is still actively growing [34] [26].

Up to present, the anticlinal fold is the only indication for recent or active tectonic deformation in the lake basin. Ongoing studies are currently investigating the relationship of this fold with a series of recent strike-slip and reverse fault structures that have recently been discovered along the southern shores of the lake. Similar anticlinal structures are also well expressed in the onshore part of the basin-floor topography towards the eastern termination of the basin.

11. Conclusions

Integrated interpretation of existing bathymetry data, new high-resolution reflection seismic profiles and new short sediment cores that were collected from Lake Issyk-Kul over the past years allows us to define and characterise the main sedimentary environments in the lake. They consist of proximal and distal delta deposits on the eastern and western shallow platforms, glacial outwash deposits along the northern margin, mass-flow deposits, basin-floor turbidites/hemipelagites and basin-floor current-controlled deposits on the flat deep basin floor. A good understanding of the distribution of these sedimentary environments is necessary before collecting sediment cores for paleoclimate studies or environmental investigations.

The seismic data confirm that the lake has been subjected to severe fluctuations in lake level during most of the Quaternary (up to several hundreds of meters below the present level), and that active compressive tectonic deformation is locally affecting the central part of the Issyk-kul Basin. Our data, however, do not bring any evidence in favour of the hypothesis of a mid-Holocene central basin collapse, which is often invoked in order to explain the peculiar lake-floor morphology.

12. Acknowledgements

We thank the captain and crew of the R/V Møltur from which the seismic data and sediment cores were collected in 1997 and 1998. The work on Lake Issyk-Kul was partially funded by the Belgian Science Policy Office. YI is funded by the Flemish IWT.

13. References

1. Korotaev, V.N. (1967) Coastal zone of Lake Issyk-Kul. Frunze, 151 pp.
2. Kodyaev, T.V. (1973) Morphometrical characteristic of Issyk-Kul Lake. *Izv. Vses. Geogr. Obshch.*, **105(4)**, 48-57.
3. Zabirov, R.D. & Korotaev, V.N. (1978) Location and morphometry of the lake. In: Lake Issyk-Kul. Frunze, 12-20.
4. Sevastyanov, D.V., Mamedov, E.D. & Rumyantsev, V.A. (Eds.) (1991) History of Lakes Sevan, Issyk-Kul, Balkhash, Zaisan and Aral. *Nauka*. Leningrad. [in Russian]
5. Bondarev, L.G. & Sevastyanov, D.V. (1991). Relief of shores and lake bottom. In: The history of Sevan, Issyk-Kul, Balkhash, Zaisan and Aral Lakes. *Nauka*, Leningrad, 78-86.
6. Aladin, N.V. & Plotnikov, I.S. (1993) Large saline lakes of former USSR: a summary review. *Hydrobiologia*, **267**, 1-12.
7. Atlas Kyrgyzskaja (1987) Moscow. 157 pp.
8. Sapozhnikov, D.G. & Viselkina, M.A. (1960) Recent sediments of the Lake Issyk-Kul and its bays. Publ. House Akademii Nauk, Moscow. [in Russian]
9. Giralt, S., Lignier, V., Klerkx, J., Julià, R., De Batist, M., Beck, C. & Kalugin, I. (2000) Sedimentological processes in Lake Issyk-Kul (Republic of Kyrgyzstan, Central Asia): the importance of tectonically influenced sedimentation. Abstract at the 8th International Symposium on Paleolimnology, 19-24 August 2000, Kingston, Canada.
10. Giralt, S., Riera, S., Klerkx, J., Julià, R., Lignier, V., Beck, C., De Batist, M. & Kalugin, I. (this volume) Recent paleoenvironmental evolution of Lake Issyk-Kul.
11. Markov, K.K. (Ed.) (1971) Sequence of the latest deposits of the Issyk-Kul basin. Moscow University Publishing House, Moscow, 164 pp. [in Russian]
12. Chedia, O.K. (1986) Morphology and neotectonics of the Tien Shan. Ilim Publications, Frunze, 313 pp. [in Russian]
13. Cobbold, P.R., Davy, P., Gapais, D., Rossello, E.A., Sadybakasov, E., Thomas, J.C., Tondji Biyo, J.J. & Urreiztieta, M. (1993) Sedimentary basins and crustal thickening. *Sedimentary Geology*, **86**, 77-89.
14. Abdrakhmatov, K.Ye, Aldazhanov, S.A., Hager, B. H., Hamburger, M.W., Herring, T.A., Kalabaev, K.B., Makarov, V.I., Molnar, P., Panasyuk, S.V., Prilepin, M.T.,

Reilinger, R.E., Sadybakasov, I.S., Souter, B.J., Trapeznikov, Yu.A., Tsurkov, V.Ye. & Zubovich, A.V. (1996) Relatively recent construction of the Tien Shan inferred from GPS measurements of present-day crustal deformation rates. *Nature*, **384**, 450-453.
15. Abdrakhmatov, K.E., Turdukulov, A.T. & Khristov, E.B. (1993) Detailed seismic zoning of the Issyk-Kul basin. Ilim Publications, Bishkek. X pp. [in Russian]
16. Cobbold, P.R., Sadybakasov, E. & Thomas J.C. (1996) Cenozoic transpression and basin development, Kyrgyz Tien Shan, Central Asia. In: Roure, F., Ellouz, N., Shein, V.S. & Skvortsov, I. (Eds.) Geodynamic evolution of sedimentary basins. Technip, Paris. 181-202.
17. Abdrakhmatov, K.E., Delvaux, D. & Djanuzakov, K.D. (this volume) Active tectonics and seismic hazard of the Issyk-Kul basin in the Kyrgyz Tian-Shan.
18. Trofimov, A.K. (1990) Quaternary deposits of Issyk-Kul Basin in connection to its tectonics. *Izvestia A.S. Kirghyzkoy SSR*, N1. [in Russian]
19. Bullen, M.E., Burbank, D.W., Garver, J.I. & Abdrakhmatov, K.Y. (2001) Late Cenozoic tectonic evolution of the northwestern Tien Shan: New age estimates for the initiation of mountain building. *Geol. Soc. Am. Bull.*, **113(12)**, 1544-1559.
20. Fortuna, A.B. (1993) Detailed seismic profiling of the Issyk-Kul depression. Ilim Publications, Bishkek. [in Russian]
21. Ghose, S., Mellors, R.J., Korjenkov, A.M., Hamburger, M.W., Pavlis, T.L., Pavlis, G.L., Mamyrov, E. & Muraliev, A.R. (1997) The Ms = 7.3 1992 Suusamyr, Kyrgyzstan earthquake: 2. Aftershock Focal Mechanisms and Surface Deformation. *Bull. Seism. Soc. Am.*, **87**, 23-38.
22. Delvaux, D., Abdrakhmatov, K.E., Lemzin, I.N. & Strom, A.L. (2001) Landslides and surface breaks of the Ms 8.2 Kemin earthquake, Kyrgyzstan. *Geologiya i Geophysika*, 42(10), 1667-1677. [in Russian]
23. Delvaux, D. (2001) DTM of Issyk-Kul. Internal Report and CD-Rom.
24. Stavinski, S., Romanovski, V., Tarasov, S., Imankulov, B., Zekper, I., Meskheteli, A., Kuznetsov, D. (1984) Results of complex profiling of Lake Issyk-Kul. Unpublished Report. Ministry of Geology, Bishkek.
25. Kalugin, I., Lignier, V., Giralt, S., Beck, C., Klerkx, J., Thouveny, N., Pourchet, M. & Boultefroy, P. (1999) Late Holocene to present day sedimentation in Lake Issyk-Kul (Kyrgyzstan, Tien Shan). Preliminary data from the 1998 gravity coring survey. Abstract at 2nd International Congress of Limnogeology "Lennou", 25-28 May 1999, Brest, France.
26. Lignier, V. (2001) Les sédiments lacustres et l'enregistrement de la paléosismicité. Etude comparative de différents cas dans le Quaternaire des Alpes Nord-Occidentales et du Tien-Shan Kyrghyze. Ph.D. thesis, Universite de Savoie, Chambery, 231 pp.
27. Scholz, C.A. (1995) Deltas of the Lake Malawi Rift, East Africa: Seismic Expression and Exploration Implications. AAPG Bulletin, 79(11), 1679-1697.
28. De Batist, M., Van Rensbergen, P., Back, S. & Klerkx, J. (1996) Structural Framework, Sequence Stratigraphy and Lake Level Variations in Livingstone Basin (Northern Lake Malawi): First Results of a High-Resolution Reflection Seismic

Study. In: Johnson, T. & Odada, E. (eds.) *The Limnology, Climatology and Paleoclimatology of the East African Lakes*. Gordon & Breach Pub. Inc., Newark, 509-521.

29. Chapron, E., Van Rensbergen, P., Beck, Ch., De Batist, M. & Paillet, A. (1996) Lacustrine sedimentary records of brutal events in Lake Le Bourget (Northwestern Alps-southern Jura). *Quaternaire*, **7(2-3)**, 155-168.
30. Van Rensbergen, P., De Batist, M., Beck, C. & Chapron, E. (1999) High-resolution seismic stratigraphy of glacial to interglacial fill of a deep glacigenic lake: Lake Le Bourget, Northwestern Alps, France. *Sedimentary Geology*, **128(1-2)**, 99-129.
31. Johnson, T.C., Carlson, T.W. & Evans, J.E. (1980) Contourites in Lake Superior. *Geology*, **8**, 437-441.
32. Johnson, T.C. (1996) Sedimentary processes and signals of past climatic change in the large lakes of the East African Rift Valley. In: Johnson, T. & Odada, E. (eds.) *The Limnology, Climatology and Paleoclimatology of the East African Lakes*. Gordon & Breach Pub. Inc., Newark, 367-412.
33. Ceramicola, S., Rebesco, M., De Batist, M. & Khlystov, O. (in press) Seismic evidence of small-scale lacustrine drifts in Lake Baikal (Russia). *Marine Geophyiscal Researches.*
34. Imbo, Y. (1998) Studie van de evolutie van het Issyk-Kul bekken (Tien Shan, Kirgizië) gebaseerd op hoge-resolutie reflectieseismische profielen. M.Sc. thesis, Ghent Universiteit, Ghent, 80 pp.
35. Sturm, M., Siegentahler, C. & Pickrill, R.A. (1995) Turbidites and 'homogenites'. A conceptual model of flood and slide deposits. Abstract at the 16 th Regional Meeting of Sedimentology. *IAS Spec. Publ.*, **22**, pp 140.
36. Trofimov, A.K. (1978) The history of Issuk-Kul Lake during Holocene. *Bull. Komis. po izuchen. chervert perioda*, **48**, 79-86.
37. Kvasov, D.D. & Seliverstov, J.P. (1960) Some problems of paleogeography of Issyk-Kul basin. Publications of the lakes Laboratory of the USSR Academy of Sciendes. Moscow-Leningrad, 10, 45-54. [in Russian]

RECENT PALEOENVIRONMENTAL EVOLUTION OF LAKE ISSYK-KUL

S. GIRALT and J. KLERKX
International Bureau for Environmental Studies
Audrey Hepburnstreet, 9/13, B1090 Brussels (Belgium)
E-mail: sgiralt@africamuseum.be

S. RIERA
Museu, Laboratorio e Jardim Botanico de Lisboa, Laboratorio de Paleoecologia
Rua da Escola Politecnica 58, P-1294 Lisboa (Portugal)

R. JULIA
Institute of Earth Sciences "Jaume Almera" (CSIC)
Lluís Solé i Sabarís s/n, E-08028 Barcelona (Spain)

V. LIGNIER and C. BECK
Laboratoire de Géodynamique des Chaines Alpines, CNRS UMR 5025 - Université de Savoie, Centre Interdisciplinaire des Sciences de la Montagne
Bâtiment Belledonne - Campus Technolac, F-73376 Le Bourget du Lac Cedex (France)

M. DE BATIST
Renard Centre of Marine Geology, Geological Institute, University of Gent
Krijgslaan 281 S.8, B9000 Gent (Belgium)

I. KALUGIN
United Institute of Geology, Geophysics and Mineralogy, Laboratory of Cenozoic Geology and Paleoclimatology
Prosp. Akademika Koptyuga 3, 630090 Novosibirsk (Russia)

1. Introduction

Closed lakes located in arid and semi-arid environments are highly sensitive to climate oscillations since small variations in their environmental conditions induce dramatic changes within the lake [1]. The sensivity of these lakes is higher if they are located in the border of biomes, as Lake Issyk-Kul, located between the steppic and high mountains environments [2]. These facts make Lake Issyk-Kul an important ecosystem

J. Klerkx and B. Imanackunov (eds.), Lake Issyk-Kul: Its Natural Environment, 125–145.

in order to study its adaptability under the environmental changes occurred in the Central Asia region.
The study of climatic change on prehistorical and historical timescales provides a valuable perspective for interpreting the natural trend of climatic variability of the past, and a database for assessing present and future behaviour under intensive and extensive human impacts [3].
These databases should allow the development of accurate environmental policies in order to establish a regional model for the sustainable development of the socio-economic structures, which could constitute one the basis of the economy (e.g. tourism and natural resources).
Lake Issyk-Kul constitutes one of the most important recreational resources in the Republic of Kyrgyzstan, with more than 100 recreational centres along its shore, where some 370,000 visitors stay annually [4][5], and will be higher in the near future due to the general interest in well-preserved natural environments. Thus, a careful management of this ecosystem is needed in order to promote and sustain this value.
In this paper, we present a preliminary comprehensive paleoenvironmental reconstruction of the lake evolution during the last 2000 years based on pollen and sedimentological analyses of a core 1.80 m long.

2. Site description

Lake Issyk-Kul is an endorheic mountain lake located at 1608 m a.s.l., in the northern Tien Shan ranges, in the Republic of Kyrgyzstan, Central Asia (Figure 1). It has an area of 6236 km^2, a length of 250 km, a width of 60 km, and a maximum depth of 668 m making it the fifth deepest lake in the world [6]. The lake is monomictic, saline (6 g/l), oligotrophic to ultra-oligotrophic (2 – 3.8 μg/l of phosphorous [5]), and it has high values of dissolved oxygen (6.5 – 7.5 mg/l at the bottom of the lake).

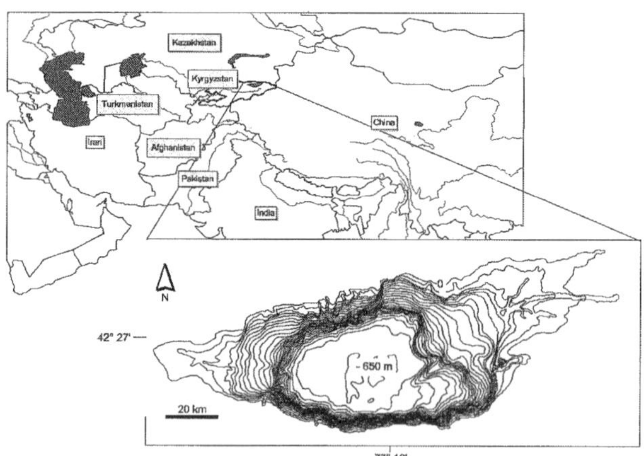

Figure 1. Geographical location of Lake Issyk-Kul.

The pH values range between 8.75 at the surface of the lake, and nearly 8 at the bottom [7]. The lake surface temperature in January is not less than 2 – 3°C, and in July it reaches 19 – 20 °C. At depths of more than 100 m, the water temperature remains constant all the year at 3.5 – 4.5 °C [8]. There are 102 streams and rivers feeding the lake. These rivers, including the two largest ones (Jergueland and Tyup rivers) are feed predominantly by melt water from glaciers and snow, located above 3,300 m a.s.l. This fact is traduced in a mean annual oscillation of the lake level of about 20 cm (Figure 2). The lake level progressively increases from February until beginning of September, decreasing afterwards progressively to the next February.

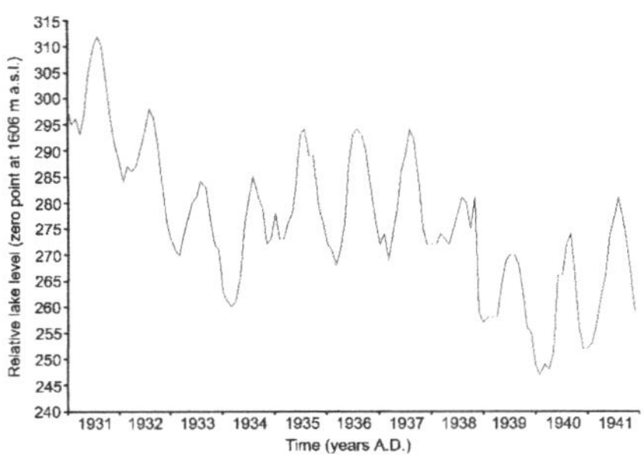

Figure 2. Monthly lake level measures for the period comprised between the years 1931 to 1941 in the town of Cholpon – Ata, located in the central northern shore of the lake.

The regional climate is highly variable from warm, temperate, and dry in the western part to slightly moist in the eastern part. Annual precipitation averages 250 mm and the annual evaporation from the surface of the lake is approximately 700 mm. The chemical water composition of the lake is dominated by sulfate among the anions and by sodium and potassium among the cations [7]. The alkalinity ranges from 310 mg/l to 330 mg/l and the calcium content is about 115 mg/l. These slight saline waters are oversaturated in calcite, monohydrocalcite and vaterite, minerals which are found in the sediments [9].
Walter and Box [10] have studied the altitudinal belts of Northern and Central Tien Shan (Figure 3), deducing that the altitudinal distribution of vegetation belts are mainly controlled by water availability.
The preliminary results of the detailed sedimentological model (seismical survey [11] and petrological studies on the recovered cores [12]) show that three main sedimentological domains have been identified [13]: 1) alternation of light and black silty-clay sediments corresponding to distal facies of deltaic lobes deposited in the eastern and western shallow shelves, apparently not affected by the regional seismicity. An interbedded coarse sandy layer in some cores from the western shelf might correspond to a debris flow, related to a punctual seismic event; 2) gravels and medium and coarse sandy sediments in a typical centimetric fining-upward sequence,

corresponding to proximal turbidite facies from the apron of the deep fans, after the bypass zone of the north and south steep borders; 3) homogenous dark muddy sediments with small lenses of very fine sands of the central deep basin, corresponding either to a muddy turbidite facies (from distal mud lobes) either to an 'homogenite' facies in the sense of Sturm et al. [14]. In both cases, those sedimentary environments indicate that all these sediments are re-worked.

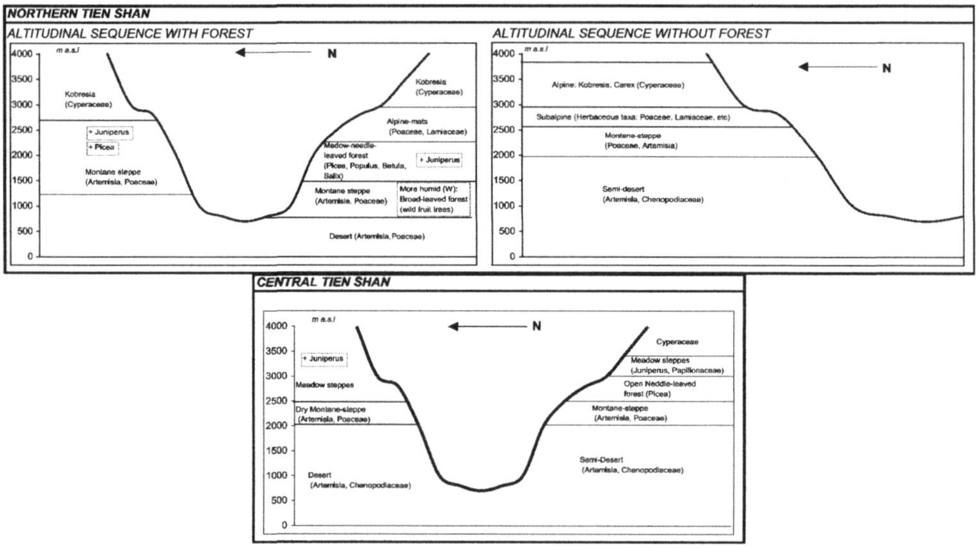

Figure 3. Altitudinal distribution of vegetation belts in Northern and Central Tien Shan ranges (modified from Walter and Box [10]).

Thus, only the sediments present in the distal part of the shallow shelves are suitable for paleoenvironmental reconstructions. For this reason, the longest core obtained in the eastern shelf (core 98i-28) has been used for this purpose.

3. Materials and methods

In September 1998, a coring program was conducted in Lake Issyk-Kul. Twenty-two gravity cores (maximum length was 1.80 m) and Ekman dredges were collected from the shallowest eastern part (23 m deep) to the deepest central part of the lake (up to 668 m deep), following a longitudinal and a latitudinal transect. The core locations were based on the results of an extensive seismic survey, carried out just before the coring [11].

Dredge sediments were used for depicting the present-day sedimentary processes while the cores were used for establishing the sedimentological model that triggers the deposition processes within the lake.
All the cores were split longitudinally, and one half were wrapped in aluminum foil and stored in a cool room at +2° C for further analyses, while the other half was used for lithological description, digitalization with a CCD camera, and sampling. These cores have been subsampled and characterized using gray-scale analysis, magnetic susceptibility, granulometry, x-ray diffraction, thin sections and pollen.
The grain size analyses were conducted using a Malvern Instruments MasterSizer/E with a 2 mW He – Ne laser (633 nm wavelength). X-ray diffractions were done with an automatic Brucker D – 5005 x – ray diffractometer: Cu radiation (K_α = 1.5405), 40 kV, 30 mA and graphite monochromator. Some sediment samples were coated with carbon or gold for scanning electron microscope (SEM) observations. SEM observations were performed in a Hitachi-3000 electron microscope. The thin sections were obtained after freeze-drying and balsam hardening the samples, and a total of 100 thin sections have been studied. The thin sections were digitalized using a CCD camera, and images were used for several measurements using Scion Image software, following the technique of Francus [15]. Samples for pollen and charcoal analysis were taken every 10 centimeters.
The pollen sequence has given us an overview of vegetation and environmental changes occurred in the lake basin during the late Holocene. Samples for pollen and charcoal analysis were prepared following standard procedures [16]. Lycopodium-tablets were added in order to obtain Pollen and Charcoal Concentration values (particles per gram of humid sediment). Minimum number of counted pollen per sample was established on 250 grains. Total pollen sum used for the calculation of percentage of each pollen taxa considered all pollen grains, excluding those from aquatic and semi-aquatic plants, that probably grow in the margins and the littoral part of the lake.

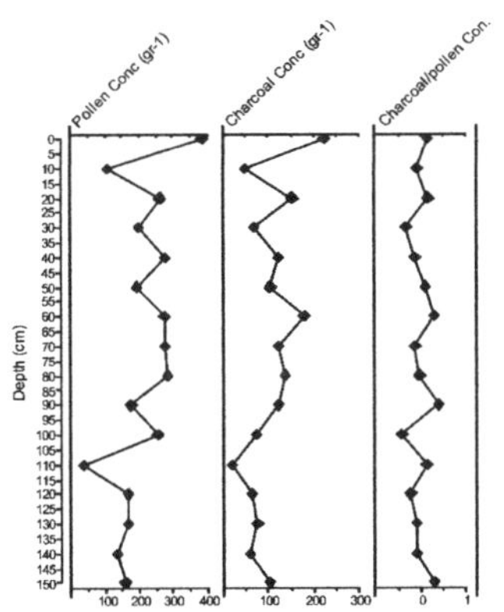

Figure 4. Pollen grains and charcoal particles concentration in the 98i-28 sequence (number of particles per gram of humid sediment).

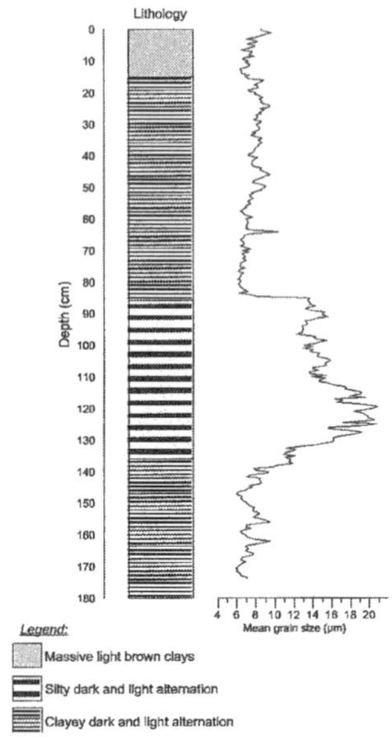

Figure 5. Main lithological units of the core 98i-28 and mean grain size of the sediments of this core.

Concentration Particle values are mainly determined by the sedimentation rate, as we can conclude from the similar concentration evolution of both, pollen and charcoal remains (Figure 4). Due to this fact, and with the objective to remove the effect of sedimentation rate, charcoal values for paleowildfires reconstruction has been calculated as follows: Charcoal Concentration x Pollen Concentration $^{-1}$ (Figure 4).

Pollen zones were established using the Edwards and Cavalli-Sforza's chord distance of the CONISS program [17].

The degree of saturation of the lake water with respect to calcite, aragonite, monohydrocalcite, and gypsum was calculated using the computer program PHREEQCI [18][19].

The chronological framework has been established by ^{210}Pb of the 10 upper centimeters and by an AMS ^{14}C date, at 15 centimeters of depth.

4. Results

4.1 SEDIMENTOLOGICAL DESCRIPTION OF THE CORE 98I-28

The core 98i-28 has a total length of 1.80 m and it is formed by two main lithological units:

1.- The upper 15 cm are mainly formed by a massive bed of light brown silty-clays.

2.- From the 15 centimeters to the bottom of the sequence the sediments are composed by a color alternation of light and dark gray clays. Between 0.86 m and 1.36 m of depth, the textural composition of this alternation is slightly coarser (Figure 5). Two types of color alternation can be differentiated: a) a centimetric (1 – 2 cm thick) alternation of light and dark gray layers. The boundaries are sharp in the top of the dark lamina and transitional in the bottom of it; b) a millimetric (1 – 2 mm thick) alternation of light and dark gray lamina with clear boundaries.

4.1.1. Microfacies

In order to characterize this alternation, thin sections and electronic microscope observations have been performed. Three microfacies have been identified:

Microfacies 1: light centimetric laminae The first microfacies is composed by massive micritic carbonate with some terrigenous grains floating in this matrix (Figure 6). These grains are mainly homogeneous in size and are mainly composed by quartz, hornblende and some elongated phyllosilicates, such as biotite, mainly disposed parallel to the surface of sedimentation. At the electronic microscope, this micrite is composed by euhedral and homogranular rhombic crystals, with a mean size of 5 μm (Figure 7). On the other hand, the detrital grains are bigger in size an anhedral. It is also noticeable the presence of fragments of gastropods, ostracods and diatoms, most of them parallel to the surface of sedimentation (Figure 8). The preservation of the ostracods is excellent, but of the diatoms very poor. This microfacies is present from the top of the core until 85 centimeters and from 136 cm to the bottom of the core, at 180 cm of depth.

Microfacies 2: dark centimetric lamina. The second microfacies is constituted by massive micritic carbonate with large fragments of plant rests, some of them with framboidal texture, and charcoal particles (Figure 9), as well as black masses, which could be interpreted as sulfide- and/or manganese- rich accumulations (Figure 10). These accumulations are laterally discontinuous, and their borders diffuse.

Microfacies 3: peletoidal lamina. The third microfacies is mainly composed by light massive micritic carbonate, rich in fecal pellets aggregates, with terrigenous minerals floating in this matrix, and large quantities of ostracods (complete shelves and fragments) and diatoms (Figure 11). In this case, the elongated particles are disposed without any preferential orientation. This facies dominates from 86 cm to 136 cm of depth.

4.1.2. Microfacies organization

The petrological and SEM observations suggest that the main part of the carbonates that compose these sediments are authigenic.

The facies of this core seems to correspond to a distal shelf sedimentation, affected by muddy contribution (evidenced by the terrigenous grains) related to the Jergueland and Tyup rivers, as well as to the eolian input.

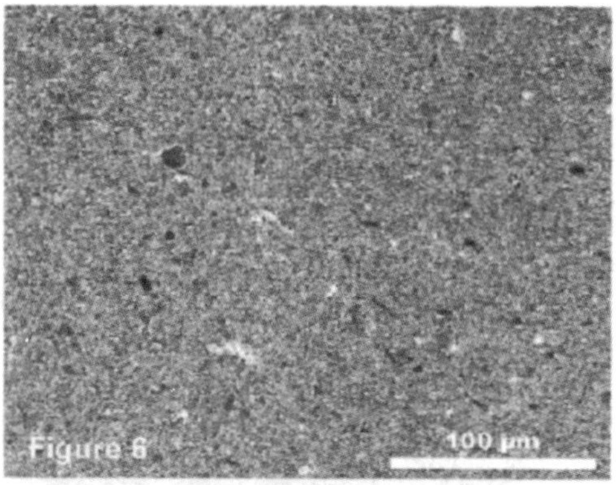

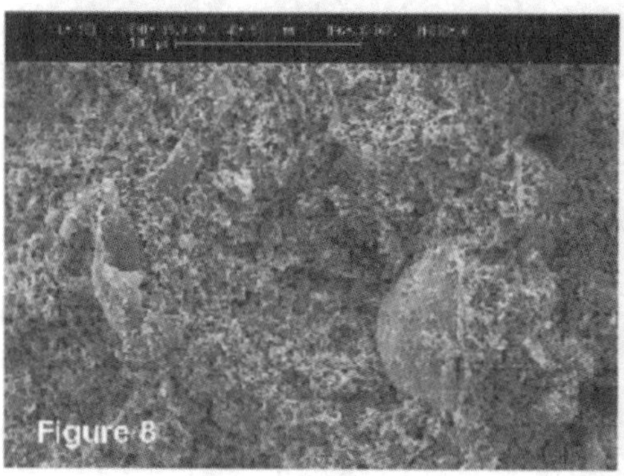

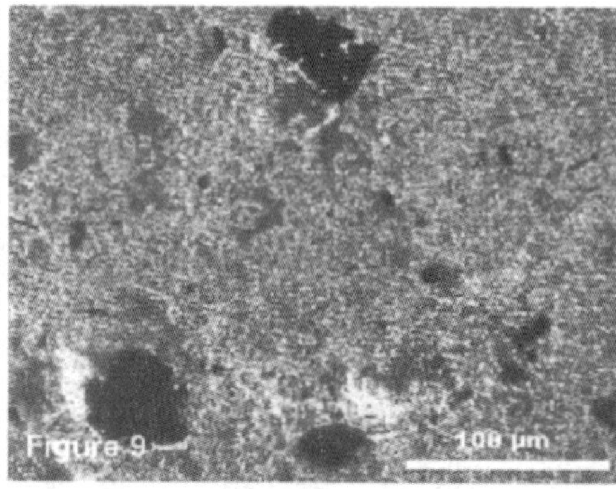

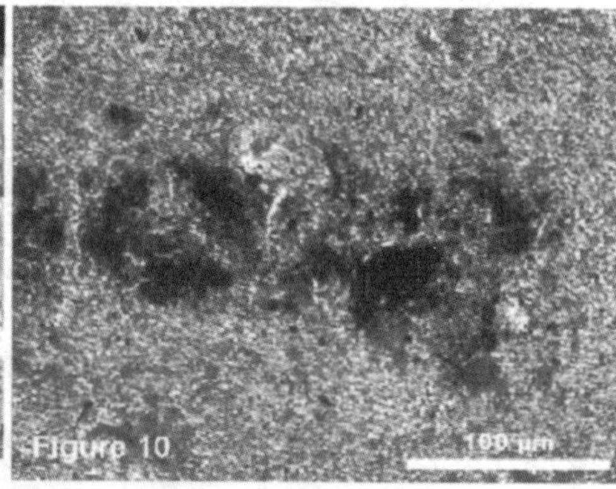

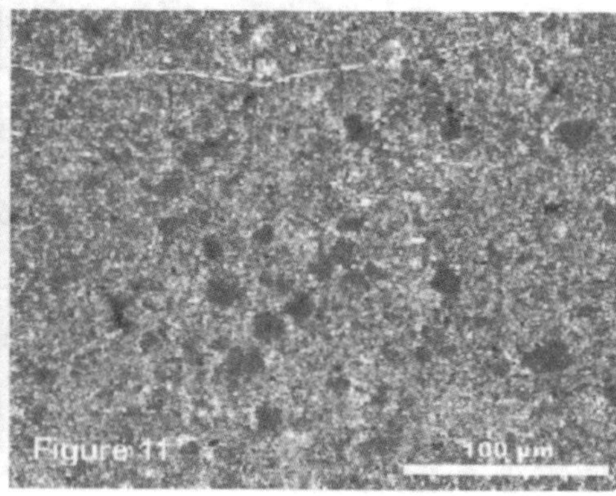

Figure 6. Common view of the microfacies 1 (82 cm of depth) in the petrographical microscope. White bar as scale.
Figure 7. Common SEM view of the crystals that compose the micrite of microfacies 1.
Figure 8. SEM aspect of the micrite composing the sediments of the core 98i-28. Note the position of the ostracods and the Lake Issyk-

Kul sequence (presence of large terrigenous particles floating in the micrite. The top of the sample is located in the left part of the photograph.
Figure 9. Common view of the dark gray lamina (99 cm of depth) at the petrographical microscope, rich in plant rests and charcoal. White bar as scale.
Figure 10. Common view of the dark gray lamina (135 cm of depth) at the petrographical microscope, rich in sulfide and/or manganese accumulations. White bar as scale.
Figure 11. Common view of the peletoidal lamina (130 cm of depth) at the petrographical microscope. White bar as scale.

These microfacies are organized forming couples of light (microfacies 1) and dark (microfacies 2) lamina. Only in the central part of the core, between 0.86 m and 1.36 m of depth, the light laminae is represented by the microfacies 3.

Origin of the millimetric alternation The ice melting, during the first half of the year, supplies large quantities of cold water, entering into the lake as interflows, over the thermocline (at 30 m of water depth, approx.) and over the chemocline (at around 50 m of water depth), providing large quantities of terrigenous sediment, as well as increasing the phosphorous and iron content in the water (Figure 12). The phosphorous maximum is achieved around May, when the first algal bloom occurs [5]. This algal bloom slightly affects the carbonate precipitation, since the water of the lake remains largely supersaturated in calcite, aragonite and monohydrocalcite during all the year (Figure 12). It is during this first half of the year that the light massive lamina would form (Figure 12).

From July to March, the temperature of the water decreases, starting the overturn at the end of November. Due to the accumulation of snow in the surrounding mountains from September the river discharge decreases. This fact is traduced in the decline of the lake level. This lake level reduction, together with the decay of the Characeae, allows a relative increase of phosphorous in the water, which induces a second algal bloom, this last in October [5]. This second algal bloom also slightly affects the inorganic precipitation of carbonates (Figure 12). During the period of overturning the organic matter is incorporated to the sediments. Therefore, during this period the dark lamina would be formed.

The presence of light peloidal microfacies in the central part of the core could indicate evidences of lake dynamic changes. In fact, this facies has usually been related with shallow and low-energy inner platform environments [20]. This fact seems to be confirmed by the large presence of rests of organisms, such as ostracods and gastropods. Therefore, the millimetric alternation of light micritic and organic-rich facies could be interpreted in terms of rhythmites related to the ice cap melting of the surrounding mountains (Figure 12).

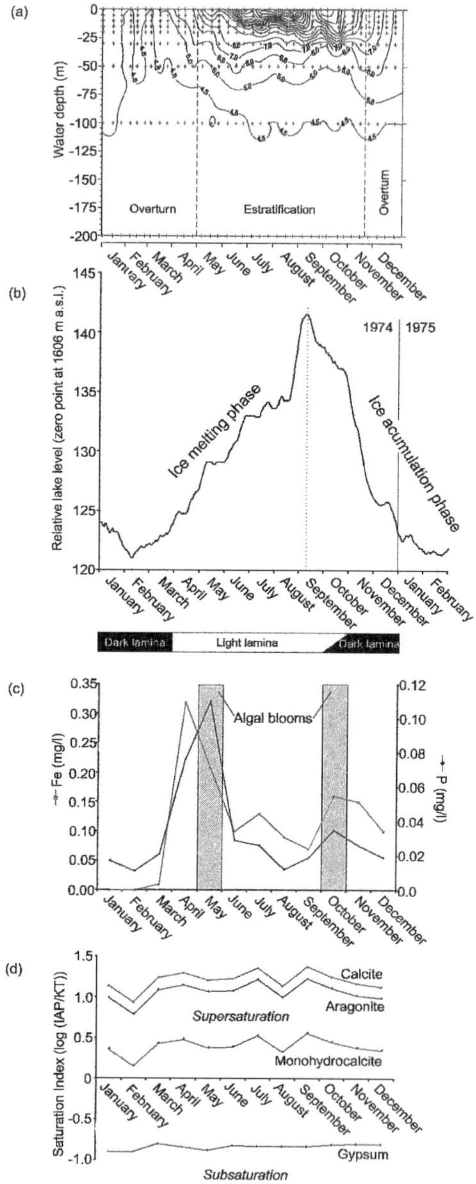

Figure 12. (a) thermal lake evolution for 1974. Note that the overturning occurs between December and April, (b) Relative daily lake level oscillation, (c) dissolved phosphorous and iron in the water, and (d) saturation indices for several carbonate species as well as for gypsum for this period.

Origin of the centimetric alternation On the other hand, the centimetric alternation of these microfacies seems to be related to early diagenetic redox mobilization of iron associated with sulfate reduction and organic matter decay.

4.1.3. Mineralogical composition of the core 98i-28

Ten mineral species have been identified: monohydrocalcite, calcite, magnesian calcite, palygorskite, clinochlorite, quartz, illite, riebeckite, microcline and albite (Figure 13). Carbonates compose up to 75% of the total sediment composition, and on basis of the dominance of the carbonate species, four main zones have been established from the bottom to the top:

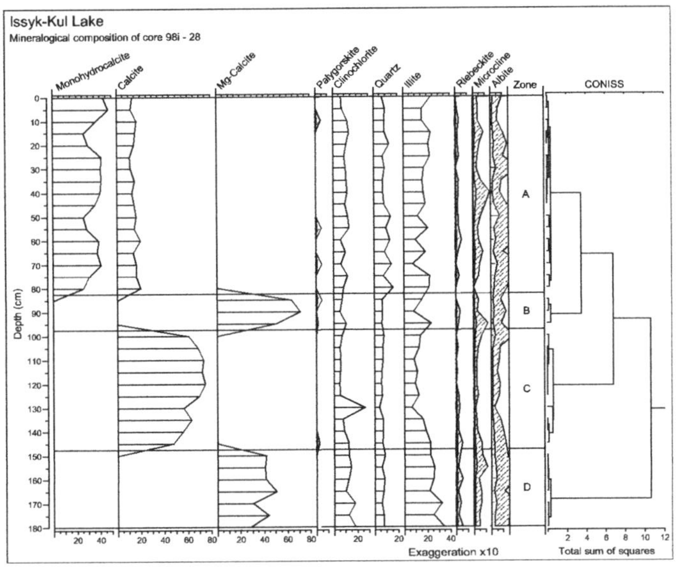

Figure 13. Mineralogical composition of the core 98i-28. Values of the different mineral species are expressed in percentage.

Zone A (83 – 0 cm deep) Monohydrocalcite is the main carbonate of this zone, together with a continuous presence of calcite. While the percentages of monohydrocalcite fluctuate (25 – 48% of the total weight), the percentages of calcite remain roughly constant (15%). Palygorskite is present in discrete levels and quartz shows the highest percentages of the sequence.

Zone B (97 – 83 cm deep) This zone is dominated again by magnesian calcite (50 – 72% of the total weight). It is also noticeable the near constant presence of palygorskite in this zone.

Zone C (147 - 97 cm deep) Calcite dominates this zone showing percentages ranging between 50 and 75% of the total weight. The other mineral species show the lowest values of the sequence.

Zone D (180 – 147 cm deep) This zone is dominated by magnesian calcite (30 – 55%), together with rough constant values of the rest of the mineral species. In this zone illite presents its highest percentages.

4.1.4. Pollen Results

With the aim to deduce and interpret the different environmental changes present in the pollen diagrams, we have grouped the different present-day pollen taxa on the basis of similar ecological requirements.
On the basis of these studies the pollen taxa have been grouped as follows (Table 1):

TABLE 1. Main ecological groups of vegetation.

ECOLOGY	TAXA
Desert	Chenopodiaceae, Ephedra Thalictrum, Caryophyllaceae
Steppe	Artemisia, Poaceae
Broad-leaved forest	Rosaceae, Acer
Needle-leaved forest	Picea, Betula, Salix Juniperus
Secondary Anthropic Indicators	Plantago lanceolata t. Plantago coronopus t. Rumex
Primary Anthropic Indicators (crops)	Cerealia t., Papaver Cannabaceae

Pollen characterization of desert and steppe vegetation has been studied with samples of modern pollen rain [21][22]. In non-forested areas of Middle East and Western Tibetan Plateau, pollen spectra are always dominated by very few taxa: *Artemisia, Chenopodiaceae, Poaceae* and *Ephedra*, while Arboreal Pollen (AP) have very low values and usually come from long distances. These studies show that high percentages of *Chenopodiaceae* are related with desertic conditions, while high *Artemisia* values have been recorded in more steppic conditions. The similar pollinisation mechanism of both taxa makes it possible to use the relationship between *Chenopodiaceae* and *Artemisia* as an aridity index (C/A), very useful for the palaeoecological and palaeoclimatic interpretation of pollen sequences located in the desert-steppe limit [23]. This index has been also applied to the Figures 14 and 15).

The use of *Poaceae* as palaeoclimatic indicator is less clear. Several authors interpreted that grassland increase in these environments as a signal of moister conditions [22][24], mainly of summer rainfall increase [24]. However, *Poaceae* can be also produced by wetland vegetation and then its variations could be related with changes in lake margins and water level. In these cases, an increase of *Poaceae* could be due to a decrease in the lake water level and so, it can be interpreted as a result of drier conditions [22].
The Lake Issyk-Kul pollen diagram is mainly characterized by the absolute dominance of Non-Arboreal Pollen (NAP). Arboreal Pollen (AP) averages along the sequence is 6.5%, with highest values of 12.2% and lowest ones of 3.5%. Mean values of NAP are about 90%. Inside the AP group, *Picea* is the best represented taxon, with mean values of 2.1%. NAP group is mainly represented by *Artemisia*, with mean values of 58% and maximum of 67%, *Chenopodiaceae* (mean values of 16% and maximum of 23.6%), *Poaceae* (with mean values of 10%), *Cyperaceae* (mean values of 2.5%) and *Ephedra* (mean of 2%).

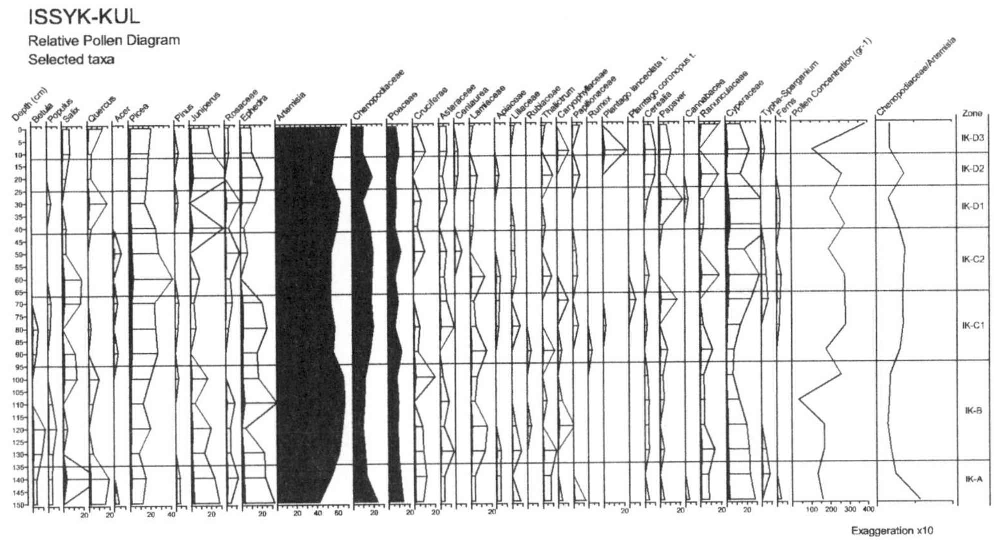

Figure 14. Relative pollen diagram of the core 98i-28. The values of the different taxa are represented in percentage.

On the basis of the CONISS results, four main pollen zones from the bottom to the top of the sequence have been established:

IK-A (150-135 cm) Besides of the high *Artemisia* pollen percentages, this zone is characterized by high values of *Chenopodiaceae* (c. 20%), Poaceae (15%), *Ephedra* and semi-aquatic herbs (*Cyperaceae*, *Typha-Sparganium* and Ferns). The synthetic pollen diagram (Figure 14) also shows the development of the desert and the retraction of

steppe communities. The expansion of AP at this moment (mainly *Salix, Quercus, Juniperus* and *Rosaceae*) can be a statistical artifact resulting from the low *Artemisia* representation, because this plant produces and disperses great amounts of pollen. Pollen and Charcoal Concentration values are low in this zone and in the next one, as a result of high sedimentation rate (Figure 4).
This pollen spectrum together with the high values of the aridity index (C/A) indicates a period of dry conditions and low lacustrine water level (Figure 14).

IK-B (135-95 cm) First major change in pollen spectra occurs at 135 cm depth. Now, *Artemisia* values increase (65%) while *Chenopodiaceae* decreases (10%). The reduction of semi-aquatic herbs is progressive and this group attains its lower values later, during the transition between zones IK-B and IK-C. The aridity index C/A is very low. All this data allow us to point out a period of expansion of steppe and retraction of desert, that means an increase in regional humidity, which preludes a later water lake raising. Lower AP relative values must be seen as the result of the increase of anemophilous taxa, such as *Artemisia*.

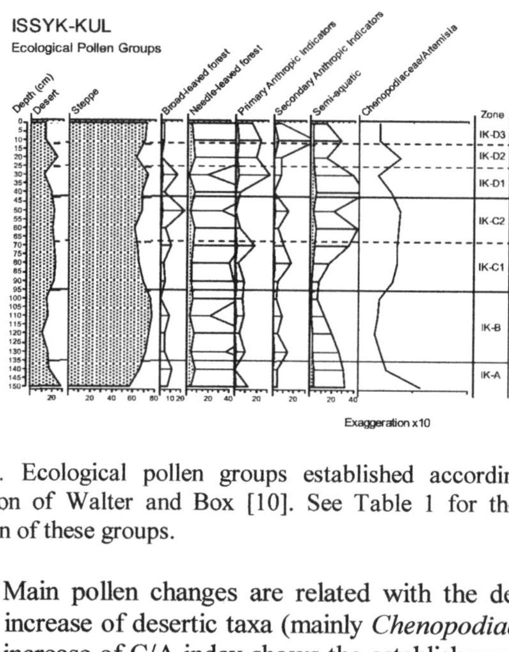

Figure 15. Ecological pollen groups established according to the classification of Walter and Box [10]. See Table 1 for the detailed composition of these groups.

IK-C (95-42 cm) Main pollen changes are related with the decrease of *Artemisia* pollen (55%) and the increase of desertic taxa (mainly *Chenopodiaceae* and *Ephedra* in subzone IK-C1). The increase of C/A index shows the establishment of drier conditions. However, semi-aquatic plants react later to this environmental change, and the subsequent decrease in water lake level occurs during zone IK-C2. Important changes occurred in the arboreal vegetation: *Juniperus* forest reduces and *Picea* expands at the same moment that broad-leaved forest (mainly formed by *Rosaceae*) develops. During this zone appear the first signals of anthropogenic activity in the region, pointed out by

the first important peak of *Papaver* during zone C1 as well as by the expansion of the secondary anthropogenic indicators (*Plantago sp.* and *Rumex*).
The increase of Pollen Concentration can be interpreted as a decrease in the sedimentation rate, a tendency that will continue during most part of he zone IK-D.

IK-D (42-0 cm) This zone is mainly characterized by the expansion of primary and secondary anthropogenic indicators related to grazing and farming activities. This fact suggests that the anthropic management of the territory plays now an important role in the landscape morphology. *Cereal* and *Papaver* crops expand now around the lake, mainly during the sub-zone IK-D1. During this zone, and related with the configuration of forests of Tien Shan slopes, *Picea* and broad-leaved forests decrease while *Juniperus* (a tree growing at higher altitude) expands (Figure 3). This forest dynamic implies an altitudinal lowering of the vegetational mountain belts towards the plains, suggesting a new climatic deterioration or man deforestation in the highest parts of the mountains.

New climatic shifts can be also observed. During the zone IK-D1, *Artemisia* develops as a result of a new increase in the moister conditions, as can be observed from the decrease of the C/A aridity index. In spite of this, higher values of semi-aquatic plants occur at the same moment showing low lake levels related to the previous aridity maximum of zone IK-C2. In the subzone D2, a new expansion of desert taxa occurs (*Chenopodiaceae, Ephedra*) showing the prevalence of dry climatic conditions during a short period of time. This short climatic deterioration could be also related to changes in the forest belts of the Tien Shan ranges (broad-leaved forest disappears and *Picea* communities reduce at the same moment that *Juniperus* develops at higher altitudes).
During the last subzone IK-D3, regional climatic moisture increases as can be deduced from high values of *Artemisia*. The reduction of desert taxa (*Ephedra, Chenopodiaceae*), the decrease of C/A index, and the low values of semi-aquatic plants can be also related with a water lake rise. This sub-zone is mainly characterized by the increase of secondary anthropogenic indicators showing again the intensification of human activities (mainly grazing), and by the increase of Pollen Concentration as a result of a low sedimentation rate.

5. Discussion

5.1. POLLEN DISCUSSION

Changes in pollen sequences entirely dominated by non-arboreal elements (NAP) are difficult to interpret in terms of environmental and/or climatic changes.
Globally, this pollen spectra allow us to establish a good relation between the Lake Issyk-Kul pollen diagram and those coming from several lakes studied in Central Asia, such as Sumxi Co [25] and Bangong Co in Western Tibetan Plateau [22], or Mans Lake in Northern Xinjiang [23]. However, in these lakes, AP values (average of c. 2%) are always lower than in Lake Issyk-Kul (6.5%). Pollen sequences from the west (e.g.

Caspian Sea) show a diverse spectra, mainly dominated by *Chenopodiaceae*, with low *Artemisia* percentages and with higher percentages of AP (c. 20%) [26].

5.1.1. Temporal framework

The chronological framework of the sequences has been established using two approaches.

The first approach has been carried out for the upper 15 cm of the sequence, where the ^{210}Pb profile and the AMS ^{14}C radiometric date has been used. A plant macrorest located at 15 cm of depth gave an uncalibrated age of 990 ±40 years B.P.

The profile of ^{210}Pb for the upper 10 cm of sequence shows the typical exponential decay with depth (Bobrov, comm. pers.). The ^{210}Pb values are roughly constant at 8 cm of depth, which means an age of 150 – 200 years at this depth. Thus, the sedimentation rate deduced from the first method is 0.5 ± 0.05 mm per year.

The sedimentation rate calculated from the second method is 0.15 mm per year. The dated plant remain was recovered from the contact between the two main lithological units. This fact, together with the disagreement with the ^{210}Pb sedimentation rate suggest that this plant remain could have a detrital origin, giving an older date.

Thus, the lack of reliable material suitable for radiocarbonic dating has not allowed to perform this technique. Then, the second approach has been applied, and a preliminary chronological framework for the rest of the sequence has been established using the lamination. Assuming the annual character of the millimetric rythmytes (as can be deduced from the annual lake limnological evolution represented in Figure 12), the mean thickness of a couple of lamina has been measured in several parts of the core. The mean thickness is about 0.1 cm, which means a sedimentation rate of about 1mm per year. This calculated sedimentation rate is in accordance with this deduced from the ^{210}Pb method.

Taking into account this hypothetical sedimentation rate, the sequence only seems to record the late Holocene, may be as far as 2,000 years. The continuous presence of crop taxa in the pollen diagram also suggests a late Holocene age.

5.1.2. Lake evolution

The total amount of carbonate throughout the Lake Issyk-Kul sequence is around 53% in average. This fact implies that the carbonate production within the lake has been roughly constant during the late Holocene. Changes in the predominant carbonate species indicates variations in the lacustrine dynamic.

The lower part of the core is characterized by the alternation of calcite and magnesian calcite, indicating changes in salinity in the lake water, from saline conditions towards more fresh water and again to more saline waters (Figure 16). But, the main change in the mineralogical diagram is present at 0.85 m of depth when appears monohydrocalcite.

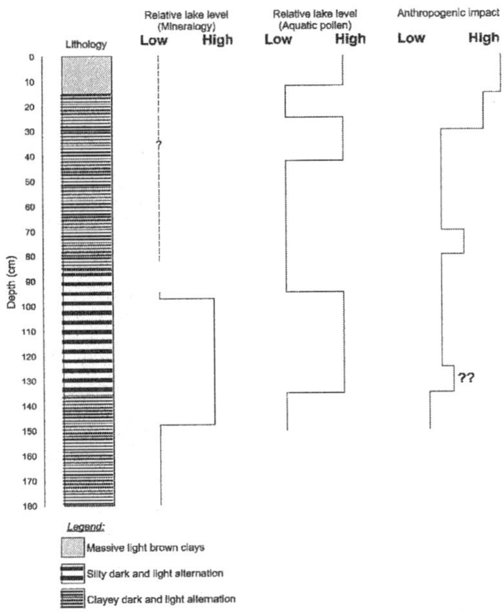

Figure 16. Lake level oscillations inferred from mineralogical composition and percentages of aquatic pollen, and compared with the anthropogenic impact on the surroundings of the lake.

The environmental conditions for the precipitation of this hydrated carbonate in natural conditions have been not yet precisely established, although it has been described in several lakes, such as as Lake Fellmongery in South Australia [27][28] and Lake Kivu in East Africa [29].

Several thermodynamic constants from monohydrocalcite have been accurately determined [30][31]. These studies have shown that this carbonate mineral species belongs to a solid phase solution, which progressively increases its stability (ikaite < monohydrocalcite < vaterite < aragonite < calcite). Ikaite and monohydrocalcite usually convert to vaterite and to calcite, which are more stable carbonate forms.

It seems that, in experimental conditions, three main factors control the precipitation of these hydrated carbonates:

1.- the salinity: high Mg/Ca ratios inhibit the formation of calcite [32][33][34] and favors both the precipitation of high-magnesian carbonates, such as magnesian calcite and dolomite, among others [35][36], and of hydrated carbonates, as monohydrocalcite [37][28].

2.- the presence of phosphate: several works have demonstrated that the presence of phosphorous in the water (as sodium hexametaphoshate or triphosphate) inhibits the spontaneous precipitation of anhydrous carbonates, even when they are supersaturated, and stabilizes the hydrated forms (such as the monohydrocalcite, ikaite or nesquehonite)

[38][30][29][39][40][41]. It seems that the phosphate tends to be absorbed on the surface of the nuclei of calcite whose growth is thus inhibited [37]. Bischoff et al. [41] have reported that ikaite crystals persisted during 160 days in Walker Lake waters, just adding 100 μm of KH_2PO_4 solution.

3.- the water temperature: the hydrated carbonates seem to be "stabilized" at low water temperatures (in the case of ikaite, this mineral species precipitates when the temperature of the water is close to 0 °C). The experiments carried out by Stoffers and Fischbeck [29] demonstrated that the precipitation of monohydrocalcite is favored by low temperatures (around +2 °C). These mineral species became unstable with the progressive increase of the water temperature, changing to a more stable carbonate species, such as calcite. This temperature dependence seems to be overcome by the presence of phosphate in the water [40][41].

Thus, the presence of monohydrocalcite in Lake Issyk-Kul can be interpreted as an increase of phosphorous in the lake water. This increase could be related to the closure of the output of the lake. This closure could be caused due to both tectonic activity and to increasing dryness. Bondarev and Sevastyanov [42] suggest that the tectonic activity led to changes in the flow of the Chu river and its separation of the lake. This activity seemed to be important during the end of the Late Pleistocene, but not during the Holocene. These authors also suggest that during the last 10,000 years the lake level oscillations seemed to be mainly triggered by changes in moisture.

In any case, both processes imply a reduction of the lake water volume, increasing the phosphorous concentration present in the water. This lake level decrease is visible in the western and southern part of the lake, where recent lacustrine sediments outcrop.

A third factor that could result in an increase of phosphorous in the water of the lake is the anthropogenic activity on the shores of the lake. In fact, the beginning of the monohydrocalcite formation broadly coincides with the first signals of anthropic management of the territory (clearly pointed out by the presence of crop taxa such as *Plantago sp.* and the first important expansion of *Papaver*). According to the preliminary chronological framework this change took place around 1100 A.D.

Since then, this anthropic activity has become progressively more important towards the present-day.

There are no clear evidences in order to decipher which factor influences the increase of phosphorous in the lake waters, but the increase of values of aquatic taxa during this moment (Figure 16) seems to indicate that the phosphorous increase was due to natural causes (climatic and /or tectonic) rather than due to human activities. These last ones, if they have affected the lake, seem to only have been acting for amplifying this phenomena.

Although the lake level has changed during this last period, the phosphorous concentration in the water has remained high enough for favoring the precipitation of monohydrocalcite until the present-day.

6. References

1. Fritz, S. C. (1996) Paleolimnological records of climatic change in North America. *Limnology and Oceanography* **41**, 882 – 889.
2. Walter, H. (1977) *Zonas de vegetación y clima*. Ediciones Omega, Barcelona.
3. Feng, Z., Thompson, L.G., Mosley-Thompson, E. and Yao, T. (1996) Temporal and spatial variations of climate in China during the last 10 000 years. *The Holocene* **3**, 174 – 180.
4. Romanovsky, V.V. (1990) *Lake Issyk-Kul – a natural system*. Ilim., Frunze, 168 pp.
5. Savvaitova, K. and Petr, T. (1992) Lake Issyk-Kul, Kirgizia. *International Journal of Salt Lake Research* **1**, 21 - 46.
6. Sevastyanov, D.V., Mamedov, E.D. and Rumyantsev, V.A. (Eds) (1991) *History of Lakes Sevan, Issyk-Kul, Balkhash, Zaisan and Aral*. Nauka, Leningrad. (in Russian).
7. Sapozhnikov, D.G. and Viselkina, M.A. (1960) *Recent sediments of the Lake Issyk-Kul and its bays*. Publ. House Akademii Nauk, Moscow. (in Russian).
8. Aladin, N.V. and Plotnikov, I.S. (1993) Large saline lakes of former USSR: a summary review. *Hydrobiologia* **267**, 1-12.
9. Giralt, S., Julià, R. and Klerkx, J. (2001) Microbial biscuits of vaterite in Lake Issyk-Kul (Republic of Kyrgyzstan). *Journal of Sedimentary Research* **71** (3).
10. Walter, H. and Box, E.O. (1983) The orobiomes of Middle Asia. In: West, N.E. (Ed.) *Ecosystems of the world: temperate deserts and semi-deserts*. Elsevier Scientific Publishing Company, Amsterdam.
11. Imbo, Y., De Batist, M., Klerkx, J., Lignier, V., Giralt, S., Vermeesch, P. and Beck, C. (1999) High-resolution reflection seismic study of the sedimentary infill of Lake Issyk-Kul, Kirghyzstan, Central Asia. Lennou Conference, Brest.
12. Giralt, S., Lignier, V., Klerkx, J., Julià, R., De Batist, M., Beck, C. and Kalugin, I. (2000) Sedimentological processes in Lake Issyk-Kul (Republic of Kyrgyzstan, Central Asia): the importance of tectonically influenced sedimentation. 8th International Symposium on Paleolimnology. Kingston, Canada.
13. Giralt, S., Lignier, V., Klerkx, J., Beck, C., Kalugin, I. and De Batist, M. (in preparation) General sedimentary pattern in Lake Issyk-Kul (Republic of Kyrgyzstan). Sedimentology.
14. Sturm, M., Siegentahler, C. and Pickrill, R.A. (1995) Turbidites and 'homogenites'. A conceptual model of flood and slide deposits. *Publ. IAS – 16th Regional Meeting Sedimentology* **22**, 140.
15. Francus, P. (1998) An image-analysis technique to measure grain-size variation in thin sections of soft clastic sediments. *Sedimentary Geology* **121**, 289 – 298.
16. Moore, P.D., Webb, J.A. and Collinson, M.E. (1991) *Pollen analysis*, Blackwell scientific publications, Oxford.
17. Grimm, E.C. (1987) Coniss: A Fortran 77 Program for Stratigraphically Constrained Cluster Analysis by the Method of Incremental Sum of Squares. *Computers & Geosciences* **13**, 13 – 35.

18. Parkhurst, D.L. (1995) User's guide to PHREEQC – a computer program for speciation, reaction-path, advective-transport, and inverse geochemical calculations. *U.S. Geological Survey. Water-Resources Investigations Report* **95 – 4227,** 143 p.
19. Charlton, S.R., Macklin, C.L. and Parkhurst, D.L. 1997. PHREEQCI – a graphical user interface for the geochemical computer program PHREEQC. *U.S. Geological Survey. Water-Resources Investigations Report* **97 – 4222**, 9 p.
20. Wright, V.P. and Burchette, T.P. (1996) Shallow-water carbonate environments. In: Reading, H.G. (Ed.) *Sedimentary environments: processes, facies and stratigraphy*. Blackwell science, Oxford, 325 – 394.
21. El-Moslimany, A.P. (1990) Ecological significance of common non-arboreal pollen: examples from drylands of the Middle East. *Review of Paleobotany and Palynology* **64**, 343 – 350.
22. Van Campo, E., Cour, P. and Sixuan, H. (1996) Holocene environmental changes in Bangong Co basin (western Tibet). Part 2: the pollen record. *Palaeogeogr., Palaeoclim., Palaeoecol.* **120,** 49 – 63.
23. Rhodes, T.E., Gasse, F., Ruifern, L., Fontes, J.C., Keqin, W., Bertrand, P., Gibert, E., Mélières, F., Tucholka, P., Zhixiang, W. and Zhi-Yuan, C. (1996) Late Pleistocene – Holocene lacustrine record from Lake Manas, Zunggar (northern Xinjiang, Western China). *Palaeogeogr., Palaeoclim., Palaeoecol.* **120**, 105 – 121.
24. Zang, H.C., Ma, Y.Z., Wünnemann, B. and Pachur, H.J. (2000) A Holocene climatic record from arid northwestern China. *Palaeogeogr., Palaeoclim., Palaeoecol.* **162,** 389 – 401.
25. Van Campo, E. and Gasse, E. (1993) Pollen- and diatom-inferred climatic and hydrological changes in Sumxi Co basin (Western Tibet) since 13,000 yr B.P. *Quaternary Research* **39**, 300 – 313.
26. Leroy, S.A.G., Marret, F., Chalié, F. and Gasse, F. (2000) Understanding the Caspian Sea erratic fluctuations: palynological results from the south basin. *Terra Nostra* **2000/7**, 45 – 49.
27. Taylor, G.F. (1975) The occurrence of monohydrocalcite in two small lakes in the south-east of South Australia. *American Mineralogist* **60**, 690 – 697.
28. Davies, P.J., Bubela, B. and Ferguson, J. (1977) Simulation of carbonate diagenetic processes: formation of dolomite, huntite and monohydrocalcite by the reactions between nesquehonite and brine. *Chemical Geology* **19**, 187 – 214.
29. Stoffers, P. and Fischbeck, R. (1974) Monohydrocalcite in the sediments of Lake Kivu (East Africa). *Sedimentology* **21**, 163 – 170.
30. Hull, H. and Turnbull, A.J. (1973) A thermodynamical study of monohydrocalcite. *Geochimica et Cosmochimica Acta* **37**, 685 – 694.
31. Königsberger, E., Königsberger, L.C. and Gamsjäger, H. (1999) Low-temperature thermodynamic model for the system Na_2CO_3 – $MgCO_3$ – $CaCO_3$ – H_2O. *Geochimica et Cosmochimica Acta* **63**, 3105 - 3119.
32. McConnell, J.D.C. (1960) Vaterite from Ballycraigy, Larne, Northern Ireland. *Mineralogical Magazine* **32**, 535 – 545.

33. Reddy, M.M. (1986) Effect of magnesium ions on calcium carbonate nucleation and crystal growth in dilute aqueous solutions at 25 °C. In: Mumpton, F.A. (Ed) Studies in diagenesis. *U.S. Geological Survey Bulletin* **1578**, 169 – 182.
34. Sawada, K. 1997. The mechanisms of crystallization and transformation of calcium carbonates: *Pure and Applied Chemistry* **69**, 921 - 928.
35. Folk, R.L. Land, L.S. (1975) Mg/Ca Ratio and Salinity: Two Controls over Crystallization of Dolomite. *The American Association of Petroleum Geologists*: 60 – 68.
36. Last, W. M. and De Deckker, P. (1990) Modern and Holocene carbonate sedimentology of two saline volcanic maar lakes, southern Australia. *Sedimentology* **37**, 967 – 981.
37. Lippmann, F. (1973) *Sedimentary Carbonate Minerals*. Springer-Verlag, Berlin.
Marschner, H. (1969) Hydrocalcite ($CaCO_3 \cdot H_2O$) and nesquehonite ($MgCO_3 \cdot 3H_2O$) in carbonate scales. *Science* **165**, 1119 – 1121.
38. Shaikh, A.M. and Shearman, D.J. (1986) On ikaite and the morphology of its pseudomorphs. In: Rodríguez-Clemente, R. and Tardy, Y. (Eds) *Geochemistry and mineral formation in the earth surface*. Proceedings of the International Meeting "Geochemistry of the earth surface and processes of mineral formation", 791 – 803.
39. Clarkson, J.R., Price, T.J. and Adams, C.J. (1992) Role of metastable phases in the spontaneous precipitation of calcium carbonate. *Journal of Chemical Society, Faraday Trans.* **88**, 243 – 249.
40. Bischoff, J.L., Fitzpatrick, J.A. and Rosenbauer, R.J. (1993) The solubility and stabilization of ikaite ($CaCO_3.6H_2O$) from 0° to 25 °C: environmental and paleoclimatic implications for thinolite tufa. *The Journal of Geology* **101**, 21-33.
41. Bondarev, L.G. and Sevastyanov, D.V. (1991) *The history of Sevan, Issyk-Kul, Balkhash, Zaisan and Aral Lakes*, Nauka, Leningrad. (In Russian).

ACTIVE TECTONICS AND SEISMIC HAZARD OF THE ISSYK-KUL BASIN IN THE KYRGYZ TIAN-SHAN

K.E. ABDRACHMATOV and K.D. DJANUZAKOV
Kyrgyz Institute of Seismology, Laboratory of Seismotectonics
720060 Bishkek, Kyrgyzstan

and

D. DELVAUX
Royal Museum for Central Africa, Dept of Geology-Mineralogy
B-3080 Tervuren, Belgium

1. Introduction

The Issyk-Kul basin is a large intramountain depression located in the northeastern part of the Kyrgyz Tian-Shan (Fig. 1). It is bordered on both northern and southern sides by E-W trending mountain ranges rising up to 5000 meters in altitude (Fig. 2). During last 100 years, these two ranges were subjected to an unusual series of major earthquakes, some of them raking among the strongest ever felt in continental areas (Table 1; Fig. 3).

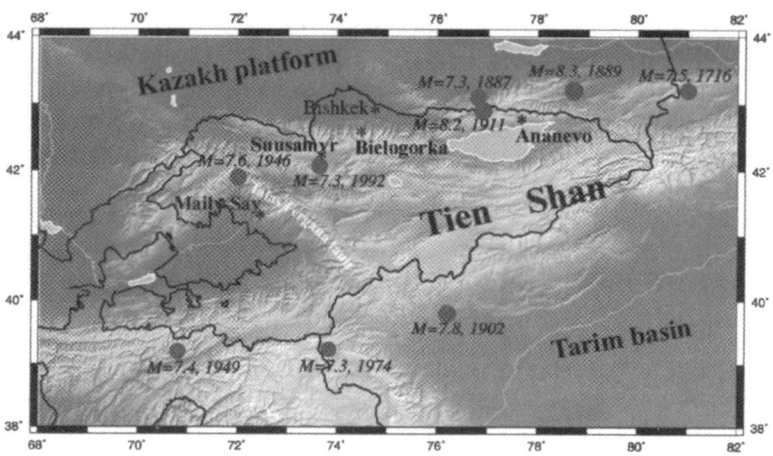

Figure 1. The Kyrgyz Tian-Shan in Central Asia, with shaded relief and location of recent strong earthquakes.

J. Klerkx and B. Imanackunov (eds.), Lake Issyk-Kul: Its Natural Environment, 147–160.

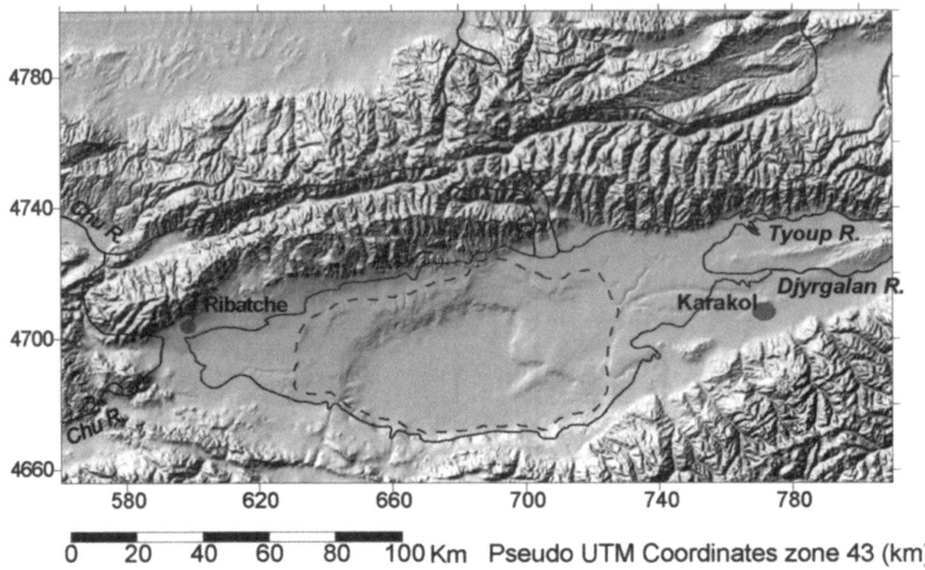

Figure 2. Digital elevation model of the Issyk-kul depression and surrounding areas. Constructed from digitised 1/200.000 topographic maps and 1/100.000 bathymetric chart of Lake Issyk-Kul. The dotted line shows the altitude of the lowest underwater terrace, highlighting a past low lake level (112 m under the present lake level).

This series started in 1887 by the (Ms=7.3) Verny earthquake near Alma-Ata. It was rapidly followed by the 1889 (Ms=8.3) Chilik, the 1911 (Ms=8.2) Kemin and the 1938 (Ms=6.9) Kemino-Chu earthquakes, in an east-to-west propagation. In reaction to this series of unusually strong earthquakes, the eastern part of the Issyk-Kul basin was later affected by the 1970 (Ms=6.8) Sarykamysh then 1978 (Ms=6.7) Djalanash-Tioup and the 1990 (Ms=6.4) Baisoorun earthquakes. Earlier seismic events in the same area are also known from historical data (1): the –250±100 Grigorevka, the 500±500 Toru-Aigir and the 1475±100 Balasogun earthquakes.

According to data of the Institute of seismology of the National Academy of Sciences Kyrgyz Republic, the level of seismic activity in the southern part of the basin is quite high but the intensity of earthquakes remained moderate, no more then 7 balls (intensity force) according to the MSK-64 scale (equivalent to Ms=6.0). A large number of small earthquakes occurred near the source of the Sarykamysh earthquake (Fig. 4). Near Kadji-Say (location on Fig. 4), earthquakes with intensity equivalent to 5-6 balls periodically took place in a cluster. The area occupied by the lake itself is surprisingly almost aseismic, with only a small number of weak events.

As a consequence of the high seismic activity in the mountain ranges bordering the lake, all the Issyk-Kul area is particularly exposed to direct and indirect effects of strong earthquakes.

TABLE 1: Strong earthquakes of Issyk-Kul basin and surrounding areas

Date Year Month Day	lat dec.deg	long dec. deg.	Ms	Shaking intensity (Balls)	Energy class K (log E)	Usual name of event	UTM X	UTM Y
-250 +/- 100	42,80	77,50	6,7	8-9	16,7	Grigorevka	704	4730
500 +/-500	42,80	76,40	6,7	8-9	15,7	Toru-Aigir	623	4730
1475 +/- 100	42,60	75,20	6,4	8-9	15,6	Balasogun	520	4716
1807 +/- 10	43,10	76,90	6,7	8-9	16,0		657	4774
1869 08 29	42,70	75,90	6,4	8	15,3		578	4730
1874 10 18	42,90	77,40	5,9	7	14,6	Aksu-Chilik	695	4751
1880 12 01	43,10	76,90	5,7	7-7	14,3		657	4774
1884 03 13	42,70	78,20	5,8	7	14,4	Tioup	762	4731
1887 06 08	43,10	76,80	7,3	9-10	16,6	Verny	646	4774
1889 07 11	43,20	78,70	8,3	10	18,0	Chilik	800	4790
1893 11 04	42,70	75,80	6	8	15,0	Chu	565	4730
1911 01 03	42,90	76,90	8,2	10-11	17,8	Kemin	659	4756
1938 06 20	42,70	75,80	6,9	8-9	16,0	Kemino-Chu	566	4728
1970 06 05	42,52	78,73	6,8	8-9	15,6	Sarykamysh	807	4712
1978 03 24	42,87	78,58	6,7	8-9	15,6	Djalanash-Tioup	794	4748
1990 11 12	42,98	77,92	6,4	8	15,0	Baisoorun	739	4775

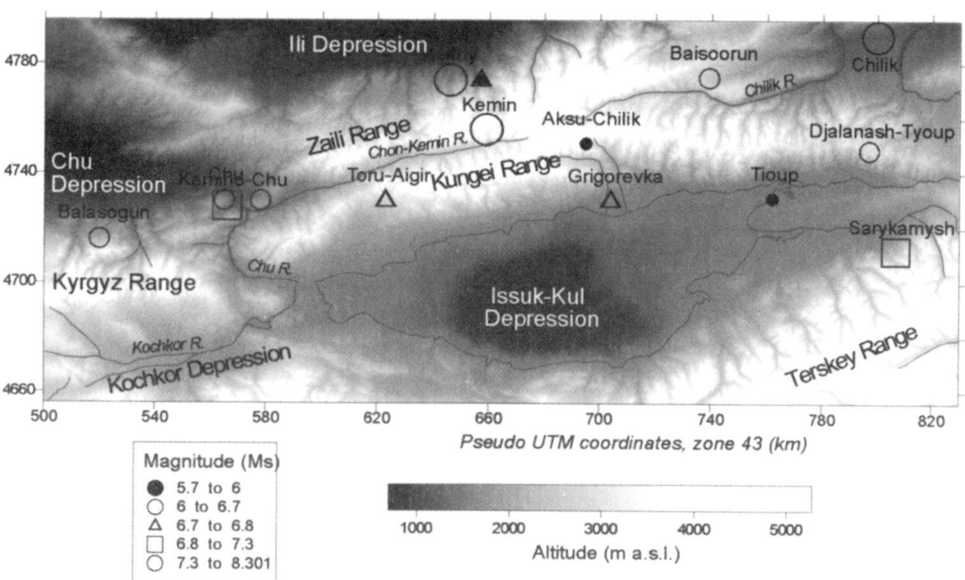

Figure 3. Issyk-Kul depression and location of strong earthquakes reported in Table 1.

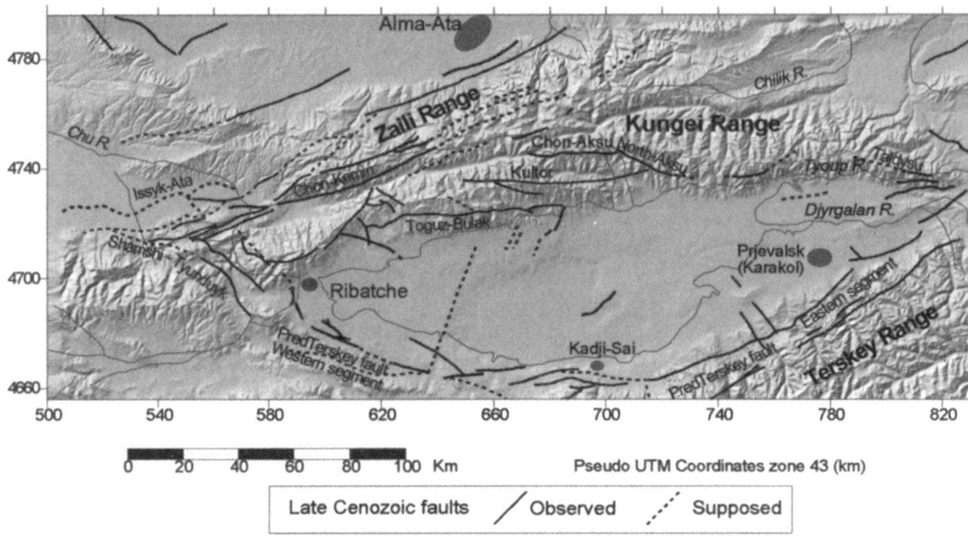

Figure 4. Undifferentiated late Cenozoic faults of the Issyk-Kul depression and surrounding area (faults active in the late Pliocene-Quaternary).

Most of this area is already characterized by critical slope stability due to the generally steep relief. Therefore strong seismic shaking can easily destabilize large rock masses and cause numerous gravity mass movements like landslides, rockfalls and rock avalanches. This aspect was the topic of an Inco-Copernicus project funded by the European Community: "Landslide Risk Triggered by Earthquakes in Kyrgyzstan, Tian-Shan".

In this article, we will first revise the recent structure of the Issyk-Kul depression, map the major faults of Late Tertiary-Quaternary age (Fig. 4) and identify those who present late Quaternary activity (Fig. 5). These late Quaternary faults are determined on the basis of morphological features (they displace Late Pleistocene or Holocene river terraces or glacial deposits). These faults are therefore considered to be seismogenic, even if no seismic activity was recorded along them during the last century.

An important part is devoted to the effects of the 1911 Kemin earthquake that affected the whole range between Lake Issyk-Kul and Alma-Ata and caused large landslides and rockslides (Fig. 6A-D). The investigation of these effects was undertaken to illustrate the potential destructive effects of such strong earthquake in mountainous areas.

Finally we briefly introduce the major approaches for estimating and mapping the seismic hazard and landslide risk of this area.

TABLE 2. Focal mechanisms of strong earthquakes for the Issyk-Kul region

Date					lat	long	H	Mag.	class	P axis		N axis		T axis		Shmax	Stress
Year	mm	dd	hr	min	deg	deg	km	Ms	K	Paz	Ppl	Naz	Npl	Taz	Tpl	azim.	regime
1970	06	05	04	52	42,50	78,90	15	6,4	15,5	013	14	106	11	231	72	13	TF
1975	02	12	13	34	43,30	78,80	15	5,0	13	337	05	240	52	072	37	337	SS
1978	**03**	**24**	**08**	**29**	**42,00**	**79,80**	**30**	**5,6**	**14**	**067**	**12**	**331**	**27**	**178**	**60**	**67**	**TF**
1978	**03**	**24**	**08**	**29**	**41,92**	**79,97**	**15**	**5,1**	**CMT**	**180**	**17**	**314**	**66**	**085**	**16**	**180**	**SS**
1979	**04**	**06**	**18**	**30**	**42,00**	**77,40**	**10**	**5,0**	**13**	**005**	**46**	**209**	**41**	**108**	**12**	**198**	**NF**
1979	**04**	**06**	**18**	**30**	**41,92**	**77,74**	**19**	**4,9**	**CMT**	**008**	**28**	**272**	**12**	**160**	**60**	**8**	**TF**
1980	**07**	**05**	**20**	**25**	**41,90**	**77,50**	**15**	**5,6**	**14**	**116**	**10**	**212**	**32**	**010**	**56**	**116**	**TF**
1980	**07**	**05**	**20**	**25**	**41,94**	**77,40**	**15**	**5,2**	**CMT**	**344**	**01**	**075**	**29**	**252**	**61**	**344**	**TF**
1983	12	21	19	30	42,07	77,45	15	5,0	13	337	09	067	01	165	81	337	TF
1982	**12**	**31**	**19**	**46**	**42,75**	**77,01**	**15**	**5,1**	**CMT**	**159**	**22**	**063**	**14**	**303**	**63**	**159**	**TF**
1988	**06**	**17**	**13**	**30**	**43,00**	**77,42**	**20**	**5,0**	**13**	**330**	**68**	**149**	**22**	**59**	**00**	**149**	**NF**
1988	**06**	**17**	**13**	**30**	**43,01**	**77,51**	**15**	**5,3**	**CMT**	**354**	**39**	**093**	**11**	**196**	**49**	**354**	**TF**
1990	**11**	**12**	**12**	**28**	**42,98**	**77,91**	**5**	**6,1**	**15**	**136**	**15**	**326**	**75**	**225**	**03**	**315**	**SS**
1990	**11**	**12**	**12**	**28**	**43,18**	**78,24**	**15**	**6,3**	**CMT**	**162**	**03**	**257**	**57**	**070**	**33**	**162**	**SS**
1996	01	18	09	33	41,92	77,45	5	5,0	13	005	04	274	11	115	78	5	TF
1996	**01**	**18**	**09**	**33**	**42,10**	**77,98**	**20**	**4,6**	**CMT**	**309**	**14**	**216**	**11**	**090**	**72**	**309**	**TF**

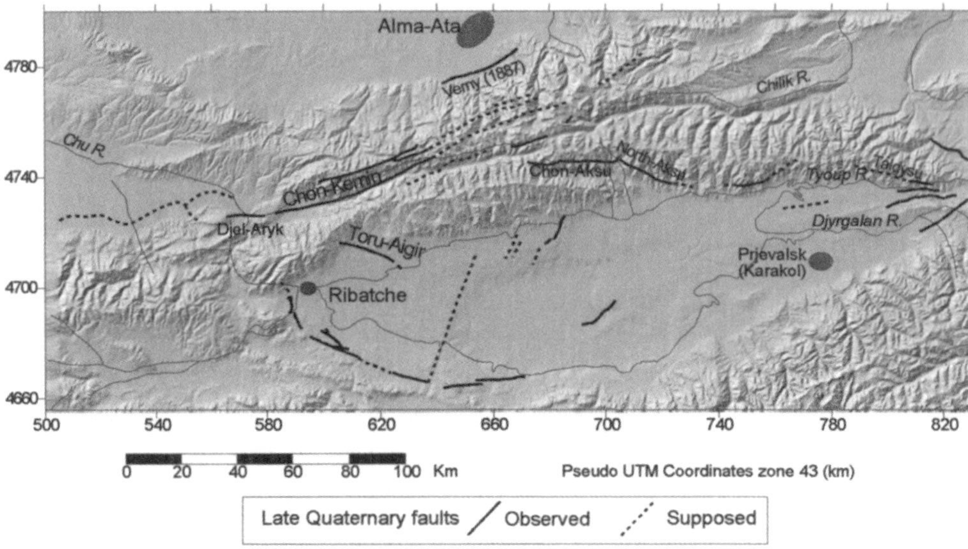

Figure 5. Late Quaternary faults of the Issyk-Kul depression and surrounding area (faults active in the Late Pleistocene - Holocene).

A/ Chon Kemin fault segment: 1911 fault scarp at Kok-Bel pass, just after the earthquake

B/ Same area as in picture A, in 1999

C/ Ananeva Rock slide along the North Aksu fault

D/ Kaindi rock slide in the lower Chon-Kemin valley

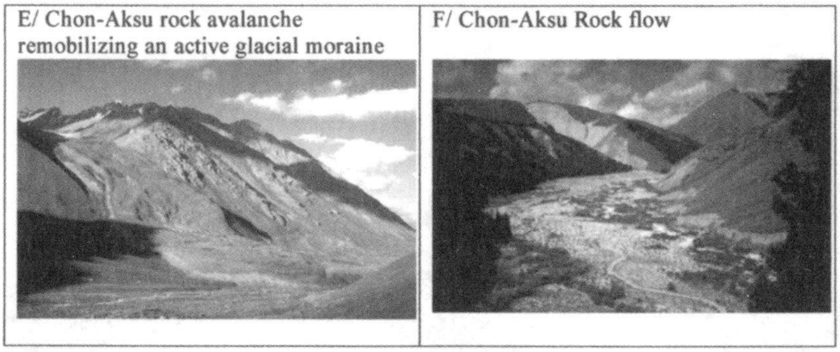

Figure 6. Photographs illustrating the effects of 1911 Kemin earthquakes (B, C, E & F: taken by D. Delvaux; A & D: reproduction of pictures published by Bogdanovitch et al. [10].

2. Recent structure of the Issyk-Kul depression and active faults

Earthquakes correspond to the sudden release of tectonic strain accumulated in rock masses by episodic movements along faults generating seismic waves. A direct implication is that the first step in the assessment of the seismic risk of an area is to identify the potential active faults. We generally consider that recent faults showing evidence for movement during late Pleistocene- Holocene (last 100-150 Ka) are active and will likely be reactivated in the nearest future. Several of such active faults are known in territory of the Issyk-Kul basin and surrounding mountain ranges (Fig. 5). They are described briefly hereafter on the background of the general Late Cenozoic structure (Fig. 4).

2.1. PREDTERSKEY SOUTH BOUNDARY FAULT ZONE

The PredTerskey boundary fault borders the Terskey Range at its foothills, along the southern margin of the Issyk-Kul basin. This fault zone is subdivided into three major segments showing different coupling between the mountain ranges and the sedimentary basin. They are referred to the Western, Central and Eastern segments.

The eastern PredTerskey fault segment is inclined 30-70 degrees to the SE, plunging under the Terskey Range. It is typically a reverse fault along which the Terskey range is transported over the Cenozoic sediments of the Issyk-Kul basin. The type section for this segment is exposed along the Dzhety-Ogyuz River and was studied recently by Cobbold et al. [2]. This segment shows no evidence for late Pleistocene and Holocene activities, but some secondary features that might be related to earthquake shaking are known. The epicenter of the Ms=6.8, 1970 Sarykamysh earthquake is located near Prjevalsk (now: Karakol) along this fault segment, at its intersection with a NW-trending transversal fault [3][4]. This earthquake caused a number of secondary

gravitational dislocations as landslides, rock avalanches and rock falls. The source mechanism indicates that this event is related to a low-angle thrusting towards the NNE.
The central segment of the PredTerskey fault is composed of two parallel faults, well expressed in the region of Kadji-Say [5] (Fig. 4). The southern fault follows closely the outcropping line of the contact between the basement and Meso-Cenozoic deposits. It is a stratigraphic contact, dipping 40-60° north, where the Mesozoic and Tertiary sedimentary rocks overlie an erosion surface over weathered granites. The northern fault runs in the middle of the Tertiary deposits, and is associated to an asymmetrical anticline. Jurassic sediments and even pieces of basement granite are injected along the fault zone. The observed microstructures show that this fault zone is dipping steeply to the south and has a transpressional character, with both reverse and right-lateral components of movement.
This deformation affected the Tertiary deposits, but did not affected the late Pleistocene terraces. The main activity of this fault zone probably occurred at the Pliocene-Pleistocene transition and in the early stages of the Quaternary.
The western segment of the PredTerskey fault zone seems to be composed of several secondary faults (Fig. 4). As opposed to the other segments of the PredTerskey fault, the fault planes are dipping north. They are generally associated to fault-propagation-folds, with brittle ruptures in the basement and monoclinal folding in the overlying Tertiary sediments. At the subsurface, such faults cut the hinge zone of the monoclines, separating the long and weakly inclined backlimb from the steeply inclined short forelimb. This deformation also occurred during the Pliocene-Pleistocene transition. However, several faults of this segment (Fig. 5) were clearly reactivated during the Late Quaternary, as indicated by the dislocation of Quaternary terraces.
According to Iudahin [7], the boundary between the central and the eastern segments of the PredTerskey fault zone corresponds to an hypothetical NW-trending Trans Issyk-Kul fault, deduced from geophysical interpretation. On a structural basis, Chedia [8] and Trofimov [24] suggested that the two segments are separated by a NE-trending transversal fault, active in late Pleistocene.

2.2. PREDKUNGEY NORTH BOUNDARY FAULT ZONE

The PredKungey fault zone delimits the Issyk-Kul basin on its northern side, separating it from the Kungey Range. It is also composed of several individual faults; from East to West: the Taldysu, North-Aksu, Kultor and Toguzbulak faults (Fig. 4).
At the extreme East, the Taldysu fault is almost vertical. It runs at the foot of the Kungey range, then it follows the Tioup valley. Along this valley, systematic morphological displacements of small streams illustrate a significant left-lateral component of movement for this fault. The 1978, Ms=6,7 Djalanash-Tioup earthquake is probably associated with this fault system. A number of gravitational dislocations were reported as a result of ground shaking during this event. Other secondary surface deformations reported by Abdrakhmatov et al. [9] evidence the high seismicity of this area during the Late Pleistocene - Holocene. Two focal mechanism solutions are available for this event: one from the Kyrgyz Institute of Seismology (KIS) and one

from the Harvard CMTS catalogue. Both are compatible with NW-SE trending fault, steeply SE-dipping (KIS solution) to subvertical (CMTS).
Further to the west, the North-Aksu fault was reactivated during Ms=8.2, 1911 Kemin earthquake, described in detail in Bogdanovich et al. [10]. Several large rockslides and rock falls with millions of cubic meters in volume were triggered by this event. One of them is situated along the Aksu fault segment, near Ananeva (Fig. 6A). This fault is dipping to the north, and the Kungey range was thrusted over the basin sediments. No lateral movements could be observed along this fault segment. To the west, the Aksu fault is relayed by the Chon-Aksu fault (Fig. 6C-D). It looks similar as the Aksu fault and runs trough the Kungey range itself, in the direction of the Chon-Kemin valley. It was also activated by the same 1911 Kemin earthquake and had also a reverse faulting mechanism. Several large rock avalanches and a debris flow (Fig. 6B) occurred in the upper reaches of the Chon-Aksu valley as the result of this earthquake.
Near Cholpon-Ata, the Kultor fault has a complicated structure. At its eastern extremity, it is reverse fault dipping under the Kungei range. To the west it changed into an en-échelon south-dipping thrust, uplifting the lower part of the range relatively to the upper one. As evidence for late Pleistocene activity, it is offsetting glacial moraines from the last glaciations.
In the Toru-Aigir area, the Toguz Bulak fault has a complex structure. It evolved in several stages during the late Tertiary and the Quaternary. The most recent activity occurred along a WNW-trending fault with a clear geomorphic expression (Figs. 4, 5).

2.3. CENTRAL PART OF ISSUK-KUL BASIN

The tectonic position of the central part of the basin occupied by Lake Issyk-Kul is not clear. The digital elevation model combining topographic and bathymetric maps illustrates the underwater morphostructure of Lake Issyk-Kul (Fig. 2). In particular, a series of steps were mapped at differed water depth. The most prominent one is at 110 meters of water depth (1495 meters above sea level). They likely correspond to underwater abrasive terraces that whiteness past low lake levels. Some investigators [6] [24] relate the lowering of lake level to a rapid tectonic collapse of the central part of the basin in the Holocene (5-7 Ka). Others believe that these terraces reflect dryer climatic conditions during or at the end of the last glacial cycle, followed by progressive lake expansion in the late Pleistocene.
Independently to this controversy, the central part of the basin appears as a fault-controlled tectonic depression with a trapezoid structure [25]. It is apparently bounded by normal faults on both its northern and southern margins [4].

2.4. CHON-KEMIN - CHILIK FAULT ZONE

The Chon-Kemin - Chilik fault zone runs in the middle of the area comprised between Lake Issyk-Kul and the Kazakhstan platform (Fig. 4). It broadly corresponds to the Chon-Kemin and the Chilik river valleys, respectively in Kyrgyzstan and in

Kazakhstan. This fault zone separates the Kungey range from the Zaili range. It mainly reactivates an important late Paleozoic fault zone [11].
The Cenozoic structure of this zone is interpreted in several ways. Some investigators suggest that the narrow Chon-Kemin valley is bounded by thrust faults. Other consider that it is bounded by normal faults [12][13]. Only recently, detailed investigations in the frame of the Copernicus project by Delvaux et al. [14] evidenced that these faults present a significant left-lateral component of displacement and that the whole structure has to be regarded as a sinistral transpressional fault zone.
Besides recent landslides and dislocations caused by the 1911 Kemin earthquake (described hereafter), several impressive rock avalanches and long-term paleoseismic fault scarps testify for repeated paleoseismic activity during the late Pleistocene, with earthquakes at least as strong as the Kemin one.

3. The 1911, Ms 8.2 Kemin earthquake

The 3 January 1911 (22 December 1910 following the Orthodox calendar), at 23H25, the Zaili-Kungei mountain ranges between Lake Issyk-Kul and Alma-Ata was hit by one of the strongest earthquake ever felt in the center of a continent: the Kemin earthquake of surface magnitude Ms=8.2 on the Richter scale. The epicentre was located near the junction between the Chon-Kemin, Chilik and Chon-Aksu valleys (Fig. 3).
The number of casualties was relatively low for such a strong earthquake, because it affected mainly mountainous areas and people used to live in yurts. However landslides, rock avalanches and rock falls triggered by the earthquake caused a lot of destructions. The following destructions were reported by Bogdanovitch et al. [10]: 452 persons killed, 740 persons injured, 1094 houses destroyed, 4545 yurts destroyed and 12962 cattle head killed.
The surface effects of the 1911 Kemin earthquake have been investigated only a few months after the event by an expedition led by K.I. Bogdanovitch in spring 1911 [10]. Since then, no further detailed investigations were performed on the seismotectonic expression of the Kemin earthquake, except by Kuchai [26], who revisited briefly the area. In the frame of the European Community Inco-Copernicus project "Landslide Risk Triggered by Earthquakes in Kyrgyzstan, Tian-Shan)", the surface effects of the Kemin earthquake were re-investigated along the Chon-Kemin, Chon-Aksu and Aksu valleys in the Kyrgyz territory only.
Combining the observations of Bogdanovitch et al. [10] and the recent observations (Delvaux et al., 2001) it can be precised that the 1911 Kemin earthquake generated nearly 190 km of surface ruptures along six different fault segments totalling 260-300 km long. The surface rupture created new lakes by damming rivers and diverting the water flows.
Along the E-W trending Chon-Aksu fault segment, paleoseismic trenching across a 18 meter-high multi-event fault scarp allowed to recognise the effects of at least 4-5 past strong earthquakes. It confirmed the reverse faulting character of that segment.

The maximum height of the fault scarp associated to the 1911 event along the Chon-Aksu and Aksu valleys ranges between 6 and 10 meter high (Fig. 6 C-D). A left-lateral component of movement between 5 and 8 meter was observed along the Chon-Kemin valley.

The earthquake caused also numerous landslides, rockslides, rock avalanches, debris flows and secondary soil dislocations over a total area of 10.000 square kilometres (200 km long, 70 km wide), elongated parallel to the fault zone. The largest rock avalanches reached 15 million of cubic meters in volume (the Ananeva rockslide, Fif. 6A). The Kaindi rockslide itself killed 38 persons in a yurt camp in the lower Chon-Kemin valley.

Several rock avalanches occurred in high-mountain areas, along the upper reaches of the Chon-Kemin and Chon-Aksu valleys. They were caused by the collapsing of steep walls of glacial valleys at high altitude. The huge mass of debris covered active glaciers and their moraines, destabilizing the whole system and moved rapidly down to the main valley, more than 1000 meter lower. They partly blocked the main valleys.

An impressive delayed effect of the Kemin earthquake is the Chon-Aksu debris-flow (Fig. 6B). It happened several months to several years after the earthquake, because it was not seen on a photograph taken by the team of Bogdanovitch and published in [10]. This was formed by a gigantic collapse of an old large glacial moraine accumulation and the debris flow submerged the Chon-Aksu valley along 2,5 km. It occurred right along the trace of the active fault. The following mechanism of formation is suggested. The earthquake destabilised the moraine accumulation and water infiltrated along open fractures. It saturated progressively the fluvio-glacial material. Suddenly the whole mass moved in a catastrophic way and rushed to the alluvial plain as a large mudflow supporting large boulders of granite. It washed out the bottom of the valley over 2,5 km long, stripped away all the trees along its way and partly dammed the valley. The mass of the debris flow is estimated at 20 million of cubic meters.

4. Seismic hazard of Issyk-Kul basin

Several approaches were tested for the evaluation of seismic hazard of the Kyrgyz territory (Abdrakhmatov et al. [9] and earlier works).

The traditional approach is an expert evaluation scheme. It takes into account the history of geological development of the territory and its block structure (Petrushevskiy, 1968), the boundary conditions [18] and the influence of Cenozoic activation expressed in a quantitative way [19]. In addition, characteristic data derived from the investigation of paleoseismic dislocations are also taken into account [20][21].

Seismogenic zones (zones prone to earthquake activity) are delimited on the basis of combined geological and geophysical approach. The data on active faults are considered with the highest attention, as they might concentrate future seismic activity. The maximum possible earthquake magnitude can be evaluated from the historical data and instrumental seismicity. Such approach was used by Djanuzakov et al. [22] to produce the seismic hazard map of Kyrgyzstan compiled in 1995. Seismic hazard zones are classified in terms of balls - intensities. This map shows that the Kungei and Zaili ranges between Lake Issyk-Kul and Alma-Ata and the Terskey range south and east of Prjevalsk (Karakol) have the highest intensity, with expected magnitudes larger than 7.5. This is a direct consequence of the strong historical and instrumental seismicity recorded in these area.

A more modern approach for the seismic hazard assessment is to construct probabilistic seismic hazard and Arias Intensity maps. This was performed in the frame of the Copernicus project, and the results are being published (Havenith et al., 2000; Havenith et al., in preparation). On the basis of seismicity (from the Kyrgyz and world seismic catalogues) and active tectonics, seismic zones are defined for the whole Kyrgyz Tian-Shan. Gutenberg-Richter laws are defined mainly from instrumental data and empirical attenuation laws are used. The probabilistic maps obtained present results in terms peak ground acceleration (PGA, in g), for defined periods (50 or 100 years) and with 90% probability of non-exedence.

This approach shows that large earthquakes can be expected in the whole region comprised between Lake Issyk-Kul and Alma-Ata (Chilik-Kemin region), and that the town of Alma-Ata is particularly at risk. Similarly, the Arias Intensity approach shows that the Chilik-Kemin region has the highest hazard for landslides triggered by earthquakes, due to the combination of strong Arias Intensities and steepness of the topography.

5. Acknowledgements

This work has been realised in the frame of the European Community Inco-Copernicus project "Landslide Risk Triggered by Earthquakes in Kyrgyzstan, Tian-Shan)". It has been presented at the 25-29 September 2000 NATO workshop at Cholpon-Ata, on environmental problems related to the Lake Issyk-Kul and surrounding areas.

6. References

1. Mushketov, I.V and Orlov, A.P. (1893) The catalog of earthquakes of Russian empire, 1893. *Proceedings of Russian Geographic Society* **26** (in Russian).
2. Cobbold, P.R., Sadybakasov, E. and Thomas, J.C. (1996) Cenozoic transpression and basin development, Kyrgyz Tien Shan, Central Asia. In F. Roure, N. Ellouz, V.S. Shein, and I. Skvortsov: (Eds.), *Geodynamics evolution of sedimentary basins,* Technip, Paris. pp.181-202.
3. Grigorenko, P.G., Mamirov, E., Sadybakazov, I.S., Talupov, M.A. and Turdukulov, A.T. (1973) *Geological structure and geotechnic conditions in the region of the Sarykamychsk earthquake*, Ilim Press,. Frunze (in Russian).
4. Trofimov, A.K. (1973) Cenozoic tectonics of Sarykamysh earthquake area, *Izvestia Academy of Science of Kyrgyz SSR* **1973(5)** (in Russian).

5. Chedia, O.K., Djumadylova, Ch.K. and Trunilin, S.I. (1988) Structure of the Predterskey boundary fault (western part). *Izv. Academy of Science of Kyrgyz SSR*, **1988(3)** (in Russian).
6. Zabirov, R.D., Korotaev, V.N. and Nikiforov, L.G. (1973) Some questions of Quaternary history of Issyk-Kul.*Geomorphology*, **1973(4)** (in Russian).
7. Iudahin, Ph.N. (1983) *The geophysical fields, structure of depth and seismisity of Tien-Shan,* Ilim press, Frunze. (in Russian).
8. Chedia, O.K. (1986) *Morphostructure and neotectonics of the Tien-Shan*, Ilim Press, Frunze (in Russian).
9. Abdrakhmatov, K.E., Chedia, O.K. and Lemzin, I.N. (1988) Seismites, In *Detailed seismic zoning of Eastern Kyrgyzia,* Ilim press, Frunze (in Russian).
10. Bogdanovitch, K. I., Kark, I. M., Korolkov, B. Y. and Musketov, D.I. (1914) *The earthquake in the northern district of the Tien shan, 22/12/1910 (4/01/1911)*, Proceedings of the Geology Committee, New series, **89**, St. Petersburg (in Russian, extended abstract in French).
11. Bakirov, A.B. and Korolev, V.G. (1979) On the age of the oldest bedrock of Tien Shan. *Izv. AN SSSR, ser.geol*, **1979(7)** (in Russian).
12. Nesmeianov, S.A. and Barhatov, I.I. (1982) The structure of cenozoic faults of Issyk-Kul basin and Kungey-Zaily district. In: *Complex scientific investigation in ingeneering research for construction*, Nauka Press, Moskow (in Russian).
13. Patalaha, E.I., and Chabdarov, N.M. (1976) Condition of relief formation of Northern Tien Shan in light of new geological observation and possible mechanism. In *Seismotectonics of some areas of South SSSR*, Nauka Press, Moskow (in Russian).
14. Delvaux, D., Abdrakhmatov, K., Jacques, E., Trefois, P. and Buslov, M. (1999) Late Cenozoic and active sinistral transpression in the Northern Tian Shan of Kyrgyzstan. EGS 24 General Assembly, The Hague, The Netherlands 19-23 April, 1999. *Geophysical Research Abstracts*, **1(1)**, 38.
15. Delvaux, D., Abdrakhmatov, K.E. and Strom, A. L. (2001) Landslides and surface breaks of the Ms 8.2 Kemin earthquake, Kyrgyzstan, *Russian Geology and Geophysics* (in press).
16. Abdrakhmatov, K.E., Turdukulov, A.T. and Khristov, E.B. (1993) *Detailed seismic zoning of the Issyk-Kul basin*, Ilim press, Bishkek (in Russian).
17. Petrushevskiy, B.A. (1968) The use of geological data for seismic zoning. In *Seismic zoning in the SSSR*, Nauka Press, Moskow (in Russian).
18. Gubin, I.E. (1974) Seismic faults and their significance for seismic zoning. *Geotectonics* **1974(6)** (in Russian).
19. Gzovskiy, M.V. (1970) The development of new direction in tectonophysics. *Izv. AN SSSR, The physics of Earth*, **1970(5)** (in Russian).
20. Solonenko, V.P. (1973) The earthquakes and relief. *Geomorphology*, **1973(4)** (in Russian).
21. Niconov, A.A. (1977) *Recent and Holocene movements of eath crust*, Nauka Press, Moscow (in Russian).

22. Djanuzakov, K.D., Chedia, O.K., Abdrakhmatov, K.E. and Turdukulov, A.T. (1995) Map of seismic zoning of the territory Kyrgyz Republic, 1:1,000,000 scale, Ilim press press, Bishkek (in Russian).
23. Havenith, H.B., Trefois, P., Jongmans, D. and Abdrakhmatov, K. and Delvaux, D. (2000) Probabilistic seismic map of Kyrgyzstan (Central Asia): application to the regional hazard of seismically induced landslides. *Proceedings of the International scientific-technical conference on geomechanical problems and geotechnical investigations of mountainous areas*, Bishkek, 4-6 October 2000.
24. Trofimov, A.K. (1990) Quaternary deposits of Issyk-Kul Basin in connection to its tectonics. *Izvestia A.S. Kirghyzskoy SSR*, **1990(1)** (in Russian).
25. Meshetely, A.V., Kuznetsov, D.V. and Buchkina, N.A. (1985) *Investigations of groundwater flow in Issyk-Kul lake. In: Hydrological investigation of intermountain basin*, Ilim press, Frunze (in Russian).
26. Kuchai, V.K. (1981) *Regional orogenesis and siesmicity*, Nauka Press, Moscow (in Russian).

THE ENVIRONMENTAL RISK RESULTING FROM GOLD MINING

Beishen IMANACKUNOV
National Academy of Sciences of the Kyrghyz Republic
Chui Prospekt, 265a , Bishkek, Kyrgyzstan

While discussing the technical-economical feasibility of gold extraction factories, the environmental problems are often considered in a simplified way. Members of the high commissions agree with elaborators' proposals and the factory starts to work, afterwards many unpredicted problems appear. The same happened to the "Kumtor" operating company. Indisputably, Kumtor is a construction of the century, promising much to the people of Kyrghyzstan. However, it is not provided with all necessary measures for protection of the environment of our mountainous country.

There were and there are problems related to the environmental consequences of tailings, cyanide-containing drains of Makmal, Kumtor factories and future integrated activity of Djerui.

We, the experts, involved in research on environmental safe methods of deactivation of cyanic drains of gold-extracting factories, want to express our opinions on that point.

In the past, it was stated that on gold-extracting factories "water and air pollution is impossible, this is admitted even by opponents". Real life has disproved all these prophetic words. The main danger of gold extraction technology and tailings, containing huge quantities of cyanide-containing drains, is resulting from earthquakes, with catastrophic consequences for people. Besides this, accidents can occur during cyanide delivery to factories as it was in 1998 in Barskoon, as well as accidents in pulp wires. Is there an expert, saying that in conditions of high altitude, in particular, at Kumtor, seismic cataclysms are impossible?

Next endangering point is the fatal effect of a hydrocyanic acid on the environment inasmuch at Kumtor factory the total deactivation of cyanide-containing drains is not envisaged. According to the project, deactivation of tailings at Kumtor is reached by natural decomposition of cyanides, the end product of which is the gaseous hydrocyanic acid and separate cyanic complexes insoluble in water.

As we can see, hydrocyanic acid gets into the air as a gaseous substance. As the experts say the construction of the tailings practically excludes the outflow of harmful substances by drainage (leakage). But there is another way for hydrocyanic acid to get into the water sources. The evaporating hydrocyanic acid with a rain stream quickly gets into the water sources as it is well-dissolved in water. As is known, up in the mountains the intensity of rains is very high.

J. Klerkx and B. Imanackunov (eds.), Lake Issyk-Kul: Its Natural Environment, 161–166.

The quantity of the harmful gases formed on the surface of tailings of the projected Djerui gold extracting factory is less than at the other working GEF; we have taken part in the expertise of the technical-economical feasibility of Djerui factory. According to their technical-economical feasibility about 200 kg of hydrocyanic acid per day are formed and educed into the air from the pond area of 16 hectares. It is easy to imagine what such an amount of hydrocyanic acid educed into the air environment can do if the maximum permissible concentration (MPC) of hydrocyanic acid in the air is 0,01 mg / m^3. Actually, some part of the hydrocyanic acid is dissipated in space, but it does not give us the right to neglect the pernicious influence of the hydrocyanic acid on our nature. We even have some fears that part of the hydrocyanic acid probably reaches the ozone layer and destroys it. But it is in the sphere of assumptions, which should be studied.

Thus, while drawing up the Technical Economical Feasibility of these factories (Kumtor and Djerui) foreign experts do not take into consideration air and water pollution with hydrocyanic acid. They lean on the experience of gold-extracting enterprises of North America and New Zealand, which probably are not seismic-dangerous zones. In these regions, the population is much lower, than here in the valleys, and the climatic conditions are strongly different.

Because of the low temperature at Kumtor, the speed of formation of hydrocyanic acid in tailings will be a little bit lower than in other places, but it does not mean that the hydrocyanic acid will not be formed.

The technology of gold extraction, excluding deactivation of cyanic drains, is applied at gold extracting factories of the neighboring states and has started to provoke a big anxiety and concern of the population and experts.

We are surprised by the fact, that in the mass media there appear declarations of "sorry - experts", saying that " on the sun cyanic compounds decompose to harmless components right away, so that they will not cause any danger to the environment". If it is so, there appears a question – for what reason such expensive researches on the deactivation of the cyanide containing drains of gold extracting enterprises are conducted all over the world? Hence, such a statement is at least irresponsible as all the data about the mechanisms of cyanide decay, including our researches, show, that during the decomposition of the cyanide in natural conditions very toxic hydrocyanic acid is formed first. There is a possibility of formation of harmless substances alongside with it, but their quota is insignificant in comparison with a hydrocyanic acid.

The scheme of cyanide decomposition proceeds by the hydrolysis mechanism: in neutral environment pH 7 cyanides of metals hydrolyzes with formation of the base and acid:

$$MeC \xrightarrow{H_2O} MeOH + HC \qquad \text{or}$$

$$Me(C)_2 \xrightarrow{H_2O} Me(OH)_2 + 2HC$$

where Me is Na^+, K^+, Co^{2+}, C^{2+}, Ni^{2+} and others

As is apparent from the reaction, the result of hydrolysis is the hydrocyanic acid.
All the cyanides are divided into toxic and non-toxic ones. Toxic ones are the cyanides of all metals, except iron. Ferrocyanides are not toxic. However, ferrocyanides under the influence of solar rays partly decay with the formation of hydrocyanic acid, so ferrocyanides can also be dangerous.
As it was stated above, at Kumtor and Makmal Gold Extracting Factories, technology of gold extraction is based on a method of cyaniding with the subsequent utilization of cyanides by chemical method and creates an extreme danger to the environment.
At Kumtor factory, deactivation is carried out by sulphurous gas O_2, as a reagent sodium metabisulfate is used, as a catalyst blue copperas, and as coagulant iron sulfate. Accumulation of these chemical reagents in big amounts in mountains is ecologically dangerous. As for Makmalgold, deactivation of cyanides is made there by liquid chlorine, which is a not less toxic substance itself. That is why one of the most actual problems of the environment protection and protection of sources of water formation in mountains is the development of highly effective environmentally safe bio-technological method of cleaning of drains of Gold Extracting Factories from cyanide.
Such a microbiological method of cyanides decomposition was developed and tested by us at the repository of "Makmalgold" factory and has shown good results.
Inculcation of the method of cyanides bio-destruction in drains of the gold-mining enterprises will allow to exclude environmentally harmful chemical methods of cyanides utilization.
Tasks on the complex use of mineral resources might be successfully solved by the development and inculcation of progressive technological processes of combined manufactures, excluding or considerably reducing dangerous influences on the environment. Mining production, in particular gold extracting factories, is one of such areas for Kyrghyzstan.
Kyrghyzstan is one of the worlds leading countries in auriferous ore deposits. They are concentrated in the highlands, in glacial zones, where drinking water sources are mainly formed. The technology of ore processing comprises the environmentally dangerous cyanide method of gold extraction.
Acting and being built gold extracting factories (GEF) of Kyrghyzstan are not provided with all necessary measures on the environment protection of our mountainous region. We, the specialists engaged in research of environmentally safe methods of extraction and rendering harmless of cyanide GEF drains, we would like to state some reasons on this point.
The main danger of gold extracting technology and tailings, containing amounts of cyanide-containing drains, is a possibility of earthquake, which catastrophic consequences are often suffered by our people. Damages in pulp-wires, accidents during the delivery of chemicals to the factory – all these are of a great danger under conditions of high altitude, where seismic cataclysms should not be excluded.
The use of biogeotechnology's achievements is one of the approaches to the solution of such environmental matters.

Biotechnology is a study of metals' extraction from ores, concentrates, rocks and solutions under the influence of microorganisms and their metabolites. A number of works [2-4] are also devoted to the processes of metals biosorbtion from the solutions. One of the new tendencies in the development of biogeotechnology is the bioenrichment by microorganisms, in particular by using bioflotation.

Group of scientists has ascertained a new type of colloidal-biochemical influence – selective adhesion of colloidal particles on the surfaces of living cells. These scientists have shown that unlike to well-known facts of microorganisms' adhesion to the different kinds of sorbents-bearers, which is unspecific and fulfilled by molecular adhesion, selective interaction of living microorganisms with colloidal particles of metals and their combinations is fulfilled by the appearance of coordination and covalent connections. Thus, selective adhesion is characterized by a variety of powers and mechanisms of colloidal-biochemical interaction: molecular, electrostatic and chemical.

The use of microorganisms while enriching the fine-dispersive gold allows deciding not only technological, but also environment protective tasks related with the purification of sewage.

According to the data of geologists, about 30 % of gold and polymetals of the ores of Kyrghyzstan are in the colloid – dispersion condition and this requires the elaboration of special technology of extraction.

For the purpose of the full extraction of ultra-disperse gold and the creation of environmentally safe technology of the gold-containing ores processing, we are developing the microbiological method of gold extraction and rending harmless the drains of GEF currently acting in the Republic.

Tasks of this development were as follows:

1. Study of applying possibilities of the μmicrobiological fine-disperse gold extraction method from the gold-containing ores of Kyrghyzstan.
2. Elaboration of cyanides bio-destruction technology for the use in the drains of gold-mining factories.
3. Determination of mineral compositions and forms of fine-dispersive gold content in the local ore.

The chemical and mineralogical analyses of the tests have shown that the investigated ore of Kumtor is related to the primary pyrite gold-containing ones.

The gold existing in the test is presented in two forms. Its high content – 10 grams per ton – has been found in graphite, which is contained in the fine fraction (up to 1%) and found out while processing. Besides, the gold is present in pyrite as a structural mixture. In what form does the gold exist in the carbon fractions? For this purpose, the X-ray phase and spectral analyses have been carried out and it was found out that there was no gold in its free display in a carbon fraction.

It was necessary to determine a quantitative correlation of gold particles according to their size. There has been carried out the sedimentation analysis of the ore. The obtained data testify that in the ore, ground up to 74 μ, about 25 % of particles have the size of 20 μ and less, the major quantity of gold (more than 75%) being in the fine-dispersive part, and more than 20 % of gold being contained in the ultra-dispersive particles (5 μ).

If one takes into account that the consolidated gold particles on the surface of microorganisms must be commensurable with their sizes, conclusions can be made about the microbiological enrichment perspectives in relation to Kyrghyzstan's ores.
The gold-containing ore of Kumtor and its concentrate have been used as the initial materials. Microorganisms were excreted from the gold-containing ore and studied by appropriate methods.
While studying the morphological-physiological properties of microorganisms' strains, grown in the meat-peptone agar (MPA), a wide spectrum of morphological forms of microorganisms' strains has been revealed, that differ from each other by size, colour, form and consistence.
The excreted cultures have been checked on the capacity of interaction with colloidal gold. Obtained results have shown that the fine-disperse particles were aggregated both from the cells' surfaces and in the intercellular spaces, in which metabolisms are concentrated. This witnesses that the excreted bacteria have a property to accumulate the fine-disperse gold by both monocell and associates or follicles of the cell.
There have been also found bacterial cells with weak interaction, being completely indifferent to colloidal gold as well.
Availability of active metalphilic microorganisms in the observed gold deposit provides a reason to say that the microorganisms played definite role in its forming.
There were made tests on the flotation of gold-containing ores of the "Kumtor" deposit by excreted microbes at the Central Lab of the State Geological Committee of the Kyrghyz Republic. For this test fresh, 16 – 18 hour's culture, excreted from ore and cultivated at the environment of 5 m, was used.
The scheme of ore processing included preliminary separation of carbon fraction, flotation with the use of reagents recommended for the "Kumtor" ore processing. During the operation of experimental flotation there has been given the biomass of microorganisms in amount of 30 g/t (in recalculation of absolutely dry matter (ADM) – of 0,4 g/t). While using the microbe mass in the form of water suspension, the output of experimental flotation concentrate increases twice, as the extraction does (Table 1). However, this does not effect on the gold content in the concentrate of the experimental flotation. Nevertheless, the content of gold in flotation tails is being considerably reduced in comparison to the idle tests.
This discrepancy can be explained by the high foam-forming ability of microbe mass. For the elimination of this discrepancy in the process of bioflotation, it was necessary to use non-water suspension of microorganisms. The appropriate researches are being conducted now in order to solve these problems.
Thus, with the help of microorganisms it becomes possible to extract fine-disperse gold, which remains in the tails while using other methods of processing. The method of bioflotation enrichment of gold-containing ores is simple in its implementation as it is easily correlated with the traditional scheme of gravitation-flotation enrichment.
For its actualization it is necessary only to create a section for microorganisms' biomass increasing. The use of bioflotation method requires the elaboration of fine-disperse gold revelation methods.

TABLE 1. The output of the concentrate of the experimental flotation

Product	Output	Content of Au,g/t	Extraction of Au,%	Conditions of experiment
Carbon concentrate	0,06	45,2	0,5	
Concentrate of major flotation	16,08	24,0	64,0	Without bio-mass
Concentrate of experimental flotation	6,23	9,4	9,7	
Tails of flotation	77,63	2,0	25,8	Expense T-80 40 g/t
Ore	100,00	6,0	100,00	
Carbon concentrate	0,08	4,2	0,5	
Concentrate of major flotation	12,20	27,0	51,8	Without bio-mass
Concentrate of the experimental flotation	6,05	12,4	11,8	Expense T-80 25 g/t
Tails of flotation	81,67	2,8	35,9	
Ore	100,00	6,4	100,00	
Carbon concentrate	0,08	45,0	0,6	With biomass
Concentrate of major flotation	13,73	26,4	58,4	Expense T-80 25 g/t
Concentrate of experimental flotation	12,39	7,4	14,8	
Tails of flotation	73,80	2,2	26,2	
Ore	100,00	6,2	100,00	
Carbon concentrate	0,11	48,0	0,8	
Concentrate of major flotation	12,32	25,6	52,3	
Concentrate of experimental flotation	14,19	8,6	20,2	
Tails of flotation	73,38	2,2	26,7	
Ore	100,00	6,0	100,0	

Part II

CASE STUDIES IN OTHER REGIONS

HYDROCHEMISTRY OF LAKES OF TAJIKISTAN

D.N. PACHAJANOV and D.L. PATINA †
Nikitin's Institute of Chemistry, Tajik Academy of Sciences
Aini Avenue, 299/2, 734063 Dushanbe, Tajikistan

N.I. VOLKOVA
Institute of Geology, Geophysics and Mineralogy, SBRAS
Akademik Koptyug Avenue, 3, 630090 Novosibirsk, Russia

1. Introduction

There are about 1500 lakes in the Pamirs [2, 5], including small ($\leq$1 km^2) lakes with total water surface area over 680 km^2, which make 97.5 % of the total number but only 9 % of their total area. The water chemistry of lakes in the Pamirs has very limited literature. An overview of Pamir research through 100 years from 1871-1971 [1] presents data on the geomorphology, geology, meteorology, soil science, flora, and fauna of this high-altitude land, but no data on this subject. At the same time knowledge of water chemistry allows to estimate the fresh water resources of Tajikistan as a potential source of drinking and irrigation water.

We analysed water samples from a number of lakes in the Pamirs and elsewhere in Tajikistan (Table 1, after [2, 3, 4]) for major ions and trace elements, including metals, and investigated the formation mechanisms of the observed chemical compounds, mineralisation, water exchange, and physiographic conditions.

Sampling, chemical analyses of lake water, and analytical material treatment was carried out in collaboration with specialists from the Pavlovsky Institute of Zoology and Parasitology of the Tajikistan Academy of Sciences and from the Hydrographic Department of the Dushanbe Hydrometeorological Observatory.

2. Methods

The chemical analysis of water samples was made mainly according to guidelines in [6], except sulphates, which were determined by the volumetric trilonometric method [7], and trace elements determined by VITR-LTI spectral chemistry [8]. Some lakes were sampled for temperature, oxygen content, dissolved oxygen and pH. Trace element abundances in lake water were determined by X–ray chemical spectroscopy at the Institute of Chemistry (Dushanbe, Tajikistan) to an accuracy of 20 %. The detection

J. Klerkx and B. Imanackunov (eds.), Lake Issyk-Kul: Its Natural Environment, 169–179.

limits were (in mg/l) 0.00003 for Ag, 0.0001 for Bi, 0.0002 for Mo, 0.0003 for Pb, 0.002 for Cu and Mn, 0.003 for Zn, 0.005 for As and Nb, 0.010 for V and W.
Concentrations of some trace elements in sediments of the Sassyk-Kul depression were determined by INAA at the Institute of Nuclear Physics (Tashkent, Uzbekistan), with standard deviations from 5 to 20% for different elements. The detection limits were 100 ppm for Ba, 25 ppm for Sr, 10 ppm for Cr and Ni, 5 for As, Rb, and Zn, 2 ppm for Ce, 0.5 for La, Mo, Cs, Se, 1 ppm for W and U, 0.2 ppm for Hf and Sm, and 0.1 ppm for Sc. Quantitative AAS was performed at the Institute of Chemistry (Dushanbe, Tajikistan) with detection limits of 100 ppm for Ti, 10 ppm for B, Zr, and Zn, and 5 ppm for Pb, Cs, Mo.

TABLE1. General information on studied lakes

Lake	Altitude m a.s.l.	Area km^2	Volume Mln. m^3	Length km	Width km	Depth m	Genetic type
Zorkul	4125	41,7	88,2	20	2,08	2,1	Tectonic
Chakankul	4126-4127	2,2	2,25	8,6	0,3	1,0	Id.
Kukdjigiti	4262	6,7	6,0	5,0	1,5	0,8	Moraine
Karadong	4268	1,5		2,3	1,2		Id.
Yashilkul	3734	36,1	450	24,6	1,5	12,5	Tectonic.
Bulunkul	3767	3,5	6,55	3,64	0,96	1,3	Id.
Drumkul	3335	1,96	69,0	3,15	0,62	35,2	Id.
Khargush	4253	0,8	3,0	1,9	0,42	3,7	Id.
Rangkul	3784	7,8	5,61	5,8	1,34	0,7	Tectonic.
Shorkul	3780	9,8	18,5	6,9	1,41	1,9	Id.
Salangur	4172	2,4		2,3	1,6		
Turumtaykul	4213	9,82	134	6,99	1,40	13,6	Moraine
Azdarkul	4382,7	0,8					
Chukurkul	3952	0,5	0,8	1,3	0,4	1,6	Tectonic.
Sasikkul	3825	8,92	30,5	4,68	1,91	3,41	Id.
Tuzkul	3798	1,84	1,95	2,98	0,62	1,04	Id.
Karakul	3914	380	26500	28,3	13,4	69,7	Id.
Sarezskoe	3260,8	79,4	15850	55,3	1,44	200	Id.

3. Fresh and saline lakes

The data in Tables 2, 3, 4, represent the chemistry of water in summer when the lakes are most actively fed from the surface, subsoil, and groundwater. The analysed major ion composition and salinity reflect the composition of leached rocks and high altitude desert and desert-steppe soils in the catchment basins of freshwater lakes, and high-altitude meadow, meadow-marsh, peatbog, glaciated, and salt-marsh (hydromorphic) soils [9] in the drainage basins of saline lakes [10]. The chemical composition and salinity of lake water form under the influence of direct (cited above) and indirect factors [10]. The latter include the type and rate of water exchange (Table 1) and climate. The arid and cold windy climate of the high-altitude Pamirs with low mean annual air temperatures and low atmospheric pressure (463 mercury mm at 4000 m above the sea level at 0^0 C [11]) is favourable for strong evaporation. The effect of predominant evaporation in the Eastern Pamirs is to increase the total content of major ions (i.e., salinity) in overflowing Lake Karakul and in closed saline lakes. Water exchange, the principal indirect factor [10], is in turn controlled by the size and rock composition (high contents of low-soluble minerals in rocks and soils) of lakes' catchment basins.

All lakes are fed from less mineralised water from source areas and remain fresh if they are overflowing (lakes Zorkul, Kukjighit, Karadong, Yashilkul, Bulunkul, Saretskoe and Drumkul in the Pamirs, and lakes Iskendeer, Kulikurdak, Harkush, Rankul, Ajarkul, and Turumtaykul outside the Pamirs). Lakes become saline in the conditions of weak groundwater circulation (lakes Salongur, unnamed lake above it, Shorkul, Karakul) or strong evaporation (times the amount of precipitation) in isolated closed depressions, often swampy, fed from groundwater vents (Sasykkul, Chukurkul, Tuzkul). Occasional mineralisation changes in wet years do not change the salinity class of lakes.

The chemical composition of dissolved compounds in saline lakes with low water exchange rates (e.g., lake Karakul) in the Northern and Eastern Pamirs results from the effect of a cold and dry climate of high-mountain desert. The major ion composition of water is controlled by composition changes (metamorphization) of the feeding surface and soil waters associated with evaporation and freezing and reflects genetic relationship with minerals leached in the catchment basins.

The salinity of Lake Karakul reaches 12 g/l as a result of the existing equilibrium in water and salt annual balance, probably, with minor deviations on the long-term scale. The annual inflow of less mineralised water to the lake of about 1 % of the total volume and almost ~0.05 % of the ion reserve cause only minor changes to the composition of the main water body. Annual ion accumulation from rivers, ground and soil waters apparently does not increase the mineralisation above 12 g/l, for several reasons, including the formation of dry salts near the coast and on islands, evaporation through groundwater outflow during water-level rise, ion loss with wind sprays, etc.

The concentrations of the detected trace elements in Lake Karakul are slightly higher than in riverine and marine water. Their significant variations with depth are associated with the biological turnover (accumulation of bound forms near and on the bottom, release of free forms, participation of bound forms in biological mineralisation of organic matter) which is responsible for the Cu, Zn, Mn, Pb, and Ag enrichment and almost invariable concentrations of Mo in groundwater.

TABLE 2. Major ions and mineralisation of water in Iskanderkul and Zmeinoe lakes (mg. l. / equ. l.).

Station	Depth m	Ca^{++}	Mg^{++}	Na^{+}	K^{+}	Bicaroca-rbonates	Sul-phates	Chlo-rides	Total ions
1	10	23,0 / 42,2	1,8 / 5,5	1,35 / 1,8	0,58 / 0,5	70,2 / 42,1	7,87 / 6,0	1,1 / 1,8	106
Northern beach		37,0 / 32,7	10,9 / 15,9	1,8 / 1,2	0,5 / 0,2	165 / 47,7	4,9 / 1,8	1,1 / 0,1	221

4. Saline lakes of the Sasykkul depression

Lakes in the Sasykkul depression are typical saline lakes. They attract interest as potential sources for extraction of rare metals, as, for instance, lakes Silver Pak and Serls (USA) in the arid zone in the centre of fissure and subsoil water discharge areas associated with a tungsten deposit, where lithium, potassium chloride, sodium sulphate, bromine, phosphoric acid, boron, soda, and tungsten are extracted from the brines.

The Sasykkul depression (3800-4000 m asl) is situated at the junction of two large tectonic zones of Southwestern and Southeastern Pamir. The climate is very dry (40 % relative mean annual humidity), with low atmospheric precipitation (70 mm/year mean annual norm), sharp seasonal (60-65°C, from -40°C in winter to +20°C in summer) and daily (20-25°C) temperature contrasts. The depression floor is occupied by brackish lakes Sasykkull, Chukurkul, Tuzkul, Shorkul, etc, with depths as shallow as 3-4 m but water surface areas up to 7 km^2 (Sasykkul - 7.02, Tuzkul, Shorkul, and 1.63, Chukurkul - 0.25 km^2). A considerable part of the depression, especially, the southern and eastern coastal zones of lakes Tuzkul and Sasykkul are highly saline. There exist mineral springs with travertine on the depression sides and in the bottom of lakes.

According to their external water exchange rate, the saline lakes of the Sasykkul depression are not overflowing. They have chloride - hydrocarbon - sulphate - sodium water composition and abnormal mineralisation of 40-130 g/l and pH 9.4-9.6. The composition of brine and salt precipitation consists of halite, trona, burnaite, thenardite, and minor amounts of brown soil and mirabilite.

High concentrations of some trace elements (Tables 2 and 3) in the lakes of the Sasykkul depression [12,13,14] are controlled by the geochemistry of granites and the presence of tungsten and polymetallic ores in the region.

Trace elements are carried to the lakes by surface and subsoil waters and besides penetrate supplied through bottom fissure hydroterms in highly permeable active

TABLE 3. Chemical composition of water of Pamir lakes

Lake	Ca^{2+}	Mg^{2+}	$Na^{+}+K^{+}$	HCO^{-}_{3}	CO^{2-}_{3}	SO^{2-}_{4}	Cl^{-}	Mineralisation	Ionic strength	Data Base
Khargush	24,0	12,4	12,7	140,3	-	8,2	9,2	206,8	0,0039	Our data
Bulunkul	30,1	4,3	29,0	114,1	-	40,3	10,6	228,4	0,0042	Our data
Yashilkul	32,1	6,1	11,8	125,1	-	12,5	9,5	196,8	0,0037	Our data
Sarezskoe I(0,5-100)	37,9	12,8	11,2	116,7	-	62,9	5,7	247,2	0,0046	Our data
Sarezskoe II (100-300)	193,0	63,5	18,3	126,5	-	619,3	21,7	1042,3	0,0247	Our data
Karakul	22,9	1126,0	1280,0	668,0	-	4802,0	1149,0	9512,0	0,254	Our data
Tuzkul	-	21,0	14279,0	1835,0	5506,0	8875,0	8850,0	39366,0	0,822	[3]
Sassikkul	-	10,0	51447,0	9190,0	22750,0	9530,0	40100,0	133027,0	2,153	[2]

TABLE 4. Average contents of trace elements in water of Pamir lakes, mkG/l

Lake	Sample	Ag	Cu	Pb	Zn	Mn	Be	Bi	Nb	Mo	W	As	V
Khargush	1	0,3	-	-	-	-	-	0,3	-	0,3	-	-	-
Turumtaykul	2	-	-	-	-	0,7	-	1,2	-	-	-	-	-
Bulunkul	2	1,3	1,8	0,6	-	-	-	1,1	-	0,3	-	-	-
Yashilkul	3	1,3	2,2	0,9	-	56,9	-	-	-	0,3	-	-	-
Sarezskoe I (0,5-100 m)	4	0,2	11,9	4,1	113,7	-	-	-	-	0,6	-	-	-
Sarezskoe II (100-300 m)	5	0,2	17,0	4,7	351,0	-	-	-	-	1,5	-	-	-
Karakul	4	11,3	48,4	8,8	402,0	88,4	-	-	-	6,7	-	-	-
Tuzkul	1	-	11,7	2,0	-	9,4	-	0,7	-	0,3	150	-	-
Sassikkul	2	950,0	103,0	53,0	570,0	684,0	5,7	2,8	57,0	570	1900	8000	57

Notice: dash means that concentrations are below detection limit of the VITR-LTI spectral chemistry method.

neotectonic zones. Some elements are accumulated as a result of dissolution of their compounds in the alkaline water.

Major elements are in extremely low concentrations in the water of the Pamir lakes, except for the highly mineralised lakes Sasykkul and Karakul, and their direct detection is a quite difficult task. In this respect, physical and chemical detection methods for metal ions in water and ground are of major importance.

The binding forms of trace elements in natural water are mainly determined by complexing mechanisms and physicochemical conditions (pH, Eh, water mineralisation, ligand concentrations). The main ligands in natural water that enter complex compounds are OH^-, $SO^2{}_4{}^-$, $CO^{-2}{}_3$, HCO_3^-, CL^-, F^-, humic and fulvic acids. The organic matter content in the studied lakes being very low (permanganate oxidizing, <2mg hydrogen/l), we confine ourselves to inorganic complexes, for which dissociation and hydrolysis constants are available from the literature (Table 5).

We suggest to analyse the correlation among different complexes in lake water by joint calculation of the mass balance equation and equations for hydrolysis constant and dissociation of complex compounds [3]. The ion activity coefficient estimated in second approximation as 4.0 by Debay-Hukkel's equation [16] provides satisfactory results in approximate calculations for water solutions with zero ion strength. The ion strength is measured by successive approximations with regard to complexing. The estimation of the activity coefficient of individual ions in high ion-strength solutions is especially problematic, as the theory of electrolytes remains poorly developed for concentrated brines providing no account of the departure of real systems from the ideal case.

Table 5 shows speciation of trace elements in some lakes (Sarez, Karakul, and Sasykkul) which covers almost the entire range of compositions, salinities, and pH levels in the water of Pamir lakes [17].

The geochemical significance of different binding forms of trace elements in lake water changes with its mineralization, as the activity of complexing is known to increase with mineralization. Moreover, transfer from solid phase becomes more active in presence of ligands, which form stable soluble complexes with metals, and the more stable the complex the higher element concentration in the solution may be expected.

According to their chemistry, the analysed trace elements belong to two groups:

Group I: $Ag+$, Cu^{2+}, Pb^{2+}, Zn^{2+}, Mn^{2+}, Be^{2+}, Bi^{3+}, Nb^{5+};

Group II: W, Mo, As, V.

The presence of negatively charged chlorocomplex (Ag), carbonate complexes with Cu, Pb, Zn, Mn, and hydroxide complexes (Be, Bi, Nb) maintains the dissolved state of the trace elements. The absence of calcium ions causes accumulation of Mo, W, As, V, as the molybdat, tungsten, arsenat, and vanadat, ions form highly soluble Na complexes. Therefore their content in alkaline lakes is directly proportional to mineralisation.

High concentrations of trace elements in water may also result from evaporation of hydrothermal groundwater supplied to lakes during Neogene-Quaternary postvolcanic activity along the Alichur Gurundin fault. New manifestations of this activity are observed as gas venting within lakes, hot and cold springs along the shorelines, travertine, small mud volcanoes.

TABLE 5. Relative percentages of binding forms of trace elements in water of Pamir lakes (in % of total concentration)

Element form	Sarezskoe I (μ =0,0046)		Sarezskoe II (μ =0,0247)		Karakul (μ =0,254)		Sassikkul (μ =2,15)	
	8,0	*8,5*	*8,0*	*8.5*	*8,0*	*8,4*	*9,4*	*9,6*
Ag^+	77	72	48	46	-	-	-	
$AgCl$	21	18	47	45	26	26	-	-
$AgCl_2^-$	-	-	2	2	67	67	30	30
$AgCl_3^{2-}$	-	-	-	-	7	7	68	68
$AgCl_4^{3-}$	-	-	-	-	-	-	2	2
$Ag(SO_4)^-$	1	1	3	3	-	-	-	-
$Ag(CO_3)^-$	1	9	-	4	-	-	-	-
Cu^{2+}	2	-	2	-	1	-	-	-
$Cu(OH)^+$	2	1	2	1	1	-	-	-
$Cu(SO_4)$	-	-	1	-	3	-	-	-
$CuHCO_3$	3	1	3	1	3	1	-	-
$Cu(CO_3)$	87	79	86	78	82	75	2	1
$Cu(CO_3)_2$	6	19	6	20	10	24	98	99
Pb^{2+}	1	-	1	-	-	-	-	-
$Pb(OH)^+$	5	4	5	4	2	2	-	-
$Pb(SO_4)$	-	-	1	-	3	-	-	-
$Pb(SO_4)_2^{2-}$	-	-	-	-	5	-	-	-
$Pb(CO_3)$	92	93	88	90	87	88	6	3
$Pb(CO_3)^{2-}$	2	3	2	6	3	10	94	97
Zn^{2+}	27	13	23	11	13	8	-	-
$Zn(OH)^+$	55	58	46	51	26	36	6	6
$Zn(OH)_2$	4	16	4	19	2	8	14	23
$Zn(SO_4)$	3	-	19	10	39	25	-	-
$Zn(SO_4)_2^{2-}$	-	-	1	-	9	6	-	-
$Zn(CO_3)$	11	13	7	9	11	17	80	71
Mn^{2+}	78	63	54	45	24	22	-	-
$MnCl^+$	-	-	-	-	3	1	-	-
$Mn(SO_4)$	7	5	34	28	55	48	-	-
$Mn(SO_4)4_2^{2-}$	-	-	2	1	11	10	-	-
$Mn(CO_3)$	8	25	5	20	4	14	14	10
$Mn(CO_3)_2^{2-}$	-	2	-	2	-	2	86	90
$MnHCO_3^+$	7	5	5	4	3	3	-	-
$Be(OH)^+$	63	33	63	33	61	38	5	3
$Be(OH)_2$	35	60	35	60	34	56	63	56
$Be(OH)_3^-$	1	7	1	7	1	6	31	41
BeF^+	1	-	1	-	2	-	-	-

BeF_2	-	-	-	-	1	-	-	-
$Be(CO_3)$	-	-	-	-	-	-	1	-
$Bi(OH)_4^-$	100	100	100	100	100	100	100	100
$Nb(OH)_5$	20	10	20	10	20	13	-	-
$Nb(OH)_6^-$	80	90	80	90	80	87	100	100
MoO_4^{2-}	100	100	100	100	100	100	100	100
WO_4^{2-}	100	100	100	100	100	100	100	100
AsO_4^{3-}	-	-	-	-	-	-	1	2
$HAsO_4^-$	92	96	92	96	92	95	99	98
$H_2AsO_4^-$	8	4	8	4	8	5	-	-
HVO_4^{2-}	22	48	22	48	20	36	6	6
$H_2VO_4^-$	78	52	78	52	70	44	1	1
$NaHVO_4^-$	-	-	-	-	10	19	92	92
$KHVO_4^-$	-	-	-	-	-	1	1	1

Notice: Dash is for concentrations below 1 %.

In addition to water analyses, we investigated sand-clay-salt bottom sediments of lakes in the Sasykkul depression in the context of the formation of continental reservoirs of saline water. The bottom sediments are (i) rich in detrital terrigenous clay (in small saline lakes), (ii) contain fragments of complex sodium carbonates, simple and complex salts (tenardite, mirabilite, etc), (iii) lack potassium salts owing to low absolute contents of potassium in the surface water, and (iv) are fed from salt basins, groundwaters, and volcanic exhalations in regions of recent tectonomagmatic activity, which explains high contents of trace elements in the lake brines.

The studied bottom sediments are of clay, sand, or halmeic (salt-bearing) lithologies. Clayey sediments are higher in Rb, Cs, Sr, Fe, Co, Sc, Pb, Zn, Ag, Th, La, Eu, Tb, and Yb than sandy and salt-bearing deposits. The latter are concentrators of Na, B, Ba, Mo, W, U whose contents reach anomalously high values in some samples. The elements of the first group increase with increasing aluminium silicates and decreasing salt components, whereas those of the second group form authigenic concentrations in salt-bearing sediments. This division is confirmed by correlation analysis, as there are high positive correlations between elements within the groups, and either negative or no correlation between elements that belong to different groups.

Contents of all analysed elements in sands are usually slightly lower than in pelites, but the distribution patterns are very similar. U, Th, Mo, W, Na, Zn exist in abnormal concentrations; some samples are high in Se, Au, Ag and low in Ni, Sb, Ta, Rb, Fe, Cs, Eu, Yb, Sc, and Tb. The distribution of trace elements in salts and salt-bearing pelites of the Sasykkul depression is more contrasting than in terrigenous deposits. The sediments are anomalously high in Y, B, Ba, Mo, W; some samples are also high in Se, Au, As, Ce, and Sm, which is related on the one hand, with the formation of boron, sulphates, tungsten, and molybdenum following the general halogenesis trend, and on the other hand, with Neogen-Quaternary postvolcanic hydrothermal activity (U, W, As, and others). The concentrations of Fe, Se, Co, Cr, Ti, Pb, Cu, Rb, Cs, Sb, Ag, Zr, Ta, Th, La, Bu, Tb, Yb in salts of the Sasykkul depression are usually very low and increase only with increase in terrigenous material.

TABLE 6. Chemical composition of deposits of Sasikkul depression

Element	Pelite (n=7)	Sandstone (n=6)	Halopelite (n=5)
Na, %	1,41-4,50	0,30-6,19	3,70-23,23
Rb	41-102	<5-88	<5-27
Cs	3,37-31,1	0,61-5,81	0,08-2,19
Sr	250-2131	140-1397	81-786
Ba	245-1623	<100-523	379-14357
B	34-320	10-140	10-2000
U	8,5-31,2	2,2-22,9	20,5-<45
Th	10,3-42,9	6,4-<95	0,3-8,2
Zr	30-700	30-700	10-100
Hf	<0,2-10,1	<0,2-40,4	0,5-10,3
Ta	0,34-1,76	0,03-2,62	0,02-0,16
Mo	10-32	<1-25	30-52
W	2,7-8,7	<1-6,5	<1-13,5
Pb	13-170	10-35	10-20
Zn	164-1212	49-247	12-145
Cu	20-50	13-50	13-50
As	<5-35	<5-12	<5-185
Sb	0,12-2,45	<0,05-0,15	0,05-0,40
Se	<0,5-54,6	<0,5-9,0	<0,5-26,3
Fe, %	0,68-7,12	0,14-21,4	0,01-0,47
Ti, %	0,08-0,40	0,05-0,60	0,01-0,50
Cr	<10-218	<10-88	<10-70
Ni	14-50	12-40	24-150
Co	2,0-20,3	0,6-8,3	0,3-1,9
Sc	3,0-18,3	0,7-90,6	0,1-1,5
La	7,92-27,41	2,77-294,1	1,0-8,20
Ce	6,1-82,1	10,0-673,0	11,8-288,4
Sm	0,47-6,38	<0,2-80,61	0,24-17,15
Eu	0,16-0,66	0,04-1,27	<0,03-0,12
Tb	<0,01-0,53	0,05-5,31	<0,01-0,88
Yb	0,05-3,99	<0,05-47,0	<0,05-0,44
Au,ppm	<0,5-4,2	1,0-4,9	<0,5-9,1
Ag	<0,05-8,64	<0,05-0,35	<0,05

5. Conclusion

Lakes in the region are of several main types according to their salinity. Saline lakes occupy a special place. High concentrations of some trace elements in the water and bottom sediments of lakes in the Sasykkul depression are controlled by the geochemistry of granites in the area and the presence of tungsten and polymetallic mineral deposits. Trace elements are supplied not only from surface and ground waters, but also from deep hydrotherms, as the lakes occur within highly permeable tectonic zones [14].

The effect of high salinity and pH on speciation of rare metals is to increase the proportion of carbonate, chloride and hydroxide complexes and their complexing degree in lake water. The absence of Ca^{2+} ions in lake water favours accumulation of W, Mo, As, and V , as MoO_4^{2-}, WO_4^{2-}, $HAsO_4^{-}$, HVO_4^{-} make compounds with highly soluble sodium [12, 17]. Accumulation of trace elements in lake water is also caused by evaporation in arid climate.

Contents of many elements have been determined in terrigenous and salt-bearing sediments of brine lakes in the Sassykkul depression. The evolution of deposition in salt-forming basins resulted in distinct division of chemical elements into those that are accumulated or dispersed in bottom sediments and those that bind or not with salts.

The water and bottom sediments of the studied lakes in the Sassykkul depression are a potential source for rare metal extraction.

6. References

1. Usufbekov, H.U. and Konnov, A.A. (eds.) (1973) *Pamir (general ways of Pamir nature research for 100 years and prospects of its development*), Donish, Dushanbe (in Russian).
2. Nikitin A.M. and Sharabaev V.A. (1982) Pamir lakes: distribution and genesis. *Proc. Middle Asia Scientific Research Hydro-Meteorological Institute* **168,** 67-71 (in Russian).
3. Kimmerih A.O. (1978) Hydrographics of Pamir and Pamir-Alay. *Misl, Moscow,* 220-221 (in Russian).
4. Kersonov G.A. (1979). New data on the morphology of the Sares lake. *Proc. Middle Asia Scientific Research Hydro-Meteorological Institute* **66 (147),** 82-88 (in Russian).
5. Nikitin A.M. (1977). Morphometry and morphology of Central Asian Lakes. *Proc. Middle Asia Scientific Research Hydro-Meteorological Institute. Issue* **50 (131),** 4-21 (in Russian).
6. Alekin O.A. (1953) *Manual in dry chemical analysis.* Hydrometeoizdat, Leningrad (in Russian).
7. Knipovich, Iy.N. and Morachevskii, Iy.V. (eds.) (1970) *Mineral raw material analysis*, Goschemizdat, Leningrad (in Russian).

8. N. (1961) *Determination of trace elements in natural waters in mineral prospecting. A handbook,* Gosgeoltechizdat, Moscow (in Russian).
9. Akramov U.A. (1987) *Organic matter of vertically zoned soils of Tajikistan and its role in soil forming and husbandry*, Donish, Dushanbe (in Russian).
10. Alekin O.A. (1970) *Fundamentals of hydrochemistry,* Hydrometeoizdat, Leningrad (in Russian).
11. N. (1966) *Handbook for chemists, v. 5: Chemistry*, Moscow – Leningrad (in Russian).
12. Krainov S.P., Petrova N.G. (1962) Trace elements of mineral waters of Pamir. *Geochemistry* **4**, 356-366 (in Russian).
13. Krainov S.R. (1966) Accumulation and precipitation of tungsten associated with evaporation in some alkaline lakes of Central Asia, *Doklady AN SSSR* **168,** 904 –907 (in Russian).
14. Krainov S.R., Petrova N.G., Baturinskaya I.V. (1971) Saline lakes of the Eastern Pamirs as trace element concentrators, *Doklady AN SSSR* **198,** 940-943 (in Russian).
15. Volkova N.I., Pachajanov D.H. (1986) *Report Acad. Sci. Tajik SSR,* **29**, 233 (in Russian).
16. Hodakovsky I.L. (1975) *Physicochemical problems of magmatic and hydrothermal processes*. Nauka, Moscow (in Russian).
17. Volkova N.I., Ozerov A.M., Pachajanov D.N. (1988) Binding forms of trace elements in the waters of Pamir lakes. *Geochemistry* **12**, 1773 (in Russian).
18. Volkova N.I., Pachajanov D.N., Ozerov A.M., Goffen G.I. (1993) Trace elements in saline lakes of the Sasykkul depression (western Pamir). *Geochemistry*, **1**, 114 (in Russian).

ZOOBENTHOS AND ZOOPLANKTON OF THE NORTHERN ARAL SEA AND POSSIBLE WAYS OF THE SEA ECOSYSTEM REHABILITATION

N.V. ALADIN, A.A. FILIPPOV, I.S. PLOTNIKOV
Laboratory of Brackish Water Hydrobiology, Zoological Institute of RAS, Universitetskaya naberezhnaya, 1, 199034 St Petersburg, Russia

Prior to anthropogenic desiccation, the average salinities in the Large and Small Seas differed to some extent. These differences were insignificant during the period from 1961 to 1970 because of sufficient water exchange between the two basins. From 1971 to 1985 after the Auzykokaral strait dried up, and Berg's strait became shallower, the average salinity in the Small Sea rose 1.5–3.0 g/l higher than in the Large Sea. However, in 1986 the difference in average salinity between the Small and Large Seas decreased again, primarily because of a change in the direction of flow of the Syrdarya River, which resulted in the diversion of more freshwater into the Small Sea. Since 1988, the average salinity in the Small Sea has decreased by 1.5–2.0 g/l when compared with the Large Sea. Subsequently, the complete drying of Berg's strait in spring 1990 has resulted in still sharper differences in salinity concentration between two basins (Table 1).

Sometimes between spring 1990 and July 1992, a small but strong flowing channel/canal appeared in the area of Berg's strait, and it once again allowed water flow from the Small to the Large Sea. This channel was formed as a result of rise in water level in the Small Sea during spring 1990. (In spring 1992, the flow rate from the Small to the Large Sea averaged 100 m^3/sec).

In July 1992 this channel was filled up with sand to build a 1 m high dam/causeway, as part of an attempt to control the water level in the Small Sea, and as an attempt to rehabilitate this basin [1][2]. The first attempt to construct a dam met with failure as the water pressure was too great, and the dam was breached. A second dam was constructed in late July/early August 1992, and it was in place for 9 months. However in March 1993, the water level in the Small Sea rose more than 1 m, and resulted in the destruction of the second dam. Later in the period from 1993 till 1999 local authorities regularly rebuild a dam many times and make it stronger and stronger. Finally in spring 1999 level of Small Aral reached level +42.6–43.0 m (according to various sources) and average salinity decreased to 18.2 g/l. 22 April 1999 when the pressure of water and influence of wave action became to be critical, this dam was completely destroyed. Due to this the present level of the Small Aral Sea is very low and oscillate at mark +40 m.

Our theoretical studies and field researches performed 1996–1999 showed opportunity of reservation and rehabilitation of northern water area of the Small Aral Sea.

J. Klerkx and B. Imanackunov (eds.), Lake Issyk-Kul: Its Natural Environment, 181–190.

TABLE 1. Average salinity (S, g/l) and level (H, m.abs.above sea level) of the Small and Large Aral Seas in 1990's (by data from different sources)

Year	the Small Sea		the Large Sea	
	S	H	S	H
1990	29.2	39.1	30.6	38.3
1991	28.1	40.0	35.1	37.9
1992	26.4	40.2	36.5	37.1
1993	20.3	41.5	37.2	36.9
1994	21.2	41.9	38.8	36.4
1995	22.4	40.8	40.5	35.8
1996	22.6	40.6	46.4 (?)	35.1
1997	20.5	41.1	56.8***	34.4
1998	19.4	42.0	59.1***	34.1
1999*	18.2	42.8	62.4***	34.0
1999**	18.2	40.1	-	-

* Before the dam destroying
** After the dam destroying
*** Average in Tshe-Bas bay
- No data

According to results of three last expeditions to the Northern Aral positive influence of construction of experimental dam in Berg's strait was proved. Construction of this dam in July, 1992 and its constant strengthening has allowed level to rise of the Small Aral Sea by 2.5–3 meters (to April, 1999) and salinity to decrease on the average by 8–10 g/l. This increase of level and decrease of salinity had positive effect on rehabilitation both Syrdarya delta and the Small Aral. Increase of brackish water species contribution both in zooplankton and zoobenthos was found. In zooplankton number of Caspian cladocerans from family Podonidae has increased again. Few years ago representatives of this crustacean were about to disappear. Now, due to decrease of salinity, their number was restored practically up to the former level. *Podonevadne camptonyx* were observed in large number and single individuals of *Evadne anonyx* were met again since more than 10 years. In zoobenthos larval chironomids have appeared again. Their number has not reached former level, however, due to this food value of benthos has increased. By data obtained from Aral Fisheries Laboratory (Z. Ermakhanov, Aralsk city, personal communication) and from local fishermen catches of flounder has increased almost twice. Unfortunately, as it was already mentioned above at the end of April, 1999 the dam in Berg's strait was completely destroyed by flood waters. Fortunately, the quick fall of level was not accompanied by considerable salinity increase. However, in future this increase could be quite significant if the dam will not be restored. Probably complete separation of Butakov and Shevchenko bays will be possible with the later complete drying up of both bays about 2005–2007. The break of dam already has resulted in separation of Bolshoy Sarycheganak bay, which could dry up completely in 1 year. It is clear that dam should be restored in the nearest future. Unhampered discharge of the Small Aral Sea into the Large will result in gradual increase of its salinity accompanying with extinction of brackishwater and marine species of hydrobionts. This, in turn, will result in complete loss of fishery importance of the Small Aral Sea. The absence of a dam in Berg's strait in the distant perspective can cause complete drying up of the Small Aral and further aridization of regional climate.

Zoobenthos studies was undertaken in September 1997 at the Bay of Butakov and in September 1999 in the Central part of the Small Aral Sea. In the Bay of Butakov benthos was collected from nine stations more or less evenly distributed down the Bay area. At each station five bottom samples were taken using Petersen Bottom sampler. In the Central part of the Small Aral Sea 40 quantitative samples were taken from 16 stations. In 1999 for the investigation of vertical zoobenthos distribution in the inshore zone of the Tastubek village, hydrobiological transect was made. Down the transect four stations were explored with the depth of 0.7, 1.5, 3 and 6 meters, five benthic samples were taken at each of the stations using Tube bottom sampler or Petersen Bottom sampler. At all the other stations 1 to 2 samples were collected using Petersen bottom sampler. The samples collected were processed using standard methods.

In the Bay of Butakov zoobenthos comprised all the benthic taxa characteristic for the modern Aral Sea: bivalves *Abra ovata* and *Cerastoderma isthmicum*, gastropods *Caspiohydrobia* and polychaetes *Nereis diversicolor*. Average zoobenthos density and biomass were rather high and reached 646 g/m^2 and 42.3 th.ind./m^2 (Tab. 2). Bivalves *A. ovata* dominated in terms of both density and biomass at all the stations. Other species contribution to the total zoobenthos biomass was much lower. Zoobenthos spatial distribution was homogenous, the same set of species was recorded at each station and their relative biomass varied between stations insignificantly [3]

The character of vertical distribution of individual species of benthic invertebrates, macrophytes and coarse organic debris in sediments has his own peculiarities in this region and differed a great deal from that at the major part of the Aral Sea. Abundance of both macrophytes and detritus as well as *S. segmentum* and *Caspiohydrobia* spp. biomass were maximum at shallower and decreased with depth increase. *C. isthmicum* vertical distribution had opposite character. Polychaete abundance did not changed much with depth.

Comparison of the data obtained with that of previous observations at the same region [4] showed that there were no appreciable year to year dynamics in the Bay of Butakov benthic assemblages. Average benthos biomass and density in 1997 did not differed much from the values registered in 1990 in the same region (501 g/m^2 and 24.5 th.ind./m^2 correspondingly). Zoobenthos structure and vertical distribution were essentially similar in both periods of observation, which confirm the nonrandom character of structural zoobenthos traits observed. Some distinctions in benthic invertebrates abundance recorded in 1990 and 1997 were probably connected with lack of coincidence of sampling stations framework in the periods of observation.

TABLE 2. Abundance (N, ind/m2) and biomass (B, g/m2) of zoobenthos at the explored sites of the Bay of Butakov in 1997.

Taxa	N/B	Station N									Среднее
		1	2	3	4	6	7	9	10	11	
A.ovata	N	18208	22528	16128	16128	20192	5632	27800	17640	11392	17294.22
	B	682	349	248	248	428	350	591	574	223	410.16
C.isthmicum	N	48	128	392	392	56	224	64	192	48	171.56
	B	65	177	382	382	73	70	45	181	64	159.98
Caspiohydrobia	N	21504	42112	4960	4960	21896	9568	50512	30104	17856	22608.00
	B	81	85	8	8	52	18	159	111	27	61.17
N.diversicolor	N	2800	3672	2496	2496	2024	872	1704	2232	1664	2217.78
	B	29	17	13	13	14	6	14	25	7	15.31
Total	N	42560	68440	23976	23976	44168	16296	80080	50168	30960	42291.56
	B	857	628	651	651	567	444	809	891	321	646.62
Macrophytes	B	588	264	0	0	268	0	1224	1232	0	397.33
Detritus	B	292	84	0	0	360	1128	448	396	8	301.78

TABLE 3. Abundance (N, ind/m2) and biomass (B, g/m2) of zoobenthos at the explored sites of the Small Aral Sea in 1999.

Taxa	N/B	Statio n N																Average
		M-1	M-2	M-5	M-7	1	2	3	4	8	9	13	11	10	21	22	23	
A.ovata	N	12679	7094	8453	8128	6792	8774	6038	14151	11520	0	0	560	4800	0	20	1760	4534.56
	B	872	597	654	399	588	313	318	308	799	0	0	12	213	0	5	151	225.60
C.isthmicum	N	264	75	151	24	943	943	189	94	400	0	0	40	600	0	0	0	267.48
	B	256	39	80	25	1179	303	98	117	196	0	0	1	718	0	0	0	217.59
Caspiohydrobia	N	5736	14189	26415	4736	36792	2642	10000	17453	1440	0	0	360	21653	0	0	1067	7617.23
	B	23	47	67	12	96	8	35	43	4	0	0	1	36	0	0	3	18.75
N.diversicolor	N	1660	1245	792	1128	1698	472	1038	3019	800	0	0	1280	440	20	0	107	739.42
	B	37	32	19	20	33	2	36	31	2	0	0	2	4	0	0	1	9.29
Chironomidae	N	302	75	302	64	0	0	0	0	0	0	0	0	0	0	60	213	22.78
	B	1	0	1	0	0	0	0	0	0	0	0	0	0	0	0	0	0.04
Total	N	20642	22679	36113	14080	46226	12830	17264	34717	14160	0	0	2240	27493	20	80	3147	13181
	B	1188	714	821	456	1897	626	486	499	1000	0	0	16	970	0	5	155	471
Detritus	B	1459	857	1887	5	0	0	0	0	0	56	0	0	512	0	0	0	47.33

In 1999 in benthos samples from the central part of the Small Aral Sea all the benthic taxa characteristic for the modern Aral Sea, bivalves *Abra ovata* and *Cerastoderma isthmicum*, gastropods *Caspiohydrobia* and polichaetes *Nereis diversicolor* were recorded. Shrimps *Palaemon elegans* were numerous in nectobenthos, though they were not caught by bottom sampler. In contrast to previous observations at the Aral Sea in 1990 [5][6], chironomids *Chironomus halophilus* larvae were recorded in benthic assemblages as well. The chironomids were met in relative abundance, their density was 64 ind/m^2 in average for the whole Sea and 186 ind/m^2 in the inshore zone near the Tastubek village (Tab. 3). The chironomids presence seemed to be related to both excessive outflows of Syrdarya River and appreciable decrease in salinity in the Small Aral Sea in spring 1999.

The results obtained showed extremely high productivity of zoobenthos in the Small Aral Sea inshore zone. Density and biomass values were rather high, reaching 46 th.ind/m^2 and 1900 g/m^2, which exceeded practically all the data obtained at different stations before [6]. In the inshore zone near the Tastubek village average benthos biomass was 795 g/m^2, which was much higher than any average value registered before in 1990s in any site investigated of the Aral Sea.

Zoobenthos vertical distribution in the inshore zone near the Tastubek village differed significantly from that at the open shores of the Aral Sea but resembled that in the Aral Sea bays and closed areas. Maximum zoobenthos abundance was registered at the shallow water stations while minimum biomass was met in the deepest places. As it was shown earlier [6] the zoobenthos vertical distribution was largely related to food material spatial distribution, which depended upon hydrodynamics peculiarities at the spot. In the Aral Sea bays and gulfs characterized by low wave action the largest amount of organic matter in sediments and correspondingly maximum abundance of benthic organisms were recorded at the shallow places. The situation similar to that was observed at the plot investigated near the Tastubek village. The water roughness was decreased here and maximum detritus concentration was observed in a shallow. This probably led to assemblages of maximal density forming namely in the shallow stations. The other possible reason of increased benthos concentration in the shallows was downward migration of benthic organisms from upper horizons as a result of Sea level fall, which took place after the dam destruction in the former Berg's Strait in spring 1999.

Analysis of results obtained at the sites explored some years earlier (Bay of Butakov, inshore zone of Tastubek Peninsula, see Filippov, [4][6] reveal no noticeable changes neither in the zoobenthos composition, no in benthic organism abundance in these areas. Relative biomass of zoobenthos main components did not changed much as well. The biomass of individual species as well as zoobenthos in a whole remained extremely high and revealed tendency of further increase. The results obtained showed that stable situation in the Small Aral Sea benthic assemblages retained during the 1990's. The high productivity of benthic assemblages along with relative stability of hydrological regime and the proximity of large city make this Sea part very promising object for the fishery restoration at the Aral Sea.

After interruption in 1998 studies of the Aral Sea zooplankton was continued in autumn 1999. In 1997 sampling was occurred only at Butakov bay being an atypical part of Aral. In September 1999 zooplankton samples were taken on the Small Aral Sea in the following water areas:

1. at Tastubek peninsula and Bultun mountain;
2. not far from Trekhgorka mountain;
3. central deepwater area;
4. Shevchenko bay;
5. Butakov bay.

In the samples there were found the following invertebrates:

1. Rotifers *Synchaeta vorax* and *S. cecilia*;
2. Copepods *Calanipeda aquaedulcis* and *Halicyclops rotundipes aralensis*;
3. Cladocerans *Podonevadne camptonyx* and *Evadne anonyx* (except of Butakov bay).
4. Larvae of bivalves *Abra ovata* and *Cerastoderma isthmicum*.

Data on number and biomass of zooplankton are given in Table 4.

TABLE 4. The Small Aral Sea zooplankton abundance and biomass in autumn 1999

Location	Rotatoria	Copepoda	Cladocera	Mollusca larvae	TOTAL
Near Tastubek	1578* 2.9	2497 8.7	415 2.1	1446 3.2	5934 14.4
Near Trekhgorka	22207 40.0	9494 23.5	3218 16.1	8529 18.8	43448 98.3
Central water area	8910 16.0	666 2.5	1249 6.2	666 1.5	11492 26.3
Shevchenko bay	5640 10.2	3301 5.5	688 3.4	3026 6.7	12655 25.8
Butakov bay	712 1.3	5220 9.7		1305 1.3	7237 13.9

* ind./m^3 / mg/m^3

Investigations has shown that during last 3–4 years in zooplankton of the Aral Sea occurred some changes connected with changes in hydrological and hydrochemical regimes of this lake. Water salinity decrease in the Small Sea has resulted in the growth of cladocerans abundance (up to 5–10% of the total zooplankton quantity) and biomass (3218 ind./m^3 and 16.1 mg/m^3) in comparison with the figures of the first half of the 90's (375 ind./m^3 and 1.9 mg/m^3), as it was predicted [7][8][9]. Now *P. camptonyx* is one of the main and persistent components in the zooplankton after copepod *Calanipeda aquaedulcis*, rotifers *Synchaeta* spp. and larvae of bivalve mollusks (the last is numerous only in summer). Nevertheless, Podonidae were not found in Butakov bay where salinity was higher than in the main water area of the Small Aral [7][9][10]. It is to note that in zooplankton some individuals of other Podonidae species – *Evadne anonyx* was found in the main water area. After its disappearance in the 1980's this cladoceran was absent during long time in the Aral Sea because of higher salinity and now they appeared here again.

In May 1996 we registered in the Small Aral Sea zooplankton sharp and substantial growth of number of rotifers from genus *Synchaeta* (12456–58969 ind./m^3) in comparison with the first half of 1990's (430 ind./m^3 in 1993) [7][8][9]. In September, 1999 the number of rotifers remained high (1578–22207 ind./m^3). In comparison with water area near the Tastubek peninsula, number of rotifers and their relative proportion in zooplankton were increased in the central part of the Small Sea, as it was observed in 1996. Thus, in the second half of 1990's rotifers have got in number of groups of invertebrates dominating in zooplankton of Small Aral. To all appearance, it was a consequence of increased intensity of primary producing process in the Aral Sea [11][12].

Number and biomass of bivalve mollusks larvae, one of the main components of the Aral Sea zooplankton, were low (while still did not reach its minimal observed values). It is normal for this season when their reproduction period is finished. Copepods were represented mainly by *C. aquaedulcis*. As it was observed in the last years, *H. r. aralensis* was not numerous [8][9].

There are differences between zooplankton in examined areas of the Small Aral Sea. Among them the water area in the North-East near Trekhgorka mountain stands out. Here zooplankton abundance and biomass are higher than in other localities at Kokturnak peninsula (43448 ind./m^3 and 98.3 mg/m^3 against 5934 ind./m^3 and 14.4 mg/m^3). Similar picture was observed in 1993 (Plotnikov, 1995a,b). We could suppose that it is connected with proximity of this water area to Syrdarya delta. The highest abundance and biomass of zooplankton in the past were observed (in 1992–1993) in the water area adjoining to the delta, where the water area characterized by high values of primary production [8][9].

Butakov bay is in some extent isolated from the main water area of Small Aral, but unlike the Bolshoy Sarycheganak bay is not losing communication with it. This bay is characterized by unstable state of zooplankton, what is expressed in significant fluctuations of its number and biomass. Most sharply they are expressed in *C. aquaedulcis*. Other feature of zooplankton of this bay is the constant presence of *H. r. aralensis* in it. This crustacean in the time of zooplankton depressive state even can predominate *C. aquaedulcis* [13][7][9][10].

At the end of September, 1997 in zooplankton of Butakov bay rotifers *Brachionus* sp., *Synchaeta vorax* and *S. cecilia*, copepods *Halicyclops rotundipes aralensis* and *Calanipeda aquaedulcis*, and also larvae of bivalve mollusks *Syndosmya segmentum* and *Cerastoderma isthmicum* were found. As before copepods *H. r. aralensis*, not numerous in the open water areas of Small Aral, and *Calanipeda aquaedulcis*, and also rotifers *Synchaeta* spp were prevailing. The larvae of bivalve mollusks - one of dominants in spring-autumn zooplankton - were in very low number.

In 1997 number and biomass of autumn zooplankton in Butakov bay (3400-43800 ind./m^3 and 9.5-68.7 mg/m^3) were much higher than in autumn of 1990 and 1991. [8][9]. Here, as well as in 1996 in the main water area of the Small Sea, number of fine rotifers *Synchaeta cecilia*, not numerous before, has increased in higher extent than number of *S. vorax*. Except for water area adjoining to the bay's mouth (stations 6, 7), where the number of rotifers was increased, composition, quantitative parameters of zooplankton development in different sites of Butakov bay were basically similar (Tab. 5).

In September, 1999 (Tab. 4) zooplankton of Butakov bay had its typical structure, and parameters of its quantitative development were comparable with that observed in the autumn 1997. It is to note, that plankton larvae of bivalve mollusks *Abra ovata* and *Cerastoderma isthmicum* have appeared more numerous in comparison with the same season in 1997. It shows, that variability of reproduction dynamics in these bivalves depends on inter-annual variability of environment factors.

TABLE 5. Number (ind./m^3) and biomass (mg/m^3) of Butakov bay zooplankton in September 1997

Station	*Rotatoria*	*Copepoda*	*Mollusca larvae*	Total
1	449.7 0.8	3597.5 8.7	0 0	4047.2 9.5
2	1557.7 2.8	1713.5 2.5	155.8 0.3	3427.0 5.6
3	259.5 0.5	9082.0 14.4	259.5 0.6	9600.9 15.4
4	334.5 0.6	3679.0 6.0	0 0	4013.5 6.6
6	10480.2 18.9	5240.1 7.7	0 0	15720.3 26.6
7	13570.1 24.4	29997.0 43.8	238.1 0.5	43805.1 68.7
11	3394.7 6.1	3637.2 5.2	363.7 0.8	7395.6 12.1

* $\frac{\text{ind./m}^3}{\text{mg/m}^3}$

References

1. Aladin N.V., Plotnikov I.S., Potts W.T.W. (1995) The Aral Sea desiccation and possible ways of rehabilitation and conservation of its North part, *Int. J. Environmetrics* **6**, 17-29.
2. Aladin N.V. (1995) The conservation ecology of the Podonidae from the Caspian and Aral Seas, *Hydrobiologia* **307,** 85-97.
3. Filippov A.A. (1999) Ecological studies at the Aral Sea: macrozoobenthos of the Bay of Butakov, Abstracts 7th Annual scientific conference "XXI century: Youth, Education, Ecology, Noosphere", 10-12 February 1999, St.Petersburg 1999, 112 (in Russian).

4. Filippov A.A. (1991) Zoobenthos of Butakov bay (the Aral Sea) in September 1990, *Proc. of Zool. Inst. of Acad of Sci USSR* **237,** 44-48 (in Russian).
5. Filippov A.A. (1993) Some structural features of the macrozoobenthos of the North Aral Sea coastal zone, *Proceed.of Zool.Inst.* **250**, 56-64 (in Russian).
6. Filippov A.A. (1995) Macrozoobenthos of inshore zone of the Aral Sea North in modern polyhaline conditions: quantity, biomass and spatial distribution, *Proc. of Zool Inst. of RAS* **262**, 103-106 (in Russian).
7. Plotnikov I.S. (1993) The Aral Sea zooplankton in 1992, *Proc. of Zool. Inst. of RAS* **250**, 46-51 (in Russian).
8. Plotnikov I.S. (1995) The Aral Sea (the Small Sea) zooplankton under conditions of the regime stabilization, *Proc. of Zool. Inst. of RAS* **262**, 167-173 (in Russian).
9. Plotnikov I.S (1995) Zooplankton of coastal zone of the Northern part of the Aral Sea under modern polyhaline conditions. Doctoral thesis, St.Petersburg, 24 p. (in Russian).
10. Plotnikov I.S. (1991) Zooplankton of Butakov bay (the Aral Sea) in September 1990, *Proc. of Zool. Inst. of Acad. of Sci. USSR* **237**, 34-39 (in Russian).
11. Orlova M.I. (1993) Materials for general valuation of productional-destructional processes in the coastal zone of the northern part of the Aral Sea. 1. The results of field observations and experiments in 1992 year, *Proc. of Zool. Inst. of RAS* **250**, 21-37 (in Russian)
12. Filippov A.A., M.I.Orlova (1996) Changes in plant and animal associations of Northern Aral Sea during the last 30 years. Abstracts of VII Congress of Russian Hydrobiological Association of Russian Academy of Science, Kazan October 14-20, 1996, v.1, 168-170 (in Russian).
13. Andreev N.I. (1999) Zooplankton of Butakov bay (the Aral Sea) in June 1990, *Proc. of Zool. Inst. of Acad. of Sci. USSR* **237**, 30-33 (in Russian).
14. Andreev N.I. (1999) Aquatic fauna of the Aral Sea under conditions of ecological crisis, Omsk State Pedagogical University, Omsk, 453 p. (in Russian).

EVALUATION OF THE DAMAGE TO ENVIRONMENTAL STATE OF THE NORTHERN-WESTERN BLACK SEA SHELF ECOSYSTEMS

V.N. EGOROV, G.G. POLIKARPOV, N.N. TERESHCHENKO
The A.O.Kovalevsky Institute of Biology of the Southern Seas (IBSS)
National Academy of Sciences of Ukraine,
Prospekt Nakhimova 2, 99011, Sevastopol, Ukraine

1. Introduction

Evaluation of the damage to environmental state of aquatic ecosystems requests different approaches of investigation. This paper contains approaches of study of the Black Sea ecosystems that allow to assess of the present state of its ecosystems as well as to make prognoses of future damage or state of the ecosystems.

The first part of the paper is devoted to the radioecological study of ^{90}Sr and ^{137}Cs in the total Black Sea and in the Sevastopol Bays; the second part describes the conceptual model of responses of organisms, populations and ecosystems to all possible dose rates as well as presents evaluation of the state of the Black Sea shelf ecosystems in the Northern-Western part of the Sea. The third part of the paper analyses chemoecological monitoring of *Modiolus phaseolinus* biocenoses in the Black Sea under pressure of natural chemical factors as well as man-made xenobiotics.

2. Distribution of ^{90}Sr and ^{137}Cs in the Black Sea

Importance of investigation of radioactive pollution in the marine environment is caused by necessity of evaluation and prediction of consequences of the accident at the Chernobyl nuclear power plant in 1986.

Radioecological monitoring in the Black Sea occupies a special place in scientific investigations of the IBSS, where the work on marine radioecology was started more than 40 years ago [13].

The first part of the paper summarizes of studies of ^{90}Sr and ^{137}Cs distributions in water and biota of the Black Sea after the Chernobyl accident [1].

The data were obtained during expeditions of oceanographic vessels and with help of the radioecological monitoring of the Sevastopol Bays and the Black Sea areas adjacent the Danube River and Dnieper River estuaries.

J. Klerkx and B. Imanackunov (eds.), Lake Issyk-Kul: Its Natural Environment, 191–206.

The area of the Black Sea received 100-300 TBq of ^{90}Sr and 1700-2400 TBq of ^{137}Cs in May, 1986, through atmospheric fall-out (Table 1). As the result of accident the content of ^{90}Sr in the Black Sea water was increased to 7 — 20% and such of ^{137}Cs was exceeded 200%. Later on the Black Sea was exposed to radioactive pollution from the Chernobyl NPP via the rivers releases.

It was shown that trends of influxes of ^{90}Sr and ^{137}Cs to the Black Sea with the Danube River and the Dnieper River run-off, as well as the outflow of these radionuclides from the Black Sea through the Bosporus Strait may be described by exponential equations [3]. The results of calculation of the ^{90}Sr and ^{137}Cs balance in the Black Sea in 1995, and the prediction, obtained by integration of the exponential functions, are presented in Table 1.

By 1995 the entry of ^{90}Sr to the Black Sea with the Danube River and the Dnieper River releases is estimated at 24.5 and 90.2 TBq, respectively. According to the prediction the entry of ^{90}Sr with rivers releases will continue about 75 years; the release of the Dnieper River will be approximately 150 TBq, and such of the Danube River — 60 TBq. Entry of ^{137}Cs will continue to the Black Sea with releases of rivers during 35 years; releases of the Dnieper River will contribute 2.0 TBq, and such of the Danube River — 24 TBq.

Transfer of ^{90}Sr and ^{137}Cs from the Black Sea through the Bosporus Strait during the period from 1986 to 1995 is estimated at 94 and 225 TBq respectively. Entry of ^{90}Sr from the Black Sea into seas of the Mediterranean Sea basin will continue more than 66.5 years and totally will be assessed to be equal to 225 TBq. Entry of ^{137}Cs will continue 23 years and its total transfer is estimated in amount about 300 TBq.

TABLE 1. Balance elements of post – Chernobyl ^{90}Sr and ^{137}Cs in the Black Sea, TBq [1].

Balance component		Entry assessment in 1986-1995		Prognosis of consequent entry		Total	
		^{90}Sr	^{137}Cs	^{90}Sr	^{137}Cs	^{90}Sr	^{137}Cs
	Control before 26.04.86					~ 1500	1400±300
(+)	Athmospheric fall-out (May, 1986)					100-300	1700-2400
(+)	Entry with the Dnieper River release	90.2	2.0*	57.8	0.06	148.0	2.1
(+)	Entry with the Danube River release	24.5	24.0*	32.8	9.4	57.3	37.6
(-)	Transfer through the Bosporus Strait	94.0	225.0	130.2	73.9	224.2	289.7

*- in 1986-1994 years.

^{90}Sr and ^{137}Cs concentrations in water of the Black Sea's surface layers will be decreased to the same level, which existed before the Chernobyl accident, during the period of 27 — 78 years. Period of residence time of post-Chernobyl ^{137}Cs atoms in water of the Black Sea body is equal to 85 years.
Flux of ^{137}Cs together with suspensions which sink on the bottom was assessed as 4-5 $Bq \cdot m^{-2} \cdot y^{-1}$ and especially in near the Danube River area it was increased to 700 $Bq \cdot m^{-2} \cdot y^{-1}$.
The observations in the framework of radioecological monitoring have shown that concentrations of ^{90}Sr in water, brown algae and mussels of the Sevastopol Bays decreased exponentially during 1986 — 1999 with the time constants of 6.9, 4.0 and 6.7 years, respectively. The constants of ^{137}Cs were equal to 5.9, 4.4 and 4.3 years, respectively [2]. A phenomenon of decrease of concentration factors of ^{90}Sr and ^{137}Cs in biota during time was found out. By using of mathematical models, it was shown that this phenomenon is caused by influence of dynamic characteristics (different fluxes values) of the radionuclide concentration processes by marine organisms.

3. The radioecological model and evaluation of the damage to the Black Sea ecosystems

The new approach in evaluation of the damage to environmental state of ecosystems is presented by the conceptual model proposed by Polikarpov [14] and later developed [20] as the model of responses of organisms, populations and ecosystems to all possible dose rates of ionizing irradiation in the environment (with the following zones of radiation effects: radiation well-being; physiological masking; ecological masking; and damage to ecosystems). Recently this model was extended and transformed into the radiochemoecological model, which covers not only ionizing radiation effects but also chemical pollutants action [21].
Figure 1 represents general radiochemoecological conceptual model of zones of ecological impact by chronic ionizing irradiation [21] with the special indication of ecological deleterious consequences in the Black Sea ecosystems, using the following ecological criteria: decrease of number or total mass (of populations), decrease of number of species (of communities), decrease of communities (of ecosystems).
It is obvious that dose commitments of environmental ionizing radiation to the Black Sea organisms are in the "well-being zone" however chemical pollutions, including, first of all, eutrophiers a.s.o., produce directly and indirectly severe ecological effects, which are identical in manifestation of action under higher and very high dose rates of ionizing radiation, available only in very few nuclear areas, such as Mayak, Chelyabinsk region. Chemical pollutions in the estuarian and shelf areas of the Black Sea cause heavy ecological effects, typical for the "damage to ecosystems zone" (number reduction, species reduction, impoverishment, degradation) under the action of chemical chronically-acting doses, which are obviously equivalent to the orders of, and higher than, 20 to > 1000 $Gy \cdot y^{-1}$ of ionizing radiation dose rates.

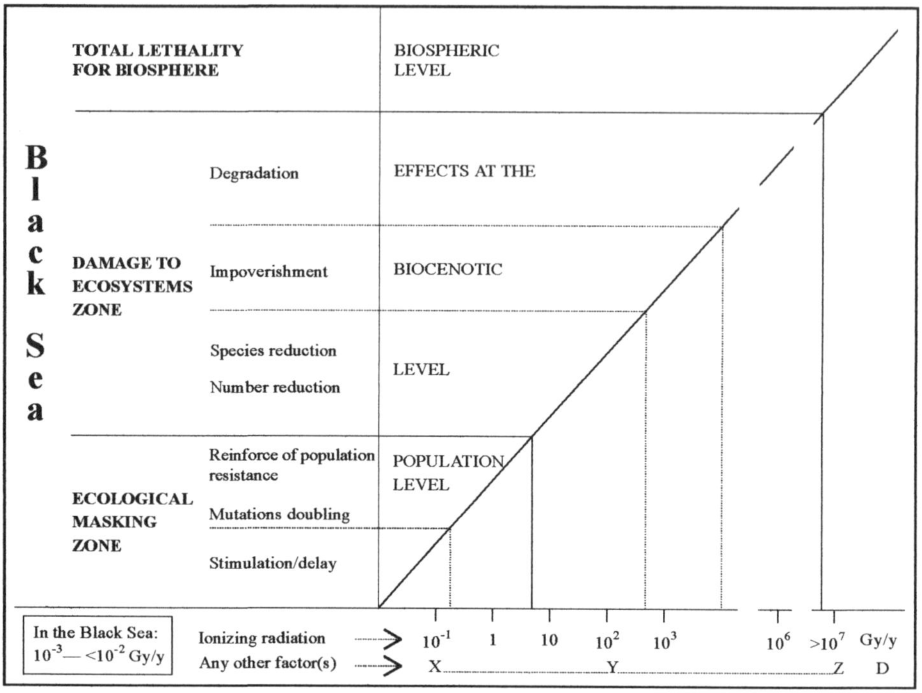

Figure 1. Long-term dependence of populations/ecosystems reactions on ionizing radiation dose rate as well as any chemical or physical factor dose in "maskin" and "damage" zones. X,Y,Z – conditional quantities of doses (D) of non-radioactive factors [23]. The Black Sea marked ecological impact is connected to non-nuclear factors because of lack of higher ionizing radiation doses.

Figure 2 demonstrates a good agreement of the microcosm data by S. Fuma et al. on nuclear/non-nuclear impacts [4] and the radiochemoecological conceptual model by Polikarpov [21].

Available measurements of metals (e.g., Cu) in the bottom sediments (1.75 to 6.17 $mg\cdot kg^{-1}$) and in mussels (3.62 to 6.89 $mg\cdot kg^{-1}$) of the Kerch Strait [29] show that, taking into account rather significant concentration factors of Cu in hydrobionts, the content of this metal in the environment (28 to 97 $nmol\cdot kg^{-1}$) and in the mussels bodies (57 to 108 $micromol\cdot kg^{-1}$) is, perhaps, a few orders of magnitudes lower in the Black Sea than the level of its severe ecological influence on the microcosm, registered by S. Fuma et al. [4].

The distribution of differently polluted water along sea-shore of Ukraine, Russia and Georgia was studied [11].

Figure 3 is a map of pollution sources at the Black Sea [19].

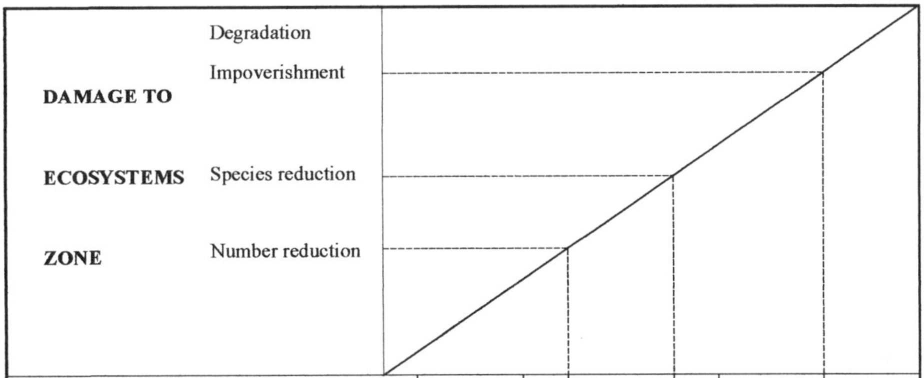

Ionizing radiation, Gy	10	50	100	500	1000	5000
Ultra-violet, kerg/mm^2				50		100
Acidity (HNO_3 + H_2SO_4), pH				4		
Cu^{2+}($CuSo_4$), µmol/L	10			100		
Gd^{3+}($GdCl_3$), µmol/L	50	100		300		1000
Ni^{2+}($NiSO_4$), µmol/L	10			100		1000
Mn^{2+}($MnCl_2$), µmol/L		100		10000		

Figure 2. Radiochemoecological model [20] combined with data on toxicants` ecological effects in the aquatic microcosm [4].

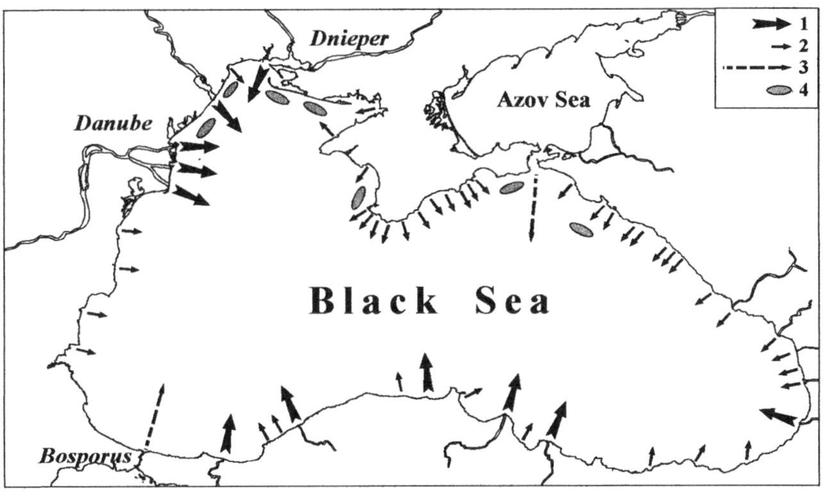

Figure 3. Scheme of location of large-scale sources of the Black Sea pollution: 1 – sewage waters input with river flows; 2 – direct discharges of sewage waters through sewage outlets; 3 – polluted waters input from adjacent seas; 4 – dumping [19].

The results of the paper describe the state of dominant and former dominant species of the Black Sea plants and animals, including some of them, now disappeared from the stage of life. Assessment of equivalency of real existing non-nuclear pollutants' after-effects in the Black Sea ecosystems is proposed to express in the units of ionizing radiation equivalent dose rates, $Gy\text{-}eq{\cdot}y^{-1}$.

The proper data on the influence upon the Black Sea ecosystems, communities and populations as well as ratios of the assessed equivalent dose rates by non-nuclear pollutants to the existing dose rates of ionizing irradiation in the ecosystems are presented in Table 2.

According to [30], the chronic lethal dose is 2 to 10 times higher than the acute lethal dose. Their mean ratio, equalled to 6, is accepted for this paper.

It is proposed to develop for the future: a detailed substantiation of equivalency approach in comparing of nuclear and non-nuclear pollutants' ecological impact as well as creation of comparative molismological theory of ecological equivalency of effective doses of all kinds of ecological-deleterious significant sources (electromagnetic, mechanical, chemical), including lack or excess of vitally important substances. It is a big, but unavoidable work in the field of ecological radiochemoprotection.

TABLE 2. Damage to ecosystems and populations of the Black Sea species caused by nuclear and non-nuclear pollutions [23]

Ecological damage assessments [30]	Lethal chronic doses of ionizing radiation [30] (Gy)	Equivalent lethal chronic doses of non-nuclear pollutions ($Gy\text{-}eg.y^{-1}$) A	Environ-mental nuclear sources [12, 15,19] ($Gy.y^{-1}$) B	Ratio A/B
1. *Cystoseira barbata* biocenosis. > 99 % of its population size were lost since 1960s on Romanian & Ukrainian shelf under eutrophication & pollution impact	For *Thallophyta* (including algae) 180-72000	180-72000	< 0.005	10^4-10^7
2. Gen. *Phyllophora* biocenosis. 97 % of population size were lost during 30 years on NW Black Sea under press of eutrophication, pollution & turbidity	For *Thallophyta* (including algae) 180-72000	180-72000	< 0.005	10^4-10^7
3. Ostrea edulis. > 95 % of its population were lost during 30 years on the Black Sea shelf under turbidity action	For *Mollusca* 600-6000	600-6000	< 0.007	10^5-10^6
4. Mytilus galloprovincialis. 60 % of its population were lost during 30 years on the Black Sea	For *Mollusca* 600-6000	600-6000	< 0.007	10^5-10^6

shelf under action of hypoxia, caused by eutrophication				
5. *Crustacea* populations. Crabs (14 species) & shrimps (> 20 species) were lost 50-70 & 60 % of their populations during 30 years under action of overharvesting of crabs & hypoxia for shrimps	For *Crustacea* 90-3600	90-3600	< 0.007	10^4-10^5
6. *Gobiidae* populations. All 20 species were lost 80 % of their populations on the shelf during 30 years under action of hypoxia on the bottom & destruction of their breeding grounds	For *Pisces* 42-360	42-360	< 0.007	10^4-10^6
7. *Marine Mammalia* populations. Dolphins (3 species, endemic subspecies) lost 90-95 % of their populations sizes during 30 years as consequences of toxicants biomagnification & killing as by catch in fisheries. Number of monk seal (*Monachus monachus*) decreased to few individuals (or to 0) because of lack of reproduction on coastal areas & toxicants biomagnification	For *Mammalia* 12-90	12-90	< 0.006	10^3-10^4

It requires wide range investigation of marine ecosystems state. Intensive radioecological study is not limited by ^{90}Sr and ^{137}Cs: our department also carries out the research of the Black Sea pollution by radionuclides of Po [9], Pu [5] as well as chemical pollutions such as mercury [8], polychlorbiphenyls [18]. Study of pollutants influence on biota is fulfilled at the level of cytogenetic effects in hydrobionts [28] and in the form of the chemoecological monitoring of *Modiolus phaseolinus* settlings [22, 26].

4. Chemoecological monitoring of *Modiolus phaseolinus* settlment

The last part of this paper is connected with chemoecological monitoring of the zone of *M. phaseolinus* habitation in the Black Sea near western coast of Crimea [22]. Chemoecological monitoring includes investigations of pollution levels in marine environment: water, bottom sediments, biota and at the same time the state and responses of biological systems under pollutants action.

Therefore it is very important to choose the proper species-monitor, which could allow (on the base of long-term observation for its settlings) to evaluate ecological quality of environment and its changes. Such a species-monitor should satisfy some requirements. Organisms-monitors should have a large natural habitat, should be mass, typical species for the reservoir. It is necessary to ensure the background investigation of spatial distribution of a species-monitor. It should be non or little movable, longstanding, it should leaves hard remains after its death, it has high concentration ability, it should be accessible for sampling and easy registered. It would be desirable that the type of feeding of the monitor is well-known as well as the size and age structure, process of reproduction and its ecological role in an ecosystem. Such a good complex of advantages in characteristics is available for the mentioned above mussel *M. phaseolinus*, because it is a benthic species of the widely spread deep-sea biocenosis in the Black Sea, which is called after the dominant species — *M. phaseolinus* biocenosis [32, 7] and which is one of the three regional biocenosis of the Black Sea. The maximum size of *M. phaseolinus* is about 14 mm, it has its hard remains after death (shells), it is an eurybiontic species and it is a mussel-filtrator. Its depth distribution, age and size structure, feeding, reproduction was studied and speed of biosedimentation and biofiltration was assessed.

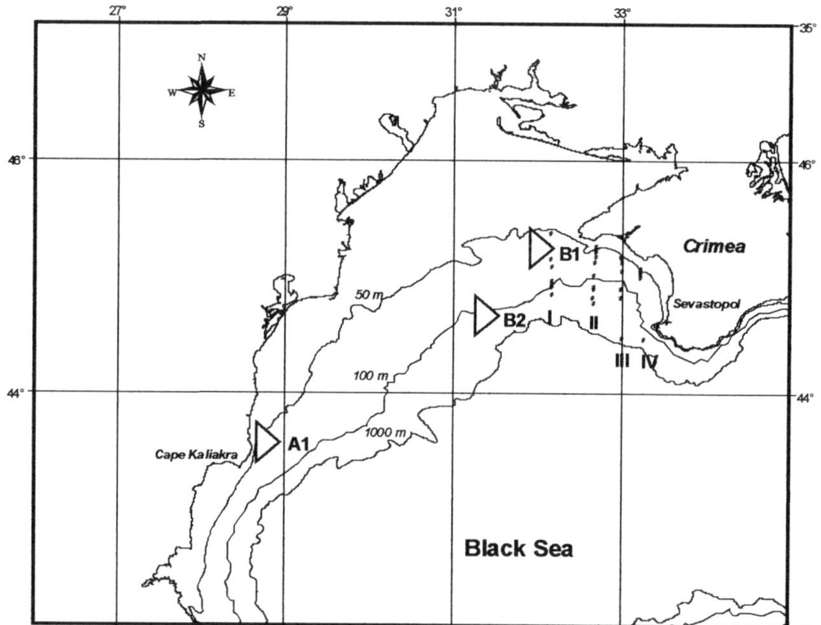

Figure 4. Working areas of *M. phaseolinus* settlings investigation in the Black Sea: • - stations at four longitude transects near western coast of Crimea [26]; A1 - area torwads the north from the Cape Kaliakra [25]; B1 – area without methane gas seeps (eastern transect) and B2 – area with gas seeps [10].

Chemoecological monitoring is fulfilled by us to determine ecological significance of natural and man-made factors.

First of all chemoecological monitoring of *M. phaseolinus* settlings was carried out to clarify the question about whether there is progressive lifting of upper boundary of the deep-sea H_2S zone in the Black Sea [16], because its settlings are located near the boundary of the oxic/anoxic interface in the Black Sea.

We had chosen the area near the western coast of Crimea (Figure 4), because depth of the bottom there grows slowly and *M. phaseolinus* distribution was studied earlier [7]. Hydrochemical and hydrophysical parameters of the environment were studied in details as well as biological characteristics of *M. phaseolinus* settlings [26]. It follows from results, which are presented in Table 3 that the diapason of 100% occuring of *M. phaseolinus* in the samples in the western part of investigated area is getting less and at the same time in the eastern part of the area the *M. phaseolinus* settlings have finished their existence.

Upper boundary of H_2S-zone (Table 3) is separated in space far enough in relation to the lower boundary of *M. phaseolinus* settlings.

TABLE 3. Depth distribution of the *M. phaseolinus* (a — average numbers of individuals, ind.m^{-2}), its shells (W — weight, kg.m^{-2}) and the depth of upper boundary of the hydrogen sulphide zone in the near-bottom water layer of the working area.

Period of observation, years	a/W				Depth, m		
	I transect	II transect	III transect	IV transect	Where investigations were carried out	100% presence of *M. phaseolinus*	Presence of hydrogen sulphide, 0.2-0.7 $mlH_2S \cdot l^{-1}$
1960-1970 [17]	1148/–	1148/–	-	-	1-200	60-125	125-130 [19]
1985	-	207/–	-	-	54-350	54-71	140
1986	-	752/496	-	-	50-100	55-57	120-140
1987	-	752/496	-	-	50-160	54-80	140
1988	92/1080	985/540	-	-	50-320	50-85	140
1989	2848/7622	1329/556	0/170	0/333	50-200	50-110*	160
1983	932/622	1982/456	-	-	50-188	50-110	125-130

*- only for I and II transect.

We observed its oscillations but no progressive lifting. It was carried out the study on the Black Sea shelf toward north of the Cape Kaliakra. According to joint programme of

investigation together with Bulgarian scientists in the same period (1986 — 1989). The observations were fulfilled in the depths 0 — 200 m at nine latitude transects [25].
M. phaseolinus was met in wide range of depths: 35 — 200 m. The number of mussels changed in the samples from some hundreds to some thousands individuals per sq. m in range of depths 50 — 130 — 170 m. *M. phaseolinus* was met in a quantity of a few individuals per sample on the edge of shallow boundary (35 m) disposal of mussels and on the deep edge of boundary (200 m). In such way we weren't observing degradation of *M. phaseolinus* biocenosis in its deep part during 1986 – 1989 years in this area. This fact confirmed absence of deep H_2S-lifting. In the area of the Kalamitsky Gulf the damage of *M. phaseolinsus* biocenosis is caused evidently by other factors, probably anthropogenic by nature such as sewage water, trawling or dumping. Particulaly the next data point on possible influence of dumping out. In the area of Kalamitsky Gulf shells were presented in samples from the western part of area in the same way as from the eastern part, but in the eastern part they were under the layer of sandy-silty bottom sediments. Its thickness was about 10 cm. There is the place of dumping in this area (Figure 3).
In such a way the chemoecological monitoring allows to clear up the question about lifting/non-lifting of H_2S in the Black Sea from its depths, and to determine the anthropogenic nature of causes of the biocenosis degradation.
When gas seeps had been discovered in the Black Sea [17] it was necessary to know what is the ecological quality of marine environment for oxibiotic macro-organisms in the area of the gas seeps. The investigations were carried out on two transects (Figure 4): in the area of the gas seeps and without gas seeps [22]. The obtained data is presented in Figure 5, 6. Analysis of depth distribution of alive *M. phaseolinus* and dead shells as well as size structure of bivalves (wide range of 100 % presence of mussels, high numbers of *M. phaseolinus* especially in the base zone of biocenoses, existence of all size classes young and adult individuals) tells about well being conditions and demonstrates the stability of the *M. phaseolinus* population in these regions. Obtained results point out absence of negative influence of methane gas seeps on the base biological characteristics of *M. phaseolinus* and show good ecological quality of marine environment for macro-oxibionts in the area of the methane gas seeps in the northern-western part of the Black Sea.
Not only natural chemical factors, which exist in the Black Sea (deep hydrogen sulphide, methane gas seeps), but also xenobiotics of anthropogenic nature appeared in the Black Sea waters and their concentrations become often dangerous in the marine waters. Polychlorbiphenyls (PCBs) belong to such class of pollutants.
In area of the Kalamytsky Gulf concentration of PCBs was equal to 10 — 100 $ng \cdot l^{-1}$ [27]. These levels of PCBs in water can reduce primary production of phytoplankton by 25 % [6], but *M. phaseolinus* being an eurybiontical organism keeps its stability state concerning to the base biological characteristics. So it can be used as the species-monitor in relation to PCBs. In 1986 — 1998 the work was carried out a lot of measurements of PCBs in water, bottom sediments and biota in the Black Sea [18]. Estimation of PCBs distribution in the main biotic and non-biotic components of the

marine ecosystem (alive *M. phaseolinus*, dead shells, bottom sediments and water) in the base zone of *M. phaseolinus* biocenosis is shown in Table 4.

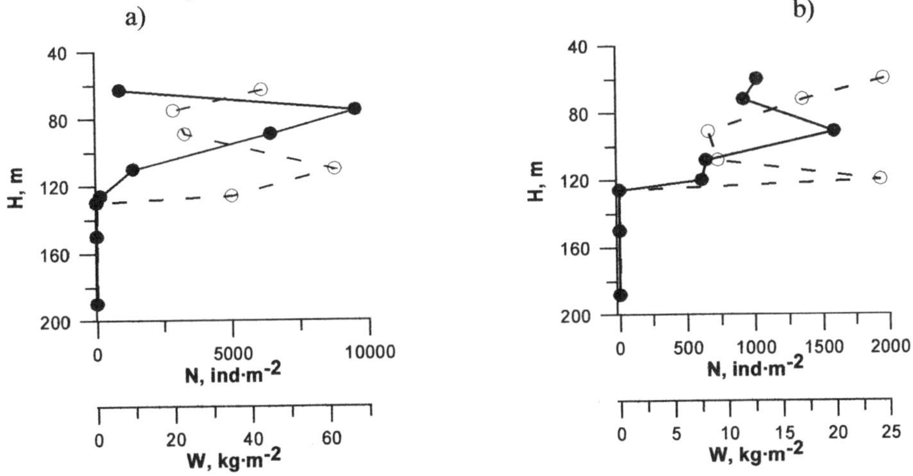

Figure 5. Distribution of the bivalve *M. phaseolinus* in the gas seep area (a) and at the eastern transect (b) where: N — number of living specimen (•), W — weight of dead shells (o), H – depth [22].

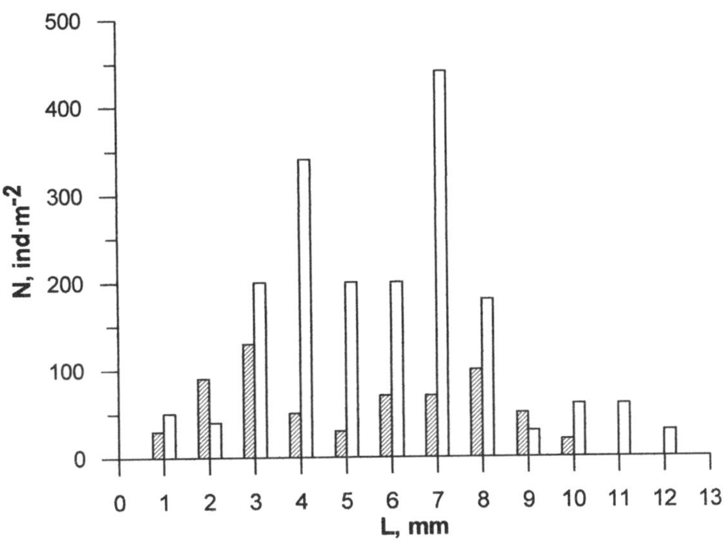

Figure 6. Shell size class distribution of alive mollusca *M. phaseolinus* at 110 m depth in the gas seep area (open bars) and at 108 m depth at the eastern transect (hatched bars) where: N — number of specimen, L — shell length [22].

The obtained results demonstrate, that the most of all PCBs are accumulated in bottom sediments (95.5%). Alive *M. phaseolinus* have the highest concentration factor (more then 16000) and accumulate 2% PCBs, almost like water (2.4%), but weight of *M. phaseolinus* is 0.004% in the ecosystem and water in the same one is 99.058% from the total weight of the ecosystem. Role of shells is not essentical (Table 4).

TABLE 4. Distribution of PCBs in main investigated biotic and non-biotic components in the base zone of *M. phaseolinus* biocenosis [27]

Components	Contents of PCBs, $ng \cdot kg^{-1}$	Concentration factor of PCBs (WW)	Weight of components, $kg \cdot m^{-2}$ (Percentage of a component in total weight of the ecosystem on 1 m^2, %)	Contents of PCBs, $\mu g \cdot m^{-2}$ (Percentage contents in a component in total contents of PCBs in the ecosystem on 1 m^2, %)
Water (0-60 m)	14		6102 (99.052)	87.3 (2.4)
Bottom sediments (0-15 cm)	62500	4371	55.900 (0.917)	393.8 (95.5)
Shells (0-15 cm)	3400	237	1.300 (0.020)	4.4 (0.1)
Alive *M. phaseolinus*	238450	16674	0.300 (0.004)	71.5 (2.0)

In such way it bear witness about the great role of biotic component in biogeochemical cycles of xenobiotics, like PCBs.

One individual of *M. phaseolinus* with length 5 mm with weight 0.02 g filters 0.6 l of water during a day. In the base zone of the *M. phaseolinus* biocenosis filters 460 $l \cdot m^{-2} \cdot day^{-1}$ and the flux of biosedimentation composes 0.7 $g \cdot m^{-2} \cdot day^{-1}$ [7]. During a year these mussels on 1 m^2 filter about 160 tons of water (in the same time the weight of water is 50 — 120 tons in the base zone of the biocenosis). It shows high environmental role of biota especially mussels-filtrators in renewal/conditioning of ecological quality of the environment. So degradation of *M. phaseolinus* biocenosis (like in the eastern part of the Kalamytsky Gulf) causes not only decreasing of food for fishes and decreasing of number of biota, but also diminishing capacity of marine ecosystems to renew and purify the environment concerning chemical and other pollution which leads to decrease reliability of marine ecosystems [15].

In our time of the global anthropogenic influence on nature, the complex chemoecological monitoring is an important tool to study ecosystems. It supplies with practical data for the conceptual model of responses of biota of different levels of organization to all possible influences of an environment and for creation united real system of assessment of the influence in ecological equivalent doses for evaluation and prediction of ecosystems state as well as for choice of regulations or adaptation tendency regarding to natural processes.

5. Acknowledgement

This work is partly supported by International Atomic Energy Agency and the European Community.

6. References

1. Egorov, V.N. (2000) Assessment and prediction of pollution of waters and biota in the Black Sea with ^{90}Sr and ^{137}Cs after the Chernobyl NPP accident, in *International Conference "Modern Problems of Radiobiology, Radioecology and Evolution" Dedicated to Centenary of N.W.Timofeeff-Ressovsky, Sept. 6-9, 2000*, Joint Institute for Nuclear Research, Dubna, pp.177 – 178.
2. Egorov, V.N., Polikarpov, G.G., Mirzoyeva, N.Yu., Kulebakina, L.G. and Artyomov, Yu.G. (2000) The tendencies of change of the ^{90}Sr and ^{137}Cs concentration in the water and hydrobionts of the Sevastopol Bays after Chernobyl NPP accident, *Ecologiya Morya* **50,** .83-87. (In Russian)
3. Egorov, V.N., Povinec, P.P., Polikarpov, G.G., Stokozov, N.A., Gulin, S.B., Kulebakina, L.G. and Osvath, I. (1999) ^{90}Sr and ^{137}Cs in the Black Sea after the Chernobyl accident: inventories, balance and tracer applications, *J. of Environmental Radioactivity* **43**, 2,137-155.
4. Fuma, S., Miyamoto, K., Takeda, H.., Yanagisawa, K., Inoue, Y., Sato, N., Hirano, M. and Kawabata, Z. (1998) Ecological effects of radiation and other environmental stress on aquatic microcosm, in J. Inaba, and Y. Nakamura (eds.), *Comparative Evaluation of Environmental Toxicants. Proceedings of the International Worksop on Comparative Effects of Health Effects on Environmental Toxicants Derived from Advanced Technologies, Chiba, January 28-30, 1998*, National Institute on Radiological Sciences, Japan, pp. 131-144.
5. Gulin, S.B., Polikarpov, G.G., Egorov, V.N. and Korotkov, A. A. (2000) Geochronological assessment radioactive pollution of the Black Sea, in *Readings in the Honour of Memory of N.W.Timofeeff-Ressovsky,* Eco-Ethics International Union, Sevastopol Chapter of the N.W. Timofeeff-Ressovsky Scientific Society 'Biosphere and Humanity', the A.O. Kovalevsky Institut of Biology of the Southern Seas of National Academy of Sciences of Ukraine, Sevastopol, pp. 88-99. (In Russian)
6. Izrael, Yu.A. and Tsiban, A.V. (1981) Ecology and problem of complex global monitoring, in *Complex Global Monitoring of World Ocean,* Hygrometeoizdat, Leningrad **1**, pp. 19-48. (In Russian)

7. Kisiljeva, M.I. (1981) *Benthos of Friable Sediments,* Naukova Dumka, Kiev. (In Russian)
8. Kostova, S.K., Egorov, V.N. and Popovichev, W.N. (2000) Ecological aspects of study of the Black Sea pollution by mercury, in *Readings in the Honour of Memory of N.W.Timofeeff-Ressovsky,* Eco-Ethics International Union, Sevastopol Chapter of the N.W. Timofeeff-Ressovsky Scientific Society 'Biosphere and Humanity', the A.O. Kovalevsky Institut of Biology of the Southern Seas of National Academy of Sciences of Ukraine, Sevastopol, pp. 216-226. (In Russian)
9. Lazorenko, G.E. (2000) ^{210}Po in hydrobionts of the Black Sea, in *Readings in the Honour of Memory of N.W.Timofeeff-Ressovsky,* Eco-Ethics International Union, Sevastopol Chapter of the N.W. Timofeeff-Ressovsky Scientific Society 'Biosphere and Humanity', the A.O. Kovalevsky Institut of Biology of the Southern Seas of National Academy of Sciences of Ukraine, Sevastopol, pp. 108-125. (In Russian)
10. Luth, U. and Luth C. (1998) Benthic meiofauna and macrofauna of a methane seep area south-west of the Crimean penincula, Black Sea, in *Berichte aus dem Zentrum für Meeres-und Klimaforschung: MEGASEEBS "Methane Gas Seep Eexplorations in the Black Sea", Project Report,* Hamburg **14**, pp. 113-126.
11. Mandych, A.F. and Shaporenko, S.I. (1992) Influence of the economic activities on coasts upon coastal waters of the Black Sea, *Geo J.* 27, **2**, 195.
12. Pertsov, L.A. (1978) *Biological Aspects of Radioactive Pollition of the Sea,* Atomizdat, Moscow. (In Russian)
13. Polikarpov, G.G. (1964) *Radioecology of Marine Organisms,* Atomizdat, Moskow. (In Russian)
14. Polikarpov, G.G. (1977) Effects of ionizing radiations on aquatic organisms (chronic irradiation), in *Alcuni Aspetti di Radioecologia, Atti, XX Congresso Nazionale,* AIFSPCR, Bologna, pp. 25-46.
15. Polikarpov, G.G. and Egorov,V.N. (1986) *Marine Dynamic Radiochemoecology,* Energoatomizdat, Moscow. (In Russian)
16. Polikarpov, G.G., Tereshchenko, N.N., Timoshchuk, V.I., Tsytsugina, V.G., Lazorenko, G.E., and Demina, N.V. (1987) Chemoecological monitoring of environment quality of lower edging zone of the mollusk *Modiolus phaseolinus* habitation in the Black Sea, *Dopovidi Academii Nauk Ukr. SSR* **8,** 77-80. (In Russian and Ukrainian)
17. Polikarpov, G.G., Egorov, V.N., Nezhdanov, A.I., Gulin, S.B., Kulev, Yu.D. and Gulin, M.B. (1989) The phenomenon of active gas escapes from mount of the slope of the western Black Sea, *Dopovidi Academii Nauk Ukr. SSR* **12**, 13-16. (In Russian and Ukrainian)
18. Polikarpov, G.G. and Zerko, N.V. (1996) Ecological aspects of study pollution in the Black Sea by chlororganic xenobiotics, *Ecologiya Morya* **45**, 92-100. (In Russian)
19. Polikarpov, G.G., Zaitsev, Y.P., Zats, V. I. and Radchenko L.A. (1994) Pollution of the Black Sea (levels and sources), in *Proc. of the Black Sea Symposium, Ecological Problems and Economical Prospects, 1991,* The Black Sea Foundation for Edication, Culture and Protection of Nature, Istanbul, pp. 15-42.

20. Polikarpov, G.G. (1998) Conceptual model of responses of organisms, populations and ecosystems to all possible dose rates of ionizing radiation in the environment, in *Radiation Protection Dosimetry*, **75**, pp. 181-185.
21. Polikarpov, G.G. (1998) Biological aspect of radioecology: objective and perspective, in J. Inaba and Y. Nakamura (eds.), *Comparative Evaluation of Environmental Toxicants, Proceedings of the International Workshop on Comparative Evaluation of Health Effects of Environmental Toxicants Derived from Advanced Technologies, Chiba, January 28-30, 1998,* National Institute on Radiological Sciences, Japan, pp. 3-15.
22. Polikarpov, G.G., Tereshchenko, N.N. and Gulin M.B. (1998) Chemoecological study of the bivalve *Modiolus phaseolinus* in habitats near the oxic/anoxic interface and near methane gas seeps in the Black Sea, in *Berichte aus dem Zentrum für Meeres-und Klimaforschung: MEGASEEBS "Methane Gas Seep Explorations in the Black Sea", Project Report*, Hamburg **14**, pp. 129-133.
23. Polikarpov, G.G. (1999) Effects of nuclear and non-nuclear pollutants on marine ecosystems, in *Marine Pollution: Proceedings of a Symposium Held in Monaco, 5-9 October 1998,* IAEA-TECDOC-1094, Vienna, pp.38-43.
24. Skopintsev, B.A. (1975) *Forming of Modern Chemical Composition of the Black Sea Water,* Hydrometeoizdat, Leningrad. (In Russian)
25. Tereshchenko, N.N., Polikarpov, G.G., Marinov, T. and Stoykov, S. (1991) Bioindication of ecologi- cal quality zones of habitation of mussels in deep zone of the Black Sea, in *Second All-Union Conference by Fishery Toxicology*, Sankt-Peterburg **2**, pp. 210-211. (In Russian)
26. Tereshchenko, N.N., Polikarpov, G.G. and Kovalenko, T.P. (1992) Monitoring investigations of *Modiolus phaseolinus* zone habitation in the area of the Kalamitsky Gulf of the Black Sea, *Hygrobiological J.* **4,** 23-30. (In Russian)
27. Tereshchenko, N.N., Polikarpov, G.G. and Zerko N.V. (2000) Chemoecological monitoring of chlororganic pollution in zone of the Black Sea *M. phaseolinus* biocenosis near the western coast of Crimea in *Readings in the Honour of Memory of N.W.Timofeeff-Ressovsky*, Eco-Ethics International Union, Sevastopol Chapter of the N.W. Timofeeff-Ressovsky Scientific Society 'Biosphere and Humanity', the A.O. Kovalevsky Institut of Biology of the Southern Seas of National Academy of Sciences of Ukraine, Sevastopol, pp. 176-189. (In Russian)
28. Tsytsugina, V.G. and Polikarpov, G.G. (2000) Cytogenetic effects in natural hydrobionts populations which were caused by radioactive and chemical pollution, in *Readings in the Honour of Memory of N.W.Timofeeff-Ressovsky*, Eco-Ethics International Union, Sevastopol Chapter of the N.W. Timofeeff-Ressovsky Scientific Society 'Biosphere and Humanity', the A.O. Kovalevsky Institut of Biology of the Southern Seas of National Academy of Sciences of Ukraine, Sevastopol, pp. 70-79. (In Russian)
29. Tretiyakova, V.D., Ageeva, T.M. and Voronenko, L.S. (1991) Toxicological characterisation on the Kerchensky Strait ecosystem, in *Second All-Union Conference by Fishery Toxicology*, Sankt-Peterburg **2**, pp. 219-220. (In Russian)

30. Woodhead, D.S. (1993) Dosimetry and the assessment of environmental effects of radiation exposure, in F. Warner and R.M. Harrison (eds.), *Radioecology after Chernobyl: Biogeochemical Pathways of Artificial Radionuclides*, SCOPE 50, pp. 291-306.
31. Zaitsev, Y. and Mamaev V. (1997) *Marine Biological Diversity in the Black Sea. A Study of Change and Decline,* United Nations Publications, New York.
32. Zernov, S.A. (1913) To the question on study of the Black Sea life, *Notes of Imperial Academy of Sciences*, Sankt-Petersburg **32**, 1, 1-283. (In Russian)

DOES BIOMANIPULATION WORK FOR WARM-TEMPERATE LAKES? FIRST CASE STUDY IN AN EUTROPHIC TURKISH LAKE

Meryem BEKLIOGLU
Middle East Technical University, Biology Department, 06531, Ankara, Turkey

1. Introduction

The process of eutrophication that arises from excessive inputs of P & N to lakes leads to world-wide water quality deterioration problem with significant losses of biodiversity, and goods and services such as water supply and recreational values [1], [2], [3]. The reduction of external phosphorus loading is the first crucial step for restoration. However, in shallow lakes, responses to the successful reduction of external nutrient loading have often been resilience. In addition, a long delay in recovery is attributed to internal phosphorus loading inducing high stock of benthi-planktivorous fish associated with low water clarity, in turn, re-development of macrophytes cannot be achieved [4], [5], [6], [7], [8], [9].

Biomanipulation as a possible restoration method was introduced by Shapiro in 1975 [10]. The term refers to the manipulation of fish community to reduce predation pressure on large sized herbivorous zooplankton, supposedly followed by an increased abundance and size of the zooplankton especially *Daphnia*. This, in turn, leads to higher grazing pressure on phytoplankton and, subsequently, to the clear-water state. However, fish reduction also leads to a high recruitment of young-of-the-year-fish (YOY), which excessively feed on zooplankton, can be a major reason for less successful biomanipulation [11]. Over 20 year of experience in biomanipulation has revealed that success does not rely solely on the simple fish-zooplankton-algae food chain. Successful recovery has been achieved only when benthi-planktivorous fish removal were to trigger secondary processes, such as reestablishment of littoral zone, reduced fish-mediated resuspension, and reduced internal loading [8], [9], [11], [12], [13], [14]. Increased abundance of herbivorous zooplankton, particularly *Daphnia*, has been found to be an important triggering factor suppressing excessive phytoplankton crop to improve under water light climate for plant growth. Submerged plants promote clear-water through several stabilising buffer mechanisms including luxury nutrient uptake thereby reducing algal crop [15], suppression of wind-induced sediment resuspension [16], provision of refuge for especially *Daphnia* against fish predation [17], [18], favouring predator fish such as pike and perch [19], [20], enhancing denitrification [21], having allelopathic effects to phytoplankton [22].

J. Klerkx and B. Imanackunov (eds.), Lake Issyk-Kul: Its Natural Environment, 207–215.

In summary, applicable recommendations for successful biomanipulation leading to clear-water state and re-established littoral zone should include over 75% fish removal, which should be rapid (within 1-3 yrs), reduction in YOY recruitment, a good reduction in benthic feeding fish, and reduction in nutrient concentration as much as possible [11], [12].

The present study aims to investigate the efficiency of biomanipulation as a restoration tool to recover the eutrophic warm-temperate Lake Eymir, and to verify the validity of the recommendations yielded from the successful biomanipulation applications on cold-temperate northern lakes.

2. Material methods

Water samples were collected at weekly, fortnightly and monthly intervals from March 1997 to August 2000 for physicochemical and phyto-zooplankton analyses. Aquatic plants were surveyed in July 1998, August 1999 and July 2000. The methods for sampling and analyses can be found in Beklioglu et al. [9] [23]. Prior to this study, the lake was sampled with a limited number of occasions between 1993-1995 for water chemistry; however, this study did not include sampling of phyto-zooplankton, fish and macrophytes [24]. In addition, the water chemistry data produced by Altinbilek et al. [24] were used for evaluation of the status of the lake.

2.1 STUDY SITE

Lake Eymir is an alluvial dam lake that was formed by damming of the Imrahor River at the beginning of the last century. Lake Eymir is located in 20 km south of Ankara (area: 1.25 km^2, Z_{max}: 6-7 m, Z_{mean}: 3.1 m, flushing rate: 0.504 y^{-1}) with a long shoreline (13 km). The earliest study carried out on Lake Eymir in 1946-47 reveals that the lake was originally in a clear-water state with >4 m of summer Secchi depth, and a large littoral zone with the outer water depth of 6.5 m (25). The submerged plants included 5 angiosperm species, and large meadows of Charophytes, *Chara vulgaris* and *Nitella opaca* [25]. The latter species are typical of nutrient-poor lakes. The Middle East Technical University has owned the lake since 1958. The lake was used to be the main drinking water supply for the campus up until 1990. The lake started to be eutrophicated during 1970s as a consequence of increased sewage effluent discharge from the surrounding urban settlement. Therefore the university ceased to use the lake water. In early 1990s, two independent studies carried out in Lake Eymir [24], [26] revealed the eutrophic characters of the lake with very high concentration of nutrients and very low water clarity. To control eutrophication of the lake, the raw sewage effluent of the nearby town, which used to be discharged into the main inflow (named as Inflow 1) of Lake Eymir, was diverted in 1995. But the Inflow1 still receives the poorly treated nutrient rich effluent of the sewage treatment work of the TEAS residency (TEAS, STW) that has a smaller effect than the former since it actively operates in summers.

The Inflow2, which dries up in summer, is the second major water source reaching to the lake.

The fish stock was determined in May 1998 using multiple gill nets (mesh-sized: 18-70 mm) with a length of 100 m and width of 3.5m. The stock was estimated as 200 tonnes (1600 kg ha^{-1}) consisting of 89 % (1420 kg ha^{-1}) of tench (*Tinca tinca*), 10 % (160 kg ha^{-1}) of carp (*Cyprinus carpio*) and 1 % of pike (*Esox lucios)*. The fish stock of the lake appeared to be very high as compared to the findings from elsewhere [7], [14]. This might have been due to prohibition of fishing using all kind of nets in the lake except licensed angling which was more selective for pike. Selective removal of the benthi-planktivorous carp and tench was initiated in August 1998, and has been still continuing. So far, nearly 45% of the benthi-planktivorous fish stock were removed from the lake. Moreover, to release the selective angling pressure on the very low pike stock, pike angling has been banned since 1998 as an extra measure on fish stock manipulation.

3. Results

In Lake Eymir, the thermal stratification sets at the depth of 4.5 m, leaving the hypolimnion with a depth of 1.5-2 m; therefore, < 40% of the total lake water mass undergoes thermal stratification.

Partial effluent diversion undertaken in 1995 led to a significant decrease in the in-lake concentrations of total phosphorus (TP) and dissolved inorganic nitrogen, which includes nitrite, nitrate and ammonium (DIN) (1206±754 μg l^{-1} and 1.4±0.79 mg l^{-1}, respectively) compared to the mean concentrations (276±17 μg l^{-1} TP and 0.124±0.027 mg l^{-1} DIN) recorded between 1997-2000 ($p<0.001$ and $p<0.001$, respectively) (fig. 1 and Table 1).

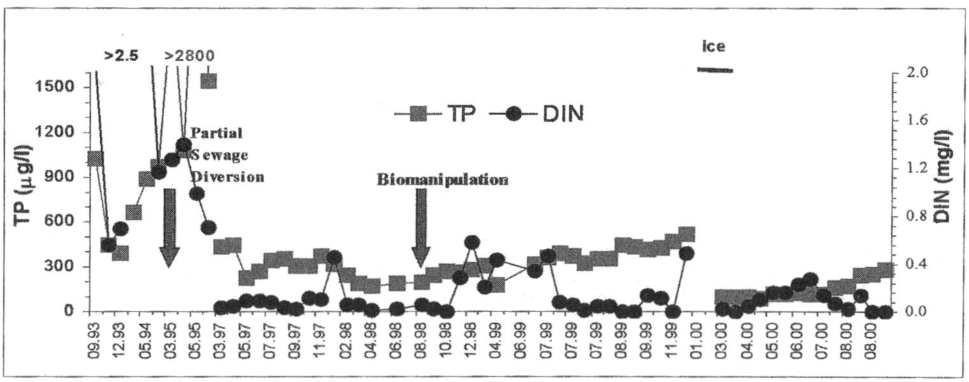

Figure 1. The changes in the concentrations of TP and DIN in Lake Eymir between the periods of 1993-1995, and 1997-2000 (the solid double lines on the axis indicates the gap between the two sampling periods; the data between 1993-95 were taken from Altinbilek et al. 1995).

The TP and DIN concentrations found in the Inflow1, which receives the poorly treated nutrient rich sewage effluent of the TEAS STW especially in summers, were very high (721±83 μg l^{-1} and 1.26±0.22 mg l^{-1}, respectively), and their concentration remained the same throughout the study period (p:0.37 and p:0.08, respectively). Although the fish removal did not lead to a statistically significant change in the overall in-lake TP and DIN concentrations, the mean TP concentration recorded in 2000 were much lower than that of pervious years regardless of the Inflow I concentration which remained as high as previous years (Fig.1 & Table 1).

Table.1. The mean values (± SD) of the variables recorded in Lake Eymir during the periods of pre-diversion (1993-1995), pre-biomanipulation (1997-1998) ,and during biomanipulation.

	1993-1995* **(n:10)**	**1997 (n:10)**	**1998 (n:9)**	**1999 (n:15)**	**2000 (n:12)**
	Pre-diver.	**Before Bioman.**			**During bioman.**
TP (μg/l)	1206±754	339±21	228±14	380±21	129±14
DIN (mg/l)	1.43±0.8	0.106±0.04	0.121±0.06	0.162±0.04	0.108±0.02
chlorophyll-a (μg/l)	27±4	19±4	17±3	13.6±2	10.8±1.2
Secchi depth (cm)	56±11	119±14	127±17	298±15	328±24
***Daphnia pulex* (Ind/l)**	-	1.3±0.03	5.7±1.5	13±3.7	39±6
Total surface coverage of submerged plants (%)	-	-	2.5	6.2	35

* taken from Altınbilek et al., 1995

Following partial sewage effluent diversion, the Secchi disk transparency doubled and the chlorophyll-a concentration has got halved compared to recorded values in the year prior to the start of the biomanipulation (119±14 cm and 19±3μg l^{-1}, respectively) (fig. 2, Table 1). The biomanipulation resulted in further progressive increase in the Secchi disk transparency (298±15 cm in 1999 and 328±24 cm in 2000) which was significantly different from the pre-biomanipulation period (p<0.001) (Fig.1 & Table 1). The chlorophyll-a concentration decreased to 13.6±2 μg l^{-1}in 1999 and to 10.8 ±1.2 μg l^{-1}in 2000 that were significantly different from that of the pre-biomanipulation (p:0.04 and p: 0.045, respectively) (fig. 2, Table 1). The recovery recorded in the Secchi disk transparency, and the chlorophyll-*a* concentration collapsed a few times during summer of 1999 that showed a sign of instability, however, the recovery in these variables has been more stable in 2000 (fig.2 & Table 1).

The several species of Chlorophyta and dinoflagellates have had the major contribution to the chlorophyll-*a* concentration throughout the study period. Though the contribution made by cyanobacteria to the chlorophyll-*a* concentration peaked before and just after the start of biomanipulation in 1998, its share was much lower in 1999 and 2000. Before the fish removal covering between 1997-98, the mean density of large-bodied *Daphnia pulex* De Geer was significantly lower (3.57±0.9 ind. l^{-1}) than that of the mean density during fish removal including 1999 and 2000 (26.6±4 ind. l^{-1}) (p< 0.001) (Table 1).

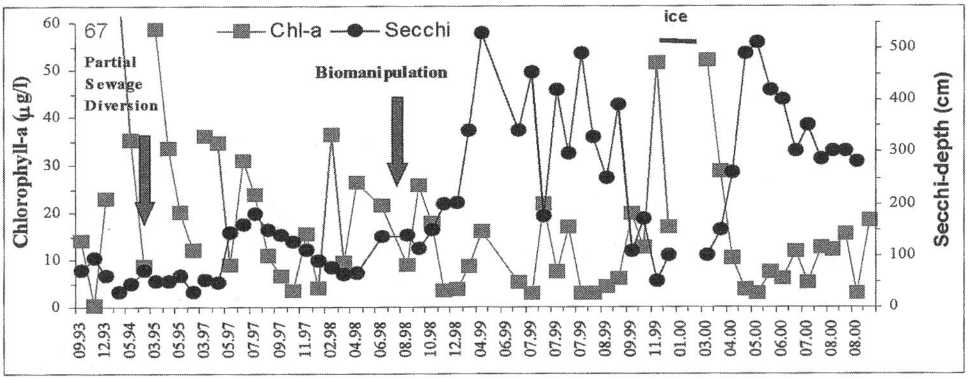

Figure 2. The changes in the concentration of chlorophyll-a and the Secchi disk transparency in Lake Eymir between the periods of 1993-995, and 1997-2000 (the solid double lines on the axis indicates the gap between the two sampling periods; the data between 1993-95 were taken from Altinbilek et al. 1995).

The density of *D. pulex* had a small spring peak, and decreased to <1 ind. l^{-1} in summer before the fish removal. Following the removal, the spring peaks of *D. pulex* recorded in 1999 and 2000 were much higher and remained high in summers. The total densities of *Arctodiaptomus bacillifer* and *Cyclops* sp., and the small-sized cladocerans, *Ceriodaphnia* and *Bosmina* species remained low throughout the study.

The surface coverage of the submerged plants was very low being 2.5% before the fish removal, and slightly increased to 6.21% in 1999 (Table 1). With a year delay in redevelopment of the littoral zone, the surface coverage of the submerged plants increase to 35% of the lake total surface area during 2000 (Table 1). The dominant species recorded in the lake were as following *Myriophyllum spicatum* L., *Ceratophyllum demersum* L., *Potamogeton pectinatus* L., *Najas marina* L., *Najas* sp. In the last plant survey, the small patches of *Chara* sp. were also recorded. This species was used to have large meadows in the lake during pre-eutrophication period [25]. The dominant emergent plant, *Phragmites austrialis* L., covers the whole shoreline as a belt with a width of 15-25 m, and with an average of 12% of the lake's total surface area.

4. Discussion

Though, partial sewage effluent diversion led to a significant decrease in the in-lake concentrations of TP and DIN, the TP concentration was still very high sustaining the eutrophic state of the lake. However, during the fish removal in 2000, the TP concentration has got halved regardless of any change in the TP loading from the Inflow 1 to the lake. The decrease in the in-lake TP concentration might have been due to the large expansion of the submerged plants with their luxury nutrient uptake which has been proven to be significant [9], [15], and as well as the reduction in the fish-induced internal phosphorus loading. This finding is in accordance with results of the biomanipulation applications in elsewhere [7], [8], [11], [12], [13], [27].

Although, partial sewage effluent diversion led to a significant increase in the overall Secchi depth, the recovery in water clarity mainly confined to the summer periods, and the spring periods had the lowest Secchi depth along with high turbidity. The clear-water attained in spring recorded elsewhere appears to be the crucial time period for submerged plants to start growing [8], [11], [13], 14], [27]. Therefore, the overall recovery in water clarity of Lake Eymir during pre-biomanipulation was not sufficient to cause substantial improvement in redevelopment of the submerged plants. The long shoreline of Lake Eymir would have made possible the redevelopment of the submerged plants. However, due to poor under water light climate especially in the springs this did not occur. In Lake Eymir, the presence of the very high benthi-planktivorous fish biomass (1580 kg ha^{-1}) and their top-down effect on the large-bodied zooplankter ought to have been the reason for the poor water clarity especially in spring. The top-down effect of the benthi-planktivorous fish on density of large daphnids appears to be an important factor for under water light climate, and in turn, submerged plant growth [7], [8], [11], [12], [13], [14], [27]. Following the fish removal, the densities of *Daphnia* increased several-fold especially in 2000 with greater spring peaks than that of the density during pre-biomanipulation period. The increased water clarity induced by increased *Daphnia* grazing especially in spring time might have explained the reappearance of submerged plants in the lake, however, this cannot be a full explanation for a delay in the growth of submerged plants in 1999 (see below for further discussion). The reduction in the benthivorous fish and their feeding-induced resuspension and turbidity should also be taken into account for the attained clarity, since benthivorous fish (e.g., carp, bream, roach) increase levels of turbidity and nutrients especially in shallow lakes [28].

The recovery was unstable in 1999 compared to that of recorded in 2000. The instability recorded in the concentration of chlorophyll-*a*, the Secchi depth and the density of *D. pulex* might have yet been due to the insufficient removal of fish. This led to an increase in the YOY recruitment, which was about 2300 (CPUE), with an expected high predation pressure on the dominant zooplankter [11]. However, in 2000, the YOY recruitment was lower than that of the previous year, this may be due to the severe winter with three months ice-cover, and the following cold spring being particularly unfavourable for the fish recruitment.

An insignificant increase in the coverage of submerged plant in the summer of 1999, maybe due partly to the high number of coots (55 ind. ha^{-1}) and their grazing on the plants. Since it has been shown that grazing can delay establishment of submerged plants after restoration [29]. The severe winter with three months of ice-cover experienced in 2000 resulted in insignificant number of coots inhabiting the lake that in turn might have decreased their suppression on the submerged plants and leading to the increase in the coverage. The effect of reduced carp biomass may also account partly for the result since this fish give a strong mechanical damage to plants [7], [30]. Through the fish removal induced improvement in Lake Eymir, the littoral zone has re-developed with reappearance of *Chara* patches as found during pre-eutrophicated period [25].

The experiences gained from biomanipulated temperate lakes suggest that the reduction in benthi-planktivorous fish removal should be 75% or more to trigger the mechanisms through which systems shift to the submerged plant dominated clear-water state. Contrary to this recommendation, in lake Eymir, the clear-water state with 35% of submerged plant coverage was achieved with only 45% of the fish removal. However, it is still unclear that if this recovery will be long lasting since severe winter conditions in 2000 was a sheer luck. However, the results presented in here suggest that the recommendations raised from the cold temperate lakes for the successful biomanipulation applications appear to be valid for the warm-temperate Lake Eymir.
Furthermore, this study is the first recorded biomanipulation application taken as a restoration measure in a warm temperate lake in Turkey. More applications are needed to develop better understanding about requirements for success in this climate. It appears that the technique is worthwhile to apply since it increases biodiversity, cheap and widely applicable as opposed to some engineering techniques such as sediment removal, chemical addition to reduce algal crop and hypolimnetic oxygenation.

5. Acknowledgement

I thank S. L. Burnak, O. Ince, C. O. Tan & C. Muluk who produced the data presented in here as a part of their MSc. projects, and to Aysegul Cetin for editorial help. The project was funded by the research grant given by METU.

6. References

1. Kristensen, P. & Hansen, H.O. (1994) European rivers and lakes. Assessment of their environmental state. European Environmental Agency, Brussels.
2. Moss, B. (1998) *Ecology of Fresh Waters: Man & Medium, Past to Future*, 3rd edition, Blackwell Science, Oxford.
3. Smith, V.H., Tilman, G.D: & Nekola, J. C. (1999) Eutrophication: impacts of excess nutrient inputs on freshwater, marine, and terrestrial ecosytems, *Environmental Pollution* **100**, 176-196.
4. Marsden, S. (1989) Lake restoration by reducing external phosphorus loading: the influence of sediment phosphorus release, *Freshwater Biol.* **21**,139-162.
5. Jeppesen, E., Kristensen, P., Jensen, P.J., Sondergaard, M., Martensen, E. & Lauridsen, T. (1991) Recovery resilience following a reduction in external phosphorus loading of shallow eutrophic Danish lakes-duration, regulating factors and methods for overcoming resilience, *Mem. Ist. Ital. Hidrobiol.* **48,** 127-148.
6. Scheffer, M., Hosper, S.H., Meijer, M-L., Moss, B.& Jeppesen, E. (1993) Alternative equilibria in shallow lakes, *TREE* **8,** 275-279.
7. Moss B., Madgwick J. & Phillips G. 1996. A guide to restoration of nutrient-enriched shallow lakes Environment Agency & Broads Authority, East Anglia.

8. Jeppesen, E., Søndergaard, M., Jensen, J.P., Mortensen, E., Hansen, A-M. & Jørgensen, T. (1998) Major perturbation in biological structure and dynamics of a shallow hypertrophic lake following a reduction in sewage loading: an 18-year study in Lake Søbygaard, Denmark, *Ecosystem* **1**, 250-267.
9. Beklioglu, M., Carvalho, L. & Moss, B. (1999) Rapid recovery of a shallow hypertrophic lake following sewage effluent diversion: lack of chemical resilience, *Hydrobiologia* **412**, 5-15.
10. Shapiro, J., Lamarra, V.& Lynch, M. (1975) Biomanipulation: an ecosystem approach to lake restoration, in Brezonik, P.L. & Fox, J.L. (eds.), *Proceedings of a symposium on Water Quality Management through Biological Control*, pp. 23-30.
11. Hansson, L.A., Annadotter, H., Bergman, E., Hamrin, S.F., Jeppesen, E., Kairesalo, T., Luokkanen, E., Nilsson, P.A., .Sondergaard, M. & Strand, J. (1998) Biomanipulation as an application of food-chain theory: costraints, synthesis, and recommendations for temperate lakes, *Ecosystems* **1,** 558-574.
12. Drenner, R. W. & Hambright, K.D. (1999) Review: Biomanipulation of fish assemblages as a lake restoration technique, *Arch. Hydrobiol.* **146**, 129-165.
13. Meijer, M-L., Boois, I., Scheffer, M., Portielje, R. & Hosper, H. (1999) Biomanipulation in shallow lakes in The Netherlands: an evaluation of 18 case studies, *Hydrobiologia* **408/409**, 13-30.
14. Meijer, M-L. (2000) Biomanipulation in the Netherlands 15 years of experience, Ph.D. thesis, Wagening Univeristy, Lelystad.
15. Ozimek, T., Gulati, R.D. & van Donk V.E. (1990) Can macrophytes be useful in biomanipulation of lakes? The lake Zwemlust example, *Hydrobiologia*, **200/201**, 399-406.
16. Hamilton, D.P. & Mitchell, S.F. (1996) An empirical model for sediment resuspension in shallow lakes, *Hydrobiologia*, **317**, 209-220
17. Timms R.M. & Moss, B. (1984) Prevention of growth of potentially dense phytoplankton by zooplankton grazing, in the presence of zooplanktivorous fish, in a shallow wetland ecosystem, *Limnol. and Oceanogr.* **29(3)**, 472-48.
18. Beklioglu, M. & B. Moss, 1996. Mesocosm experiments on the interaction of sediment influence, fish predation and aquatic plants with the structure of phytoplankton and zooplankton communities. *Freshwater Biol.* **36**, 315-325.
19. Grimm, M.P. (1989) Northern pike (*Esox lucius* L.) and aquatic vegetation tools in the management of fisheries and water quality in shallow waters, *Hydrobiol. Bull.* **23 (1)**, 59-67.
20. Persson, L. (1993) Predator-mediated competition in prey refuges: the importance of habitat dependent prey resources, *Oikos*, **68**, 12-22.
21. Weisner, J., Eriksson, G., Graneli, W. & Leonardson, L. (1994) Influence of macrophyte on nitrate removal in wetlands, *Ambio* **23**, 363-366.
22. Wium-Anderson, S. (1987) Allelopathy among aquatic plants. *Arch. Hydrobiol. Beih.*, **27**, 167-172.
23. Beklioğlu, M, Burnak, L. ve İnce, Ö., 2000, Benthi-planktivorous fish-induced low water quality of Lake Eymir before biomanipulation, *Tr. J. of Zoology*, **24(3)**, 315-326.

24. Altınbilek, D., Usul, N., Yazıcıoğlu, H., Kutoğlu, Y., Merzi, N., Göğüş, M. Doyuran, V. & Günyaktı, A. (1995) Gölbaşı-Mogan-Eymir gölleri için su kaynakları ve çevre yönetim planı projesi, Mogan ve eymir gölleri 1. Çevre Kurultayı, 13-21 (in Turkish).
25. Geldiay, R. (1949) Çubuk Baraji ve Emir Gölünün Makro ve Mikro Faunasının Mukayeseli İncelenmesi, *Ankara Üniversitesi, Fen Fakultesi Mecmuası* **2**, 146-252 (in Turkish).
26. Diker, Z. (1992) A Hydrobiological and ecological study in Lake Eymir, MSc thesis, METU, Ankara
27. Scheffer, M. (1997) Ecology of Shallow Lakes, Chapman & Hall, London. Breukelaar, W.A., Lammens, H.R.R.E., Breteler, J.G.P.K. & Tatrai, I. (1994) Effects of benthivorous bream (*Abramis* brama) and carp (Cyprinus carpio) on sediment resuspension and concentrations of nutrients and chlorophyll-*a, Freshwater Biol.* **32**, 113-121.
28. Lauridsen, T.L., Jeppesen, E. & Andersen, F. Ø. (1993) Colonisation of submerged macrophytes in shallow fish manipulated Lake Væng: Impact of sediment composition and waterfowl grazing, *Aquat. Bot.* **46**, 1-15.
29. Crivelli, A.J. (1983) The destruction of aquatic vegetation by carp, *Hydrobiologia* **106**, 37-41.

COMPUTED MEAN MONTHLY WATER BALANCE OF A LARGE LAKE: THE CASE OF LAKE TANGANYIKA

Laurent BERGONZINI
UMR 8616 "ORSAYTERRE", Université Paris-Sud
Bât 504, F-91405 Orsay (France)

1. Introduction

Freshwater is, and has always been, a primordial and strategic resource. Consequently, the large lakes of the East African Rift, which represent one of the most important surface fresh water reserves of the world, are a major wealth for the adjoining countries. Moreover, these large lakes constitute the water supply for some of the major African rivers, the Congo River, the Nile River and the Zambezi River. They also are economically important for the nutrition and the development of the lakeside populations. Beyond these functions, they constitute particular hydrological objects, which have a climatic memory making them topics for scientific research unique throughout the world. Furthermore, the climatic dependence of the region of the African lakes from the atmospheric circulation registered above oceans has been recently evidenced. This actualizes the analysis of the lake balances, considering that numerous studies confirm that the African lakes underwent important lake level oscillations during the Quaternary. However, despite their economic and scientific relevance, the hydrological balance of the Lake Tanganyika water masses has been poorly investigated.

Consequently, different reasons justify the study of the water balance of Lake Tanganyika, as well as the factors responsible of their behavior. The present study intends to investigate at a monthly scale the average water budget of Lake Tanganyika and to evaluate in a qualitative and quantitative way the role of the different components that influence the variability. Whereas the hydrology of the other great Central African lakes such as Lake Victoria, Malawi or others has received considerable attention [1][2][3][4][5], the water budget of Lake Tanganyika has been neglected. Indeed, information concerning the water balance of Lake Tanganyika is very scarce. Data on the water budget given in recent publications [6][5][7] are rough estimates based mainly on the work of Theeuws [8] and Gillman [9]. The specific work of Devroey [10], Bultot [11] and Hecky and Degens [12] are also often cited. One of the specificities of Lake Tanganyika is that - as it is the case for many large lakes - it is shared between different countries: Congo, Burundi, Tanzania and Zambia, and that its

J. Klerkx and B. Imanackunov (eds.), Lake Issyk-Kul: Its Natural Environment, 217–244.

catchment lies in Congo, Burundi, Tanzania, Zambia and Rwanda. This specificity induces supplementary difficulties for the establishment of the water budget. This study is based on a database assembling the monthly and annual scale information on pluviometry, water level variations, drainage and output associated with the lake and its catchment. This database results from the important effort of compiling dispersed information. In this note only monthly and yearly long term means values have been used, the inter-annual variability of the various components have not been approached.
Lake Tanganyika (3°20'-8°45'S; 29°00'-31°20'E; Fig. 1), the Earth's second deepest lake, covers 32 600 km^2 at about 773 m altitude. It has a maximum depth of 1470 m and a water volume of 18 800 km^3. Its catchment covers 197 400 km^2 [11], from 1° to 10°S and from 29° to 34°E, and includes the drainage area of Lake Kivu to which it is connected through the Ruzizi River. It presently has a small outflow, the Lukuga River, which flows to the Congo River basin.

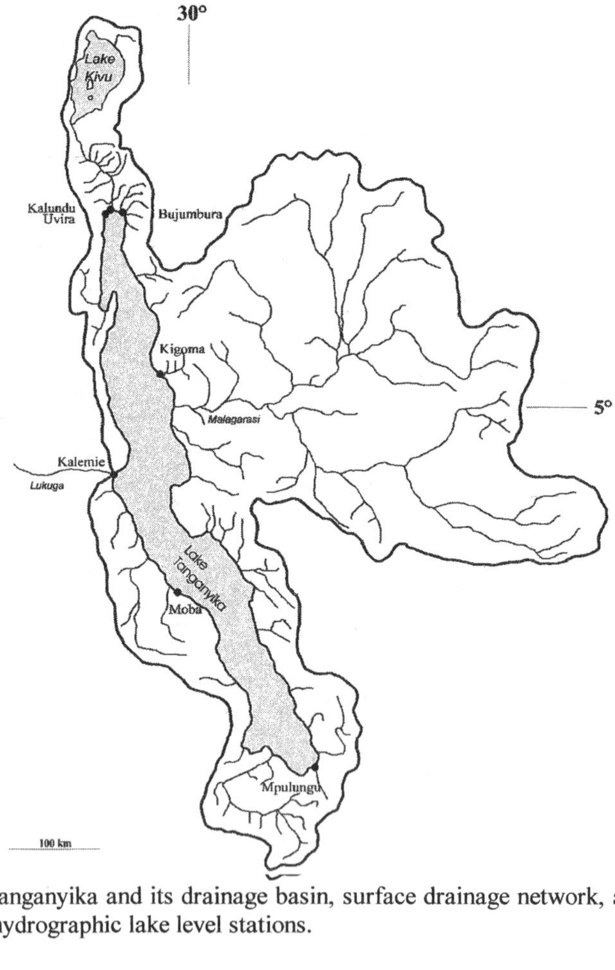

Figure 1. Lake Tanganyika and its drainage basin, surface drainage network, and localization of hydrometric and hydrographic lake level stations.

2. Control of the water budget of Lake Tanganyika

The water volume of the lake is not constant. However, according to the law of conservation of matter, there is equilibrium between inflow components, outflow components and the change of water volume for each chosen interval of time. The water budget of a lake reservoir as Lake Tanganyika can be expressed and synthesized by equation (1) which relates the stock variations with the other components of the balance for the considered time scale (in our case, monthly or annual).

$$\Delta S = P + Qi + E + Qo \qquad (1)$$

where ΔS is the variation of the water storage, P the precipitation over the lake, Qi the inflow from the drainage basin, E the loss by evaporation and Qo the outflow through the outlet. This equation, which connects the above components, is applicable only if the input and output components are correctly evaluated. In the practice there are errors involved in the estimates of the various components [13]. Indeed, for Lake Tanganyika the groundwater contribution and losses are small enough not to be considered with respect to the other components. In fact, as it will be discussed below, the groundwater contribution (cf. section 7) is integrated in the component Qi of the equation (1), which corresponds to the total input from the drainage basin.

Usually, in a lake balance study, one of the components, often the evaporation or the input from the drainage, is assimilated to an unknown parameter, and is determined with values of the other component by the equation (1). In the developed approach, all the components are estimated in an independent way, and the equation (1) is used only to test the accuracy of the different terms. For commodity reasons, as it will be shown in section 3, the values of the different components can be expressed in $mm.month^{-1}$ and in $mm.year^{-1}$ over the Lake Tanganyika surface (32 600 km^2). This choice is justified in order to easily compare the different estimations.

The next part will focus on the estimation of terms of the equation (1), the stock variations, precipitation input, evaporation output, outflow via the Lukuga outlet and then on inflows from the drainage basin. Each of these components will be quantitatively approached under mean conditions. Lastly, the accuracy of the equation (1) will be discussed and tested.

3. Monthly and annual mean stock variations

3.1 PRELIMINARY REMARKS

The accumulation term of the water balance (equation 1), ΔS, represents the change of the storage or water volume in the lake during a defined period. More often, the stock variation is estimated by a single hydrographic record. This section firstly identifies and illustrates some problems, which are linked to this component for a large lake system. Indeed, difficulties associated to the stock variation evaluations are often underestimated. However, remarks will be rapidly presented and illustrated. At least, these steps will inquire about the limits of this approach.

Firstly, for the determination of the various accumulation factors, one needs data on the morphometric characteristics to determine the relation between volumes and lake levels of the water body. The stock estimations between two lake levels have to take into consideration the basin morphology. For the tectonic Lake Tanganyika and in regard to the surface of the lake it has been demonstrated [14] that it is possible to consider a constant lake surface for the historic level fluctuations. This hypothesis leads to an error in the water storage of less than 1%. In fact, a constant surface of the lake associated to a level variation of 1 m induces an error, expressed in mm over the lake, of less than 5 mm.

Secondly, the problems of the evaluation of the stock variations are related to the chosen study period. When the estimation of the stock variation of a lake is based on daily lake level records, additional errors and inaccuracies appear if the period of investigation becomes smaller. For this reason, it is better if the studied time scale is at least equal to the one of the lake level records. Furthermore, it can be noticed that this limit is not specific to the stock variations but should be taken into account for all the different water budget components. The above comment justifies the chosen monthly and yearly time scales, presented in this note.

Lastly, to be representative for the whole lake surface a particular lake level record has to be considered in function of climatic forcing. In fact, numerous limnimetric registrations are available all around Lake Tanganyika: at Mpulungu, Moba, Kalemie, Kigoma, Bujumbura, Kalundu and Uvira (Fig. 1). The comparison of the monthly lake levels at the surrounding shores of the lake has shown significant differences between the data obtained at the different hydrometric or hydrographic stations [15]. The existence of a seasonal evolution of the relative monthly level of the lake surface is shown to be related to climatic forcing. Under influence of the South-North, Inter-Tropical Convergence Zone (ITCZ), migration and general climatic circulation over the region [16], an annual periodicity in the differences of the monthly lake levels is registered between Bujumbura and Mpulungu, which are respectively the northern and southern extremities of the lake. It appears that during the months of June, July and August, which correspond to the period of decreasing level, the relative level at Mpulungu is inferior by about 10 cm to that of Bujumbura. During September, October and November - the low level period -, this difference decreases, and is opposite during December, January, February and March - the rising level period -, when the relative lake level at Mpulungu is 10 cm higher than the one of Bujumbura. This tilting in the lake water level topography is clearly linked to the mean wind directions and intensities in the region [17], showing the importance of the local and regional surface forces on the water table topographies and on the lake level records. In fact, from March 1958 to February 1959, the monthly relative lake level records at Mpulungu, Moba, Kalemie, Kigoma, Bujumbura and Uvira stations, show the same seasonal tendencies, with a precision of the cm, and the same amplitude in the annual variation, with a precision of a few cm. But the differences between punctual monthly lake levels records can differ a few centimeters and are maximal in June when the amplitude of variations between the stations reaches 10 cm. This deviation appears to be smaller during high and low lake level periods, of the order of 2 to 3 cm, and more important, during the decreasing but also during the rising lake level periods [15]. This difference is minimal during the change of seasons and maximal during the windy period which also corresponds to the dry season. In addition, the lake topography tilting is in good agreement with the water circulation knowledge [6][7][18]. Both the identification and the quantification of this phenomenon, as well as its variability from one year to another inform about the stock variation estimation accuracy, and also about water circulations and other specific chemical, sedimentological and biological characteristics of the lake. It must be kept in mind that, for several annual periods, no significant changes appear between the record of the seven stations.

3.2 STOCK VARIATIONS

For the determinination of the monthly mean storage profile, the monthly limnimetric record of the Kalemie station, from 1932 to 1995, had been used. In fact, the monthly lake level record at Kalemie is available only from 1932 to 1959, but had been reconstructed for the remaining period, from the Bujumbura record that started in 1948 [19][20][21]. The choice of Kalemie as the reference station is justified by its location at the outlet of the lake, and this record will be necessary to estimate the outflow during the studied same period. On the contrary, the inconvenience of this choice is that the recorded level fluctuations of the Kalemie station are higher at the beginning of the decrease as well as of the rising periods compared with the other available records. This specificity is induced by the lake level regulation capacity of the outlet. It is thus possible that the reconstructed profile of the lake storage deduced from Kalemie records presents the same shift compared to the whole lake.

To determine the average monthly storage variations, the monthly level records from 1932 to 1995 had been used for initially calculating a mean monthly level profile (Fig. 2). Secondly, this mean monthly profile had been transformed in average monthly storage values. The applied method [14] transforms a mean monthly level record in a mean daily record, in order to obtain information on the lake level at the beginning of each month. The method has been tested on the different stations for the period 1950-1959 (period with published daily data). It has been shown that the method induces mean punctual gaps of less than 1 cm, which correspond to the precision of measurement. It is consequently necessary to keep in mind the above mentioned difficulties when considering the estimation of the stock variations.

The Lake Tanganyika level record shows high levels in May and low levels in October (Fig. 2). The stock variations are thus positive during six months, from November to April - corresponding almost to the beginning and to the end of the rainy period -, and then negative for the other part of the year which almost correspond to the dry season. The yearly amplitude of variation of the daily lake level record is around 80 cm [10][11]. As it has been noticed, this can be associated to stock variations between daily high and lows levels of about 26 km^3, that is 80 cm over the lake surface. Between monthly mean levels this annual amplitude of storage variation is estimated close to 70

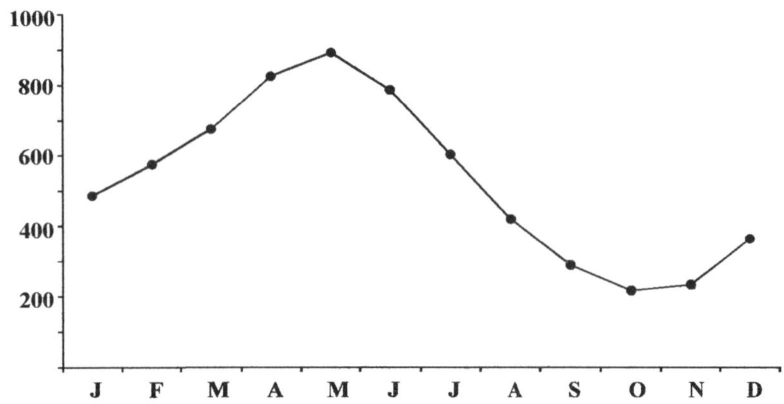

Figure 2. Mean monthly Lake Tanganyika level profiles, reconstructed for the Kalemie station from the period 1993-1995, expressed in mm with 0 = 774 m.

cm. Since 1932, the lake level presents, as many open lakes, a relatively low amplitude of variation, with less than 4 m between the low level of 1949 and the high level of 1964. This evolution of the level is in agreement with the other intertropical large African lakes of the region [22][14][23].

TABLE 1. Estimated average monthly and annual variation of the storage of water in Lake Tanganyika, during the period 1932-1995, expressed in mm over the lake surface.

	Jan	Feb	Mar	Apr	May	Jun	Jul	Aug	Sep	Oct	Nov	Dec	Ann
ΔSo	91	86	132	134	-22	-161	-193	-162	-102	-38	86	132	-17

The evaluated average storage values, established for the period 1932-1995, have been transformed and are expressed in mm with respect to the lake surface (Table 1 and Fig. 3). The average monthly profile (Fig. 3) evidences a positive water budget from November to April, with two maximum storage values occurring in December and in March-April, associated with a relative diminution of the storage rate centered in February. Negative storage values occur from May to October, with maximum losses in July. The maximum water storage change occurs in July during the lake level falling period. The standard deviation of the average monthly storage profile is fixed at 127 mm and the mean value of the absolute storage is fixed at 112 mm. The average yearly storage variation is estimated at -17 mm, and is induced by the recorded differences of levels between 1932 and 1995. This proves that, at an inter-annual scale, the storage of water in Lake Tanganyika, as in other open lakes, can be considered near or equal to zero.

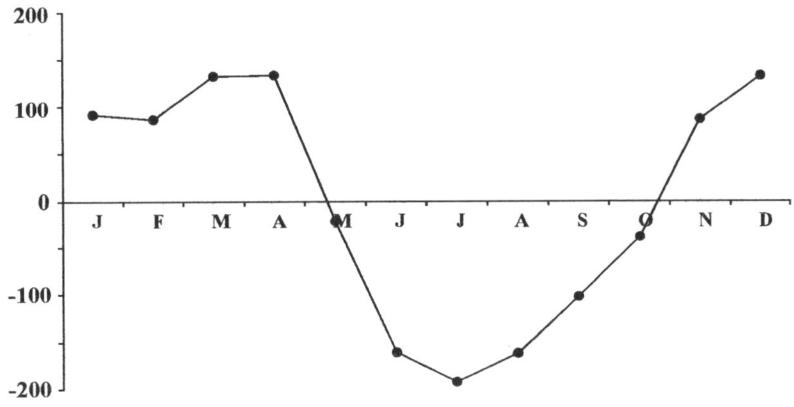

Figure 3. Mean monthly water storage profile of Lake Tanganyika obtained from Kalemie mean monthly lake level profiles (Fig. 2), expressed in mm over the lake surface (32 600 km^2).

4. Monthly and annual mean precipitation rates over the lake and its catchment

4.1 INTRODUCTION

The first interest of this section is to estimate and quantify the monthly mean precipitation rates over the lake surface, but also the precipitation over the drainage has been investigated. After a rapid presentation of the general atmospheric circulation and the pluviometric patterns, the origin of the used data is listed. Next, the spatial and temporal variabilities are crudely approached and lastly the estimation of monthly amounts are proposed and commented in regard to the literature.

4.2 REGIONAL PLUVIOMETRIC LANDSCAPE

The pluviometric regime over East Africa is primarily controlled by the migration of the Inter-Tropical Convergence Zone (ITCZ) [27]. The migration of the ITCZ is associated with intense convergence activity, which usually leads to abundant heavy rainfall. The ITCZ lies in the southern part of the catchment during the boreal winter, and is located northwards of the lake during the summer. Most of the catchment area has a well defined rainy season, from October to May, and a dry season from June till September concomitant with a fairly constant southerly wind. The monthly rainfall amount varies considerably with the maximum rates in December and in March. The boreal spring rains are usually stronger, the autumn ones, more variable [24]. Rain generally falls in heavy thunderstorms and most of it occurs on the high escarpments surrounding the lake basin. The entire region is under the prevailing influence of easterly circulation bringing moisture from the south-western Indian Ocean during the boreal spring and from the equatorial Indian Ocean during the boreal autumn and winter. The ITCZ provides an important control of the seasonal variability of the rainfall. The interannual variability of the rainfall is due to the amount of boreal autumn, and seems to be controlled by sea-surface conditions in the tropical Pacific and Indian Oceans [25][26][16].

4.3 ORIGIN OF THE DATA

The monthly variability of the precipitation has been studied at a regional scale. Mean profiles of amount of rainfall amount from 152 meteorological stations were selected over and at the proximity of the Lake Tanganyika drainage basin (Fig. 4). This set is the result of the compilation of data originating from diverse publications [28][29][30][31][32][33] and from the communications of many authorized sources (Zambian Meteorological Departments, Tanzanian MAJI service, P. Camberlin, Y. Richard, D. Plisnier, and G. Mahé). Most records last over at least 30 years. But, in order to obtain a good picture of the amount of the spatial rain variability, a few stations, which present less than 30 years of records were also selected. These last profiles, principally located in the Congo sub-basin, have been considered only if no information was available in the surrounding area. The density of the network is not uniform, most of the stations are located in the northern part of the basin, in Burundi and in Rwanda. The spatial density variability of these stations appears clearly linked with the human occupation of the region.

4.4 SPATIAL AND TEMPORAL DISTRIBUTION OF THE PRECIPITATION

In figure 4, the total annual precipitation is presented. Values have been used to define 5 classes with a gap of 250 $mm.yr^{-1}$. The global image shows a general NW-SE gradient with lower pluviometric rates in the southern and eastern parts of the basin (under 1250 $mm.yr^{-1}$), whereas the important yearly precipitation (over 1500 $mm.yr^{-1}$) occurs in the North-West. It can be noticed that the high precipitation rates often correspond to high relief zones: Nyiragongo, Karisimbi and Mitumba mounts. The monthly analysis (not shown here) shows a minimum of precipitation in July, but the months of June, August, and September are also characterized by a low rate. It also appears that the rainy season is more persistent in the northern part of the drainage, whereas it is more concentrated in southern part. The rainy season starts then preferentially in September-October in the northern-eastern part of the drainage which corresponds to high relief of the Rwanda, Burundi and Congo-Zaïre, whereas precipitation is lower and start preferentially in October-November in the Western basin corresponding to Tanzania. During January and February the precipitation are more important in the southern part of the basin (150-300 $mm.month^{-1}$) even if high monthly rate (100-200 $mm.month^{-1}$) still occurs in the North. At the end of the rainy season, in April-May, precipitation presents the particularity to be persistent on the bordering of the lake. It can be noticed that the scattering of the allowed profiles presents a great density variability over the catchment. Although, the region with the least quantity of information, corresponding to the eastern Malagarasi and Ugalla drainage (Tanzania high-tableland), represents as well the area with a most homogeneous relief and rainfall amounts distribution. Alternatively, the northern basin, which is characterized by important escarpments and important precipitation, corresponds to a well-informed basin with many observations.

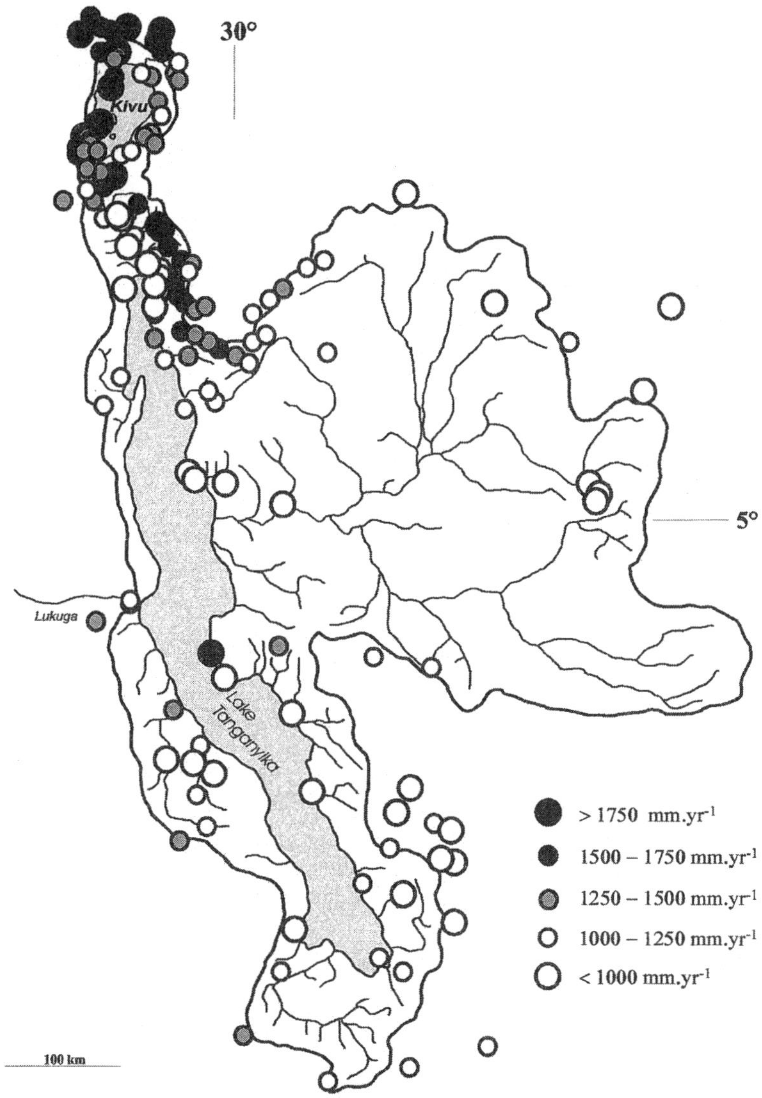

Figure 4. Distribution of the annual precipitation rates recorded at the selected stations located on or at proximity of the Lake Tanganyika drainage basin.

An analysis of the main component parameters of the monthly means identifies rainfall regimes, which are characterized by their respective importance and specificities. Four "hydro-climatic" sub-types have been determined:
(1) Equatorial areas, which have a regular high rainfall, distributed all over the year, with two maximums in April and October-December (e.g. the northern part of Zairian ranges and Nyiragongo relief).
(2) Sub-equatorial regions with two rainy seasons separated by two dry seasons (e.g. the eastern part of Lake Kivu and the southern part of Lake Victoria drainage areas).
(3) Tropical sectors with a rainy season centered in January and February (e.g. the southern part of the Lake Tanganyika catchment).
(4) Areas with relative persistent important precipitation at the end of the rainy season in May, linked to the nearby lake and by the intensity of southern trade winds and to orographic processes (e.g. the lake bordering stations).

4.5 ESTIMATED AVERAGE MONTHLY RAINFALL RATES OVER THE LAKE AND ITS CATCHMENT

The statistical spatial integration and interpolation of the records allow to quantify the variability of precipitation, both on the lake and over the drainage basin. The average monthly values have been summarized (Table 2 and Fig. 5). As no direct measurement is available, the precipitation feature over the lake itself has been gathered mainly from the shoreline station records. This approach presents drawbacks, because differences in amounts of precipitation can exist between the lake and the surrounding land surfaces. Large lakes can greatly influence the low atmospheric phenomena as well as local circulation between the lake and the shorelines [27]. Therefore it must be kept in mind that the estimated rainfalls over the lake include some uncertainties.
The monthly profiles for both lake and drainage surfaces, show a dry season between May-June and October and a pluvial rainy season between November and April, with two maximums: one in December (maximum monthly precipitation rate), one in March, and a relatively low amount in February. A similarity clearly appears between these two profiles and the storage profile (Fig. 3), confirming the strong influence of the rainfall on the lake balances. The yearly total precipitation over the lake has been estimated, under mean conditions, at about 1090 $mm.yr^{-1}$. This estimation is in agreement with other existent values (900 $mm.yr^{-1}$ [9]; 1000 $mm.yr^{-1}$ [34]; 1050 $mm.yr^{-1}$ [7]. Over the drainage basin, annual precipitation has been estimated at about 1035 $mm.yr^{-1}$ of the same order of magnitude as the Bultot reports (1000 $mm.yr^{-1}$ 1993) and higher than those of Gillman (850 $mm.yr^{-1}$ 1933). Therefore, the precipitation rate appears within the same order in both lake and drainage surfaces, even if higher rates seem to prevail over the lake at the beginning of the rainy season. The monthly scattering of the amount appears to be in agreement with other estimates obtained with less meteorological stations and with the Theisen approach [21].

TABLE 2. Estimated monthly precipitations over the Lake Tanganyika surface (Pl), and over its drainage basin (Pb), expressed in mm.

	Jan	Feb	Mar	Apr	Mai	Jun	Jul	Au	Sep	Oct	Nov	Dec	Ann
Pl	149	139	167	157	66	5	2	3	19	49	142	192	1090
Pb	152	142	168	155	49	5	1	4	15	41	124	179	1035

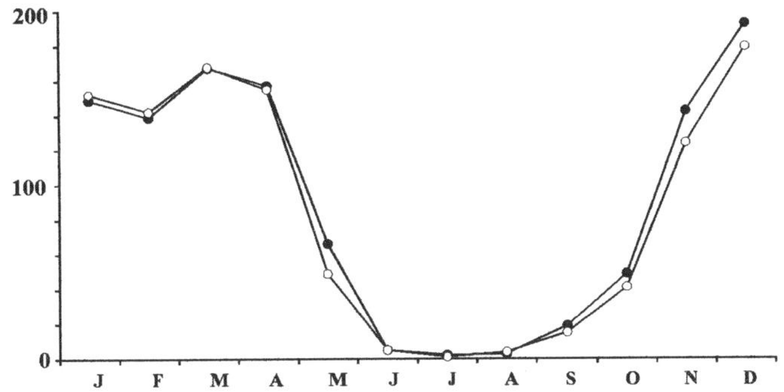

Figure 5. Mean monthly profiles of precipitation rates over both the Lake Tanganyika surface (black circles) and its catchment (open circles), expressed in mm.month^{-1}.

5. Monthly and annual mean evaporation rates from the lake surface

A previous estimation of annual evaporation at the lake surface of about 1350 mm.yr^{-1} has been obtained from the hydrological balance by Theeuws [8]. This value was used by many authors from Gillman [9] to Crul [5]. As in this approach the evaporation corresponds to the flowing deficit, it cannot be considered as an independent estimation. Furthermore this value had been obtained with a partial knowledge of the other water budget components. More recently, the energy budget method applied to the Lake Tanganyika water body has allowed to estimate the monthly and yearly mean evaporation rates over the lake surface. The annual mean value that has been then fixed at around 1696 mm.yr^{-1} [11], has been confirmed in a recent study [34]. Furthermore, the water budget equation and the lake level records developed at different time scales confirm the monthly values obtained by Bultot [11][34]. Furthermore this value appears more realistic in comparison with (1) others estimations of the Lake Tanganyika annual evaporation rate [17][35] and (2) assumed values for other East African Lakes [36][1][37]. It must also be noted, that other evaporation rates from shore lake stations had been estimated, from at least 10 years of records and with a penman approach, at 1749 mm.yr^{-1} and 1718 mm.yr^{-1} respectively at Kasanga and Kigoma [30][38]. These values are a little higher by a few mm than the Bultot ones, but present almost the same monthly profile. From the above considerations, it is considered that the Bultot's values

(Table 3), which were obtained on the basis of the Belgian data set established in the '50 and '60, can be considered as a good estimation of the monthly evaporation rates. Indeed, no changes in the climatic feature, which can induce an important shift of the evaporation from the lake surface, had been detected, and the low magnitude of variations are usually attributed to annual values of this component.
The monthly profile (Fig. 6) presents a maximum in September at the end of the dry period and a minimum in February when humidity and cloud cover are important. Even if the average monthly profile looks relatively smooth, the monthly evaporation rates present an annual cycle. The evaporation is relatively low and constant during the rainy season and increases during the dry season in order to reach its maximum value at the end of the latter, and then falls quickly with the return of the rains and associated cloud cover. This cycle shows the "direct" relation between the sun radiation and evaporation and also the "indirect" relation with the rain occurrences. The mean monthly value of evaporation is estimated at 141 mm.month^{-1}, with standard deviations of 12 mm. As the monthly values are associated to this low standard deviation, even if it appears to be an important component of the yearly hydrological budget, the evaporation cannot be considered as responsible for the intra-annual variability of the water budget.

TABLE 3. Estimated average monthly evaporation rate from the Lake Tanganyika surface, expressed in mm, from Bultot 1965 & 1993.

	Jan	Feb	Mar	Apr	Mai	Jun	Jul	Au	Sep	Oct	Nov	Dec	Ann
E	132	124	138	130	149	147	144	147	167	153	135	130	1696

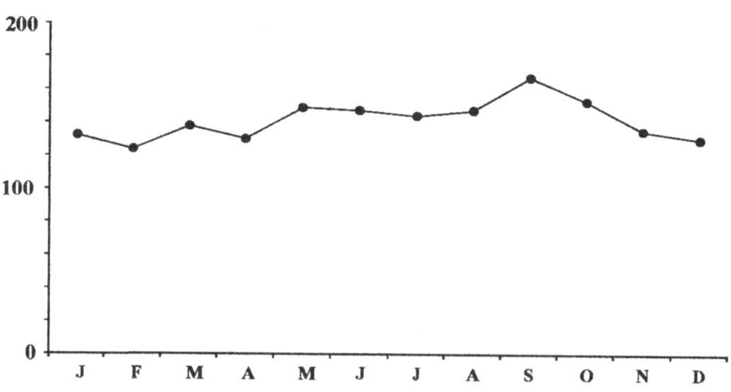

Figure 6. Mean monthly profile of evaporation rate from Lake Tanganyika surface, expressed in mm (from Bultot 1965, 1993).

6. Monthly and annual mean outflow via the Lukuga outlet

In a lake balance equation, the surface outflow can be equal to zero for a "closed" lake basin. For an "open" lake basin, the surface outflow is usually directed into a single outlet channel. In the case of Lake Tanganyika this surface outlet corresponds to the Lukuga river which ultimately joins the Atlantic Ocean via the Congo river. The lake outflow can be reconstructed from direct measurements if there is a hydrological station at the outlet. The outflow can then be reconstructed from a record of the lake levels and associated gauging curves, what has been done for the Kalemie station based on 162 gauging measurements, made between 1931 and 1995 and a lake level record [21].

TABLE 4. Estimated average monthly and annual outflow from Lake Tanganyika for the period 1932-1995, expressed in mm over the lake surface.

	Jan	Feb	Mar	Apr	Mai	Jun	Jul	Au	Sep	Oct	Nov	Dec	Ann
Qo	30	29	35	37	40	36	33	28	24	24	23	27	366

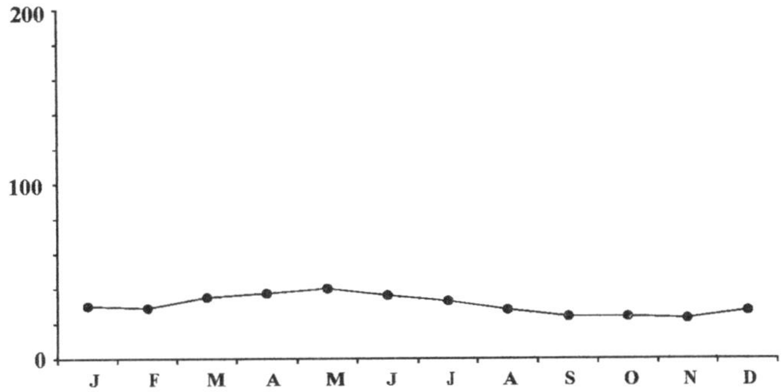

Figure 7. Mean monthly outflow profile of the Lukuga at Kalemie, established for the period 1932-1995, expressed in mm over the lake surface (32 600 km^2).

The integrated data allow to estimate the monthly mean outflow rate for the period 1932-95. The reconstructed values of the mean monthly and mean annual outflow, expressed in mm over the lake surface, are presented in table 4 and figure 7. The Lukuga regime appears pondered as shown by its average monthly value of 31 $mm.month^{-1}$ (380 $m^3.s^{-1}$) and by its standard deviation, around 6 mm. The Lukuga presents a maximum flow in May, corresponding to the highest monthly lake level and a minimum in November during low lake level period. The differences between the lake level profiles (Fig. 2) and the outflow profiles (Fig. 7), appearing as a decrease of the outflow rate whereas the average lake level grows between January and February, are induced by the considered number of days for the respective months. The mean outflow estimated at around 380 $m^3.s^{-1}$ for the period 1932-95, corresponds to a water height, reported at the lake surface, of 366 $mm.yr^{-1}$ It can be noticed that the evacuation capacity of the outlet has changed since the beginning of the century[21]. Since 1965, the outflow of Lake Tanganyika through the Lukuga river tributary into the Congo river system has a tendency to increase, whereas the general outflow of the Congo system shows a tendency to decrease [39]. The specific outflow rate of almost 1,6 $l.s^{-1}.km^{-2}$, seems to be very small due to intense evaporation at the lake surface, but appears higher than other estimations [10][11][40][6][7]. This difference is attributed to the considered period, and is due to the increase of the outflow since the beginning of the 60's. It appears that yearly outflow values present a large standard deviation induced by the lake level relative inertia and both (1) the linked inter-annual tendencies, and (2) the relatively low capacity of the outlet. But, by the low standard variations of the monthly means of this component, as for evaporation, the surface losses cannot be responsible for the intra-annual variations of the stock, which attests the influence of the input in the monthly stock profile.

7. Monthly and annual mean runoff from the drainage basin

7.1 INTRODUCTION

Rivers, which drain the lake catchment, are of different types. The Malagarasi River presents a relatively low stream, with many meanders and swampy areas. The Ruzizi River, the Lake Kivu outflow, has an upper torrential stream and a calmer lower stream in the alluvial plain. The Kalambo River, a less important river, falls for 200 meters to join the lake after a calmer stream on the high-plains. These rivers and torrents are characterized by important variations of their flowing regime, and can present extreme variations, which induce the formation of important alluvial fans. The flow regimes are clearly induced by the seasonal distribution of the precipitation.

To determine the runoff into a large lake that has many tributaries, it is necessary to estimate the contribution of the ungauged or unstudied areas. In this approach, the contribution of gauged/studied stations is first analyzed. Secondly, the contribution of ungauged areas and of the total inflow from the catchment are considered. In this section, the origin of the used data will be briefly presented, and then the monthly and annual mean inflow from diverse rivers of the lake drainage will be presented in a global view and synthesized to determine the total runoff inflow. Finally, the relation between the precipitation and runoff will be crudely approached.

7.2 ORIGIN OF THE DATA

Monthly and annual mean flows were compiled from various hydrological stations associated to rivers, that drain sub-basins of the lake catchment. The data originate from Rwanda, Burundi, Tanzania and Zambia, none has been found for the Congo adjacent drainage:

- Rwanda records originate from the Belgian and Rwanda hydrological yearbooks [19][41]
- Burundi records originate from the hydrological yearbooks of the IGEBU [42], from the publication of Pruvot [43], and from the personal communications of the IGEBU and of D. Plisnier.

- Tanzania records originate from the hydrological yearbooks [44][45] and from the NORAD publications [30][38], as well as from the personal communication of the Tanzanian Ministry.
- Zambia ones come from the Zambian Hydrological publications [46].

7.3 INFLOW FROM THE GAUGED SUB-BASIN

A database of monthly and annual mean runoff has been computed from 26 hydrological stations located upon the rivers that drain the basin. The location of the different stations is presented in figure 8. Some stations are located on the upstream and the downstream of the same river, which means that some chosen stations integrate each other. This had been taken into account to better characterize the surface behavior of the different sub-basins. Table 5 presents the monthly and annual mean flows of the 26 computed gauging stations, expressed in $mm.month^{-1}$ over the drained surface, the yearly flow expressed in $km^3.yr^{-1}$, as well as the associated basin surfaces and the recorded period used to establish the mean values. These gauging stations are associated with drainage basins of various surfaces (from 128 for the Mungonya River, up to 123 300 km^2 for the downstream of the Malagarasi River). Without taking into account the sub-basin that is integrated to another, the gauged surface represents 154 229 km^2 or 78% of the total lake catchment.
Although the recording periods are different (from 2 to 25 years), and not synchronized for the various stations, they can be considered in this approach as significant enough to qualify the behavior pattern of the considered rivers, as well as the surface characteristics. Finally, to quantify the inflow in mean conditions, it has been accepted that they can be used as an indicator of the hydraulic pattern of the various sub-basins. From table 5, it appears that the lake is supplied by three major rivers: the Ruzizi River which originates at the Lake Kivu (5,3 $km^3.yr^{-1}$), the Malagarasi River, which drains the Tanzanian highlands (4,3 $km^3.yr^{-1}$), and the Lufubu River which drains the southern extremity (2,4 $km^3.yr^{-1}$). A north-south gradient seems to characterize the monthly profiles (Table 5). Whereas the hydraulicity is minimum in the eastern part of the catchment, a north-south gradient appears along the surrounding area of the lake, with higher flowing height in the North than in the South. Maximum monthly flowing height values appear during April and sometimes in May in the northern part of the catchment, at the opposite, in the southern part, the maximum occurs in March. The lowest monthly flowing height occurs in August-September in the North and in September-October in the South. This behaviour is clearly determined by the spatial and temporal distribution of the precipitation (cf. section 3), and proves the qualitative coherence of the above data set. The annual specific flow fluctuates between 0,7 $l.s^{-1}.km^{-2}$ to 24,6 $l.s^{-1}.km^{-2}$ (or from 19 $mm.yr^{-1}$ to 768 $mm.yr^{-1}$ reported over drained surfaces) showing the great heterogeneity of the surface conditions. For the total of the gauged surface (154 229 km^2), the mean annual specific flow is estimated at about 3,3 $l.s^{-1}.km^{-2}$ (or 104 $mm.yr^{-1}$ over the catchment). Considering the gauged surface without the Malagarasi, which represents 123 300 km^2 (or 62% of the total drainage area), the mean annual specific flow is estimated at 12,1 $l.s^{-1}.km^{-2}$ (or 382 $mm.yr^{-1}$ over the residual surface), proving

the particularity of the Malagarasi basin characterized by a low specific flow. This analysis shows higher mean specific flow in the Burundi Mountains (around 20 $l.s^{-1}.km^{-2}$ or 600-650 $mm.yr^{-1}$), and also higher values for the southern escarpment surrounding the lake (around 10 $l.s^{-1}.km^{-2}$ or 300-350 $mm.yr^{-1}$).

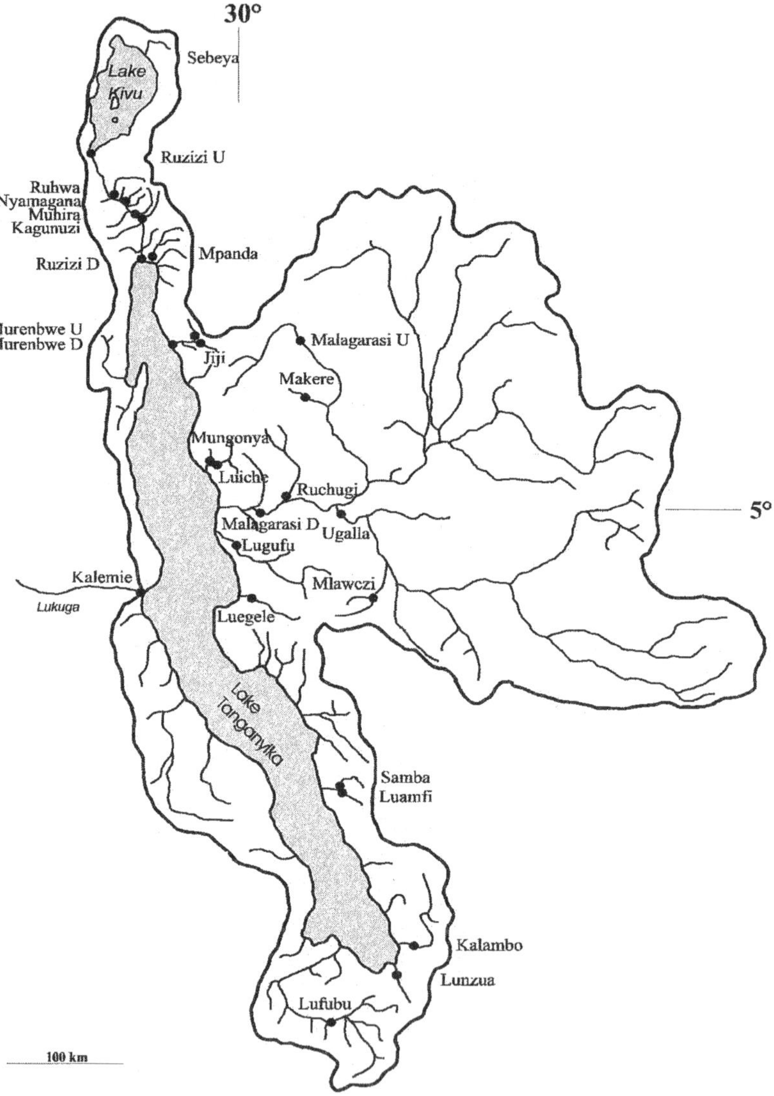

Figure 8. Localization of the selected different hydrometric and hydrographic stations located on the water surface network draining Lake Tanganyika catchment.

TABLE 5. Flow characteristic of considered stations (river name, station mane, drained surface in km^2, period of record, monthly flow values expressed in $mm.month^{-1}$ above the drained surface, annual flow values in $km^3.yr^{-1}$).

River / Station	Surf	Period	Jan	Feb	Mar	Apr	May	Jun	Jul	Au	Sep	Oct	Nov	Dec	An.
Sebeya / Gisenyi	320	1951-74	15	18	22	29	28	16	14	11	15	15	19	19	0,1
Ruzizi U/ Bukavu	7400	1951-73	30	31	31	33	36	35	32	29	28	28	29	31	2,8
Ruhwa / id	622	1983-84	54	53	62	65	47	36	33	30	33	51	55	53	0,4
Nyamagana / Murambi	231	1983-85	42	40	42	94	71	50	38	36	27	31	39	44	0,1
Muhira / Ruhagarika	243	1983-85	50	49	64	100	68	43	29	26	29	37	53	49	0,1
Kagunuzi / Ndava	432	*	59	54	52	68	67	49	33	22	27	28	50	33	0,3
Mpanba / Gatura	255	1984-85	45	63	66	120	66	43	37	33	36	46	47	46	0,2
Ruzizi D/ Gatumba	±14000	1981-89	31	32	34	41	41	33	27	23	22	24	28	30	5,2
Murenbwe U/ Rumeza	299	1984-85	49	53	75	137	85	48	36	27	20	32	42	52	0,2
Murenbwe D/ Mutambara	940	1981-89	69	82	78	93	79	40	29	22	20	24	40	57	0,6
Jiji / Ndago	197	1984-85	71	72	103	182	92	42	28	20	16	26	49	64	0,2
Malagarasi U/ Taragi	8525	1970-88	22	27	27	40	48	22	11	7	5	5	8	18	2,1
Makere / Makere	783	1964-88	17	20	23	30	24	10	6	5	4	4	7	15	0,1
Mungonya / Mungonya	128	1973-88	32	24	49	81	49	22	14	8	8	8	12	61	0,1
Luiche / Simbo	1626	1974-88	40	40	54	74	45	21	16	13	11	11	14	28	0,7
Ruchugi / Pwaga	2989	1974-88	18	18	22	29	19	19	6	5	5	5	7	16	0,5
Malagarasi D/ Mbelagule	123300	1975-88	2	3	4	5	6	5	4	2	1	1	1	2	4,3
Lugufu / Katyentye	4196	1976-88	25	25	54	67	26	9	6	3	2	2	9	17	1,0
Ugalla / Sindi	51775	1976-88	1	1	2	3	4	3	2	1	1	0	0	1	1,0
Mlawezi / Uvinza	951	1975-88	21	30	35	61	36	16	11	8	6	5	7	18	0,3
Luegele / Lubalisi	1304	1975-88	24	30	40	42	26	13	8	5	4	3	6	19	0,3
Samba / Masolo	353	1975-88	15	32	35	35	18	10	7	4	3	1	3	9	0,1
Luamfi / Masolo	531	1975-88	24	44	51	49	24	8	5	3	3	2	16	21	0,1
Kalambo / Kapozwa	3126	1975-88	22	46	88	83	44	17	8	5	3	2	2	8	1,0
Lunzua / Lunzua	386	1957-66	19	26	81	78	52	38	30	24	19	6	6	13	0,2
Lufubu / Kezo	4090 ?	1957-66	76	82	140	108	48	30	21	16	12	10	12	29	2,4

* 1971-1973 & 1983-85

These areas correspond to many short seasonal torrents coming from the border of the graben. The lowest recorded specific flows are associated with the eastern part of the drainage surface (around 1 $l.s^{-1}.km^{-2}$or 20-40 $mm.yr^{-1}$) which corresponds to the highland drained by the Malagarasi and Ugalla Rivers. These low values are explained by the relative low amount of precipitation, by the low relief of the area, and especially by the presence of many swamps and humid areas over these catchments. These observations are confirmed by a principal component analysis (PCA) statistical approach on the mean monthly flow profiles. The first eigenvector of this analyse opposes then sub-catchments located in the Rwanda and Burundi mountains and sub-catchments, which drain the Tanzanian highland. This dominant eigenvector is highly correlated with the total annual specific runoff. This disparity in the specific flows, associated with the different regions of the total catchment, can be related to the precipitation and relief variability, but underlines the geographic coherence of the selected data. Thus, regions immediately surrounding the lake have a major contribution as already suggested by Bultot [11]. The comparison of the collected data (Table 5) with the published estimates [11] confirms the spatial organization of the considered area, even though all the flows from the different basins are underestimated in the Bultot estimation. This compilation allows to characterize the specific discharge of the different surfaces and to localize regions of important water contribution.

7.4 INFLOW FROM THE TOTAL LAKE CATCHMENT

In order to determine the total inflow into the lake, the different monthly flows of the 13 downstream stations have been summarized. Considering the above mentioned characteristics, the Malagarasi River which drains about 62% (123 300 km^2) of the total basin, had to be considered separately. For the remaining surface (74 100 km^2), the available information is associated with only 42% of this surface. In order to estimate the contribution of the un-gauged area (58% of this 74 100 km^2), a same flowing height between the gauged and un-gauged areas has been supposed. The runoff values of the gauged surface then have to be multiplied by 2,4. Alternatively, in order to close the annual yearly budget (considering P = 1090, E = 1696, Q0 = 274, ΔS = -17 and Qi-Malagarasi = 132, all this components expressed in $mm.yr^{-1}$ over the lake surface), it is necessary to consider a multiplication coefficient of 2,3 between the gauged part and the total of the considered 74 100 km^2. The established accuracy between the two coefficients 2,4 and 2,3 confirms the representativeness of the flow conditions obtained with the available data (Table 5). The coefficient of 2,3 between the inflow from the gauged reduced catchment (total catchment minus Malagarisi River and un-gauged ones) and the inflow from the total reduced catchment (total catchment minus Malagarasi River ones) is used further on. The total estimated inflow profile, from both the Malagarasi River and the other part of the drainage basin, reported in $mm.month^{-1}$ at the lake surface (Table 6 and Fig. 9), leads to an annual inflow from the drainage, over the lake, of about 955 $mm.yr^{-1}$. It must be noted that using the coefficient 2,4 between gauged and total inflow leads to an annual inflow value of 995 $mm.yr^{-1}$ corresponding, in regard to the above estimation, to a difference of +4%. The estimation of 955 $mm.yr^{-1}$ appears higher than the previous ones, available in the literature: 530 $mm.yr^{-1}$ [9], 784 $mm.yr^{-1}$ [47], 440 $mm.yr^{-1}$ [6], 560 $mm.yr^{-1}$ [7]. This value of 955 $mm.yr^{-1}$ (or 31 $km^3.yr^{-1}$) is associated to a mean specific flow of 4,7 $l.s^{-1}.km^{-2}$ for the entire catchment, and to a specific flow of 10,2 $l.s^{-1}.km^{-2}$ for the reduced ones (surfaces outside the Malagarasi sub-catchment). This above value appears appropriate compared with available values (Table 5), and does not overestimate the flow of the different parts of the drainage. The average monthly flow input is about 80 $mm.month^{-1}$, but is associated to a high standard deviation, about 42 mm. The maximum flow input occurs in April - about 153 mm over the lake surface, whereas the lowest value occurs in September - about 34 mm.

TABLE 6. Estimated average monthly and annual runoff from Lake Tanganyika catchment, expressed in mm over the lake surface.

	Jan	Feb	Mar	Apr	Mai	Jun	Jul	Au	Sep	Oct	Nov	Dec	Ann
Qi	90	96	149	153	115	76	58	41	34	35	44	65	955

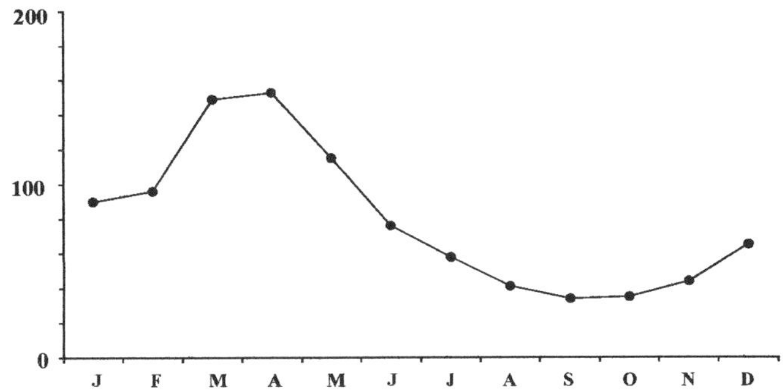

Figure 9. Estimated mean monthly total inflow profile from Tanganyika catchment to the lake, expressed in mm over the lake surface (32 600 km^2).

Even if the chosen approach, based on a K multiplying coefficient between gauged and total area in the way to close the annual budget, does not induce important changes in regard to the use of a coefficient based on the hypotheses of same runoff conditions between gauged and un-gauged area, it presents inconveniences. This approach leads one to reconsider the lesser known components of the budget. For the Lake Tanganyika the un-gauged basins correspond essentially to the Congo-Zaire sub basin. In the study of the precipitation, it has been noted that the rainy season had a tendency to start preferentially in this region. Elsewhere, this geographical area corresponds to the area closely surrounding the lake and to the part of the basin with the highest slope. It is consequently possible that these regional specificities induce a precocity in the high flow of the un-gauged basin, implying more important flow at the end of the calendar year, and more pronounced low flow values. The present estimation (Table 6 and Fig. 9) may underestimate the flow rate at the beginning, and on the other hand overestimate the flow during the second part of the mean year. This inflow profile (Fig. 9) appears to be determined by the precipitation one (Fig. 5), even if differences occur especially at the beginning of the rainy season, considering the inertia of the surface after the dry season and its higher efficiency during the second part of the rainy season. Compared with the precipitation on the catchment, only 15% joins the lake as surface inflow. Though, it had to be noticed that this characteristic is the result of a high spatial variability. This coefficient is less than 5% for the Malagarasi River catchment and at about 50% for some North-East shore catchment. At a monthly scale, the mean precipitations are well correlated with runoff (r = 0,82) but with a time lag of one month. On the other hand, compared to the stock variation profile (Fig. 3), the inflow profile seems to be a determinant factor.

8. Monthly and annual mean water balance of Lake Tanganyika

In this section, the approach is briefly discussed, and then the mean monthly and annual water budgets of the Lake Tanganyika are presented. In conclusion, the relations between the mean climatic factors and the hydrological budget are considered and the hydrologic characteristics of the lake are calculated.
In the previous sections, the methodological difficulties to obtain synchronized and representative data have been discussed for the estimation of water storage (ΔS), precipitation (P), evaporation (E), and outflow (Qo). For the inflow component (Qi) the approach has been justified and its limitations have been underlined. Assuming that the groundwater components are negligible compared to other terms of the hydrologic budget, the proposed monthly water budgets under mean conditions have been established. The monthly values of the different hydrological components of the equation (1) had been estimated considering a closed yearly water budget. This approach has conducted to balance the yearly budget without changing the gauged monthly inflows profile (Qi). Though, it has been shown that the used coefficient (2,3) is very close to the one obtained applying the same runoff rate for the gauged and un-gauging surfaces (2,4). Consequently, the approach has to be tested on the monthly time scale. To test the accuracy of the collected data set, the estimations of the monthly stock variations obtained from the observed lake level records (ΔSo, section 3) and calculated based on the other terms of the water balance (ΔSc, equation 1) have been compared. The monthly differences between observed stock variations (ΔSo) and calculated stock variations (ΔSc) show inaccurate estimates of at least one term of the water budget, and can be attributed to an inadequate methodology or to inaccurate measurements. This comparison can be obtained directly, at a monthly rate, with the absolute gap between ΔSo and ΔSc (Table 7 and Fig. 10). It can also be apprehended with linear regression between ΔSo and ΔSc (Fig. 11), which illustrates the data coherence. From a theoretical point of view the slope of the regression between ΔSo and ΔSc must be equal to one.

TABLE 7. Calculated stock variation from the other components of the hydrological budget of Lake Tanganyika ($\Delta Sc = P + Qi - E - Qo$), expressed in mm over the lake surface.

	Jan	Feb	Mar	Apr	Mai	Jun	Jul	Au	Sep	Oct	Nov	Dec	Ann
ΔSb	77	82	143	143	-8	-102	-117	-131	-138	-93	28	100	-17

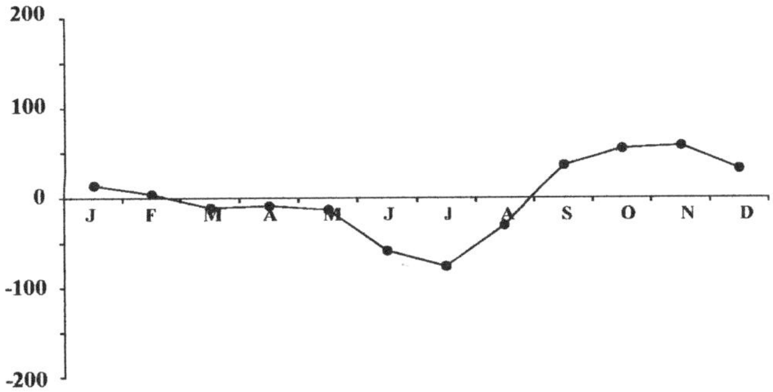

Figure 10. Mean monthly profile of the difference between observed water storage (ΔSo) and calculated water storage from equation 1 (ΔSc), expressed in mm over the lake surface (32 600 km^2).

The mean difference between the monthly absolute values of ΔSc and ΔSo is about 33 mm (Table 7), a relatively acceptable value, but some differences are still important. In fact, the differences between ΔSo and ΔSc are positive from September to February and negative from March to August. The maximum shift is observed in June and July during the important lake level decrease period, when ΔSc is lower than ΔSo. Some less important opposite differences appear during October and November (Fig. 10). The annual equilibrium hypothesis accentuates the repartition of the difference in two periods. The interpretation of the absolute values of the difference must be made with caution. Alternatively, the monthly repartitions of the differences are not conditioned by artifacts. They can be explained by (1) the difficulties to obtain an accurate estimation of ΔSo in regard to the lake topography, and (2) the uncertainties of the other components P, E, Qo and Qi. It has been already noticed (section 3), that the months of June and July correspond to the periods with the high standard deviation between the punctual level records located over the lake shore, and that these months are also characterized by important stock changes. Furthermore, during this period the decrease of the lake level recorded at Kalemie is relatively higher than at the other recording stations, and inversely during October to December, the rise at Kalemie is relatively more important than elsewhere (section 3). Inversely the months with smaller differences between the stock variation estimates (lower than 20 mm), correspond to the

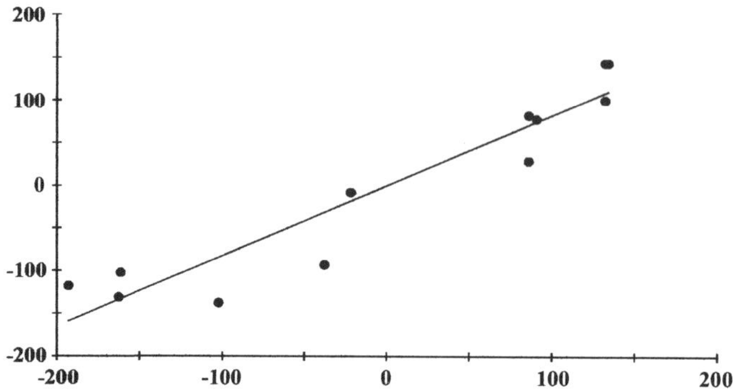

Figure 11. Relation between observed (ΔSo) and reconstructed (ΔSc) mean monthly water storage profiles, ΔSb = 0,82 ΔSo -0, 17 (r = 0,95 ; n = 12), both expressed in mm over the lake surface (32 600 km^2).

period associated to low standard deviation between the level records. The stock variations calculated from the Kalemie records present the same distortions in regard to other lake level records and values obtained by monthly water budgets (equation 1). Nevertheless, these shifts could also be induced by an under-estimation of the runoff during the beginning and an over-estimation at the end of the hydrologic years. As this observation is also in agreement with the remarks in section 7, it is possible, especially for the western part of the catchment, that the un-gauged area may contribute earlier than the gauged ones, as it can be expected from the distribution of the precipitation and from relief patterns. Furthermore it must also be noticed that this difference may be induced by an under-estimation of the evaporation rates at the lake surface during the dry seasons. Finally, if the ΔSo estimations are most probably responsible for the observed differences, the estimated runoff might probably participate in this shift. Nevertheless, figure 11 which connects ΔSo and ΔSc shows the good adequacies of the proposed budget and the coherence of the data, even if the slope of the regression is statistically different from one (for a test at a level 0,05). This observation validates the used approach, approximations and methodology, in particular for the input estimations.
The mean difference, 33 mm, is small in regard to the mean total absolute flow which is involved in monthly lake budgets, and represents an error of less than 10%. The magnitude of the lake level variations, simulated by the water budget, is calculated at about 590 mm, a value that is inferior to the one obtained from the Kalemie record 678 mm. This difference (88 mm), in the order of 13%, is still acceptable in regard to the precision of the other components, and to the representativeness and precision of the Kalemie record.
From the comparison of the mean stock variation profile with the total input and total output monthly mean values (Fig. 12), the determinant role of the input is clearly shown as responsible for the intra-annual fluctuation. During the year, with a mean value of 172 mm.month^{-1} and a standard deviation of less than 13 mm, the output profile is

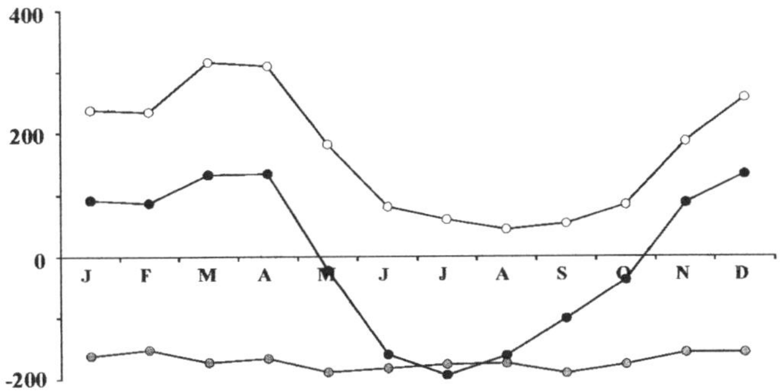

Figure 12. Mean monthly profiles of observed water storage (ΔSo; black circles), total input (P+Qi; open circles) and total output (-E-Qo; grey circles) of the reconstructed mean monthly water balance, expressed in mm over the lake surface (32 600 km^2).

relatively constant from one month to another. But because of the indirect influences of the cloud cover and the precipitation on the evaporation rates, a negative relation is shown between the evaporation and the stock variations (r = 0,70). Alternatively, the good linear relation between ΔSo and the total input (r = 0,94), clearly proves the input profile to be responsible for the monthly evolution of the stock variations. For the case of Lake Tanganyika, it then appears that, when directly considering the precipitation and indirectly the input from the catchment (section 7), the montly precipitation profile determines the stock evolution and consequently the lake level fluctuations. A positive linear relation connects the direct precipitation and stock variations values, associated with a correlation coefficient of about 0,98. It is clear that this case is specific for an equatorial lake, related to the geographical position of the lake Tanganyika, and that in a non-tropical environment the diverse components may play a different role on large lake water balances.

For the yearly budget, the annual components are then estimated at about 35-36 $km^3.yr^{-1}$ (1090 $mm.yr^{-1}$ over the lake surface) for the direct precipitation, 55 $km^3.yr^{-1}$ (1696 $mm.yr^{-1}$) for the evaporation, 12 $km^3.yr^{-1}$ (366 $mm.yr^{-1}$) for the surface outflow, and 31 $km^3.yr^{-1}$ (955 $mm.yr^{-1}$) for the catchment inputs. These values are associated with a mean storage decrease between 1932 and 1995 of about 1 to 0 $km^3.yr^{-1}$ (17 $mm.yr^{-1}$). Then, the mean yearly water balance appears primarily controlled by evaporation, and by precipitation falling on the Lake surface, which represents respectively about 82% of the losses and 53% of the inputs. From, these data and taking into account the volume of the lake (V = 18880 km^3) it is possible to calculate the mean residence times at less than 300 years, as well as the draining times (V/Qo) at 1500 years and the filling up (V/Qi) at 600 years. In any case, as Lake Tanganyika is stratified, the residence time in the epilimnion must be shorter. Modification in the water quality in the inflow must induce a modification of the lake composition. The proposed values, smaller than the ones reported in the literature, show that if Lake Tanganyika is buffered, it is nevertheless sensible to hydro-climatic conditions.

TABLE 8. Average monthly and annual difference between the stock variation observed from the lake level records(ΔSo) and calculated from the other components of the hydrological budget of Lake Tanganyika (ΔSc = P + Qi - E -Qo), expressed in mm over the lake surface.

	Jan	Feb	Mar	Apr	Mai	Jun	Jul	Au	Sep	Oct	Nov	Dec
ΔSo-ΔSb	14	4	-11	-9	-14	-59	-76	-31	36	55	58	32

9. Conclusion

In conclusion, some remarks can be made: firstly about the used approach, and secondly, about the specific case of the Lake Tanganyika water budget.
The used approach, which corresponds to a independent estimation of all the terms of the water balance equation (equation 1), corresponds to a more rigorous way, in regard to the often used one, which consists in using the water balance equation to determine an unknown components. The collected data have made allowed to obtain an estimation of the mean annual and monthly values of the different hydrological components. It has been shown that a relatively good coherency appears between the proposed independent component values. That was not so easy because only partial, punctual and not synchronized information had been used. It is possible that with more precise informations the mean stock variations could be better constrained at both monthly and perhaps inferior time scale. But even if the water balances close well, it must be kept in mind that errors involved for each independent estimate components can tend to cancel each other. So a good equality between the components of the equation 1, should therefore not be interpreted as a guarantee that the estimates of individual components are all correct. Nevertheless, the relatively well equilibrated proposed budget allow to trust the proposed values. Even, if the difficulty to obtain an estimation of the stock variations had been underlined, it is possible, with acceptable approximation, to establish a mean monthly water budget of a large lake. The above work had lead, with the examination of a mean year, to approach the direct and indirect impact of each component on the mean monthly water budget. This result can be used as a useful tool or as a base for characterizing specific years and inter-annual variability of the water budget. Even, if the differences between ΔSo and ΔSc are small enough, and if the proposed values relatively well equilibrate the water budget, a more focused study must be developed on specific components to constrain more precisely the lake water budget.
If the different estimations of the components of the Lake Tanganyika water budget proposed in the present work must be used with caution, the proposed values may lead to specific observations. It was evidenced that the Lake Tanganyika water budget is influenced by a north-south climatic gradient which as a whole consists in a decrease of the annual pluviometry, a shortening of the rain season and probably an increase of the potential evaporation. This study has also informed about the relative importance of the various components of the yearly lake water balance. The estimated values of each component, especially for surface inflow and outflow, appear higher than the one

presented in literature. It then appears that, by importance of order, the main components are evaporation from the lake surface, rainfall over the lake, inflow from the drainage area and lastly outflow via the outlet, whereas the stock variation appears insignificant at the interannual scale. It has been shown that the inputs by direct precipitation (P) and by inflow from the drainage area (Qi) contribute each for about 50% of the total input, about 53% for the precipitation and about 47% for the inflow from the drainage. For the loss, the evaporation from the lake surface, which corresponds at about 82% of the annual total loss, appears as the predominant term compared to the outflow via the Lukuga River, which represents only about 18% of the total water loss. These results underline firstly, that small variations of the precipitation or of the evaporation could rapidly induce the closure of the lake, and so major changes of the water cycle and lake proprieties, evidencing the relatively precarious equilibrium of this lake system. Secondly, as noticed above, input by runoff, which is a function of the climate, but also of the basin morphology and the soil occupation is not negligible. So important modifications in the surface occupations, of the most contributing area of the catchment, may induce sensible changes in the quantities as well as in qualities of the inflows.

At a monthly scale, the differences between observed and calculated monthly stock variations could come from an underestimation of few $mm.month^{-1}$ of the evaporation rate especially during the dry season. This hypothesis seems to be confirmed by the monthly evaporation rate proposed by the NORAD [38] which differs by a few $mm.month^{-1}$ higher from the Bultot one's [11][34] which had been considered in this study. An underestimation of the input especially at the end of the year may also be invoked to explain the observed shift. In fact, the misreading of the runoff conditions on the western part of the catchment (the Congo-Zaïre sub-basin) limits the above approach. Nevertheless, the stock variations are clearly related to the input, precipitation falling directly on the lake, and inflow coming from the catchment. In fact the lake rapidly reacts on variations of the input, and the monthly lake level profile is very sensitive to the evolution of the precipitation. It is possible that for different time scales this relation appears less important. This specificity is clearly induced by the equatorial climate specificity and by the outlet characteristics which induce relative constant monthly values of losses along the year. Indeed it appears that evaporation influences the level of the lake only in a cumulative way and during a sufficiently long period of time.

It is apparent that the water storage in lakes as reservoirs serves a variety of human activities and that the optimal us of these resources requires effective management. Extending and broadening this study will be useful only if it helps us to improve our ability to manage these important fresh water resources. As a matter of fact, the demographic growth and its consequences on the catchment basin and on the quality of water could induce at the short term considerable consequences. Finally, proposals for refining and prolongating the study must be made only on an inter-national project involving the different bordering countries.

10. References

1. UNDP/WMO (1976) *A water resources assessment of lake Malawi.* Project 71/518, 167 p.
2. Kidd, C. H. R. (1983) *A water resource of lake Malawi and the Shire river.* UNDP Project Report, MLW/77/012, 135 p.
3. Sene, K. J. and D. T. Plinston (1994) A review and update of the hydrology of Lake Victoria in East Africa. *Hydrological Sciences Journal des Sciences hydrologiques,* **39, 1,** 47-63.
4. Crul, R. (1996) *Limnology and Hydrology of Lake Victoria.* UNESCO, studies and reports in hydrology, 53, 79 p.
5. Crul, R. (1997) *Limnology and hydrology of Lakes Tanganyika and Malawi.* UNESCO, studies and reports in hydrology, 54, 111 pp.
6. Coulter, G. W. and R. H. Spigel (1991) Hydrodynamics, in G.W. Coulter (ed.), *The Lake Tanganyika and its life.* Nat. Hist. Mus. Publ. & Oxford Univ. Press., pp. 49-75.
7. Spigel, R. H. and G. W. Coulter (1996) Comparison of Hydrology and Physical Limnology of the East African Great Lakes : Tanganyika, Malawi, Victoria, Kivu and Turkana (with reference to some North American Great Lakes), in T. Johnson and Odada: (ed.), *The Limnology, Climatology and Paleoclimatology of the East African Lakes,* Gordon & Breach, Toronto, pp.103-138.
8. Theeuws, R. (1920) Note personnelle et inédite sur le problème de la Lukuga.
9. Gillman, F. (1933) Hydrology of Lake Tanganyika. *Bull. Geol. Surv. Dar-es-Salaam. Tanz.,* **5**, 1-27.
10. Devroey, E.-J. (1949) A propos de la stabilisation du niveau du lac Tanganyika et de l'amélioration de la navigabilité du fleuve Congo (Bief moyen du Lualaba, Kindu - Ponthierville). *Mém. Inst. Roy. Col. Belg., in-8°., 5, 3, 135 pp.*
11. Bultot, F. (1965) A propos de l'évaporation du lac Tanganyika. *Bull. Séan. Acad. Roy. Sci. Outre-Mer Belg,* **4**, 1226-1241.
12. Hecky, R. E. and E. T. Degens (1973) Late Pleistocene-Holocene chemical stratigraphy and paleolimnology of the Rift valley lakes of central Africa. *Woods Hole Oceanographic Institution,* **73, 28**, 93 pp.
13. Ferguson, H. L., and V. A. Znamensky (1981) Methods of computation of the water balance of large lakes and reservoirs. *UNESCO Studies and reports in hydrology,* **31**, 120 pp.
14. Bergonzini, L. (1998) Bilans hydriques des lacs (Kivu, Tanganyika, Rukwa et Nyassa) du Rift Est-Africain. *Mus. roy. Afr. centr. Tervuren Belg., Annales Sciences Géologiques,* **103**. 183 pp.
15. Bergonzini, L. (1997) Mise en évidence d'oscillations saisonnières à la surface du lac Tanganyika. *Mus. roy. Afr. centr. Tervuren Belg. Dépt. Géol. Min., Rapp. ann. 1995 & 1996,* 125-134.
16. Nicholson, S. E. (1996) A Review of Climate Dynamics and Climate Variability in Eastern Africa, in T. Johnson and Odada (eds.), *The Limnology, Climatology and*

Paleoclimatology of the East African Lakes, Gordon and Breach, Toronto, pp. 25-56.

17. Capart, A. (1952) Exploration hydrobiologique du lac Tanganyika, (1946-1947) Résultats scientifiques : Le milieu géographique et géophysique. *Inst. Roy. Sci. Nat. Belg.*, **1**, 1-27.
18. Branchu, P. (2001) Cycle du carbone et des éléments majeurs et traces dans les grands lacs de Rift Tropicaux (Lacs Tanganyika et Malawi) : Processus biogéochimiques et enregistrements sédimentaire. *Mus. roy. Afr. centr. Tervuren Belg., Annales Sciences Géologiques,* **106**, 374 pp.
19. Devroey, E.-J. (1952-60) Annuaire hydrologique du Congo belge et du Ruanda-Urundi (1951-1959). *Mém. Acad. Roy. Sci. d'Outre-Mer Belg.*
20. Bergonzini, L. (1995) Les écoulements de la Lukuga a l'exutoire du lac Tanganyika. *Mus. roy. Afr. centr. Tervuren Belg. Dépt. Géol. Min., Rapp. ann. 1993 & 1994,* 79-89.
21. Bergonzini, L. (1998) Les fluctuations du niveau du lac Tanganyika et les termes de surface de son bilan hydrique, in G. Demarée, J. Alexandre and M. De Dapper (eds.), *ITCMH*, Acad. roy. Sci. Belg, pp. 613-637.
22. Grove, A. T. (1998) Variability of African rivers discharges and lakes levels, in G. Demarée, J. Alexandre and M. De Dapper (eds.), *TCMH*, Acad. roy. Sci. Belg., pp. 470-478.
23. Bergonzini, L., P. Camberlin and Y. Richard (2001) Variations interannuelles du bilan hydrologique du lac Tanganyika: les précipitations dans l'Afrique des Hautes-Terres sont-elles seules en cause ?, in *Les montagnes tropicales : identités mutations développement* Talence, DYMSET, CRET, Espaces Tropicaux, 16, pp. 145-159.
24. Beltrando, G. and D. L. Cadet (1990) Variabilité interannuelle de la petite saison des pluies en Afrique Orientale : relations avec la circulation atmosphérique générale. *Veille climatique satellitaire,* **33**, 19-37.
25. Ropelewski, C. F. and M. S. Halpert (1996) Quantifying Southern Oscillation Precipitation Relationships. *Journal of Climate,* **9**, 1043-1059.
26. Richard, Y. (1992) Connexion entre le début de la saison des pluies en Afrique orientale méridionale et les circulations cellulaires zonales indienne et pacifique. *Publication de l'Association internationale de climatologie,* **5**, 113-123.
27. Leroux, M. (1983) *Le climat de l'Afrique Tropicale.* Champion ,Paris, 633 pp.
28. Ergo, A. B., and B. de Halleux (1979) *Catalogue mondial des données climatiques moyennes*, Vol. II, fasc. I, première partie.
29. Ergo, A. B., and B. de Halleux (1984) *Catalogue mondial des données climatiques moyennes*, Vol II, fasc. II, deuxième partie.
30. NORAD (1982): Rukwa Water Master Plan, Vol. 7 Hydrology, Final Report, Oslo, 194 pp.
31. Nicholson, S. E., J. Kim and J. Hoopingarner (1988) Atlas of African Rainfall and its interannual variability. *Florida State University,* 252 pp.
32. Vose, R. S., R. L. Schmoyer, P. M. Steurer, T. C. Peterson, R. Heim, T. R. Karl and J. K. Eischeid (1992) The Global Historical Climatology Network: Long-Term Monthly Temperature, Precipitation, Sea Level Pressure, and Station Pressure Data.

ORNL/CDIAC 53, Environmental Science Division 3912.
33. Bultot, F. and D. Gellens (1994) Sur le caractère stationnaire et cyclique des précipitations au Rwanda. *Mém. Acad. Roy. Sci. d'Outre-Mer Belg. in-8°,* **19, 1**, 53 pp.
34. Bultot, F. (1993) Evaporation from a tropical lake: comparison of theory with direct measurements, comment. *Journal of hydrology,* **143**, 513-519.
35. Bergonzini, L., F. Gasse and F. Chalié (1997) Paleoevaporation and paleoprecipitation in the Tanganyika Basin at 18 000 years B.P. Inferred from Hydrologic and Vegetation Proxies. *Quaternary Research,* **47**, 295-305.
36. Balek, J. (1977) Hydrology and water resources in tropical Africa, in V.T. Chow (ed.), *Developments in Water Science 8,* Elsevier, 208 pp.
37. Hastenrath, S. L. and J. E. Kutzbach (1983): Paleo-climate and water budget of East African lakes. *Quaternary Research,* **19**, 141-153.
38. NORAD (1989) Kigoma Water Master Plan, Hydrology updated 1989. Publ. Ministry of Water Tanza. & Norw. Agency for Development Co-opération, Oslo, 62 pp.
39. Olivry, J.-C. (1993) Evolution récente des régimes hydrologiques en Afrique intertropical, in *Au fil de l'eau, hommage à René Frecaut*, Presses universitaires de Nancy, 181-190.
40. Camus, C. (1965) Fluctuations du niveau du lac Tanganyika. *Bull. séan. Acad. Roy. Sci. Outre-Mer Belg.*, **11**, 1242-1256.
41. Bureau du Génie Rural du Rwanda (1961-74): Annuaire hydrologique du Rwanda. Ministère de l'Agriculture. République Rwandaise.
42. IGEBU (1986) Recueil de données hydrologiques: année 1984-85. *Publ. Inst. Géogr. Bur., Bujumbura.*
43. Pruvot, P. (1974) Burundi, l'hydrologie de l'Imbo. Document de travail, FAO, Rome, 50 pp
44. MAJI (1976) Hydrological Year-Book (1965-1970). Ministry of Water Development and Power. Unit. Rep. Tanz., 666 pp.
45. MAJI (1985) Hydrological Year-Book (1971-1980). Ministry of Water Development and Power. Unit. Rep. Tanz., 364 pp.
46. Zambian Ministry of Rural Development (1957-66): Hydrological Year Book. Prepared by the Branch Water Affairs Department, Lusaka, Zambia.
47. UNESCO (1978): World Water Balance and Water Resources of the Earth. UNESCO Studies and Reports in Hydrology ISBN 92-3-101497-8.

RADON STUDIES FOR INVESTIGATION OF NUCLEAR WASTE DEPOSITS AND NATURAL EMISSIONS

S.E. BEAUBIEN, S. LOMBARDI, N. VOLTATTORNI
Earth Science Department, University of Rome "La Sapienza"
Piazzale Aldo Moro, 5, I 00185 Roma, Italy

1. Introduction

In underground repositories for radioactive waste, significant quantities of gases may be generated due to several processes (i.e., corrosion of metals in waste and packaging, radiolysis of water, microbial degradation of various organic waste, etc). These gases may migrate through the engineered barrier system and the natural geological barrier. It is therefore recommended that the potential impact of gas accumulation and migration on the performance of the various barriers should be addressed and assessed in the development of safety cases for radioactive waste repositories.

Gas geochemistry has been proven to be a reliable tool to investigate the origin and processes involving fluids during their ascent to surface. The wider knowledge of the nature and behaviour of deep fluids establishes a fundamental basis for hydrogeological surveys, mineral exploration (ore deposits, petroleum and geothermal reservoirs) and studies for environmental protection and seismic zonation. The soil gas techniques (soil gases sampling, injection test, sampling and analysis of gases dissolved in groundwaters, etc.) provide a useful tool for gas permeable faults detection (secondary permeability) and enhance the study of gas migration conditions. Soil-gas methods allows an easy acquisition of geochemical data in areas where traditional geochemical methods, based on trace-element migration in groundwater, may fail because of the thickness and low permeability of certain lithotypes. One of the innovative aspects of this method is the attempt to overcome the limitations imposed by the presence of thick and/or low permeable sediments over buried mineral resources and hidden structural features [9, 2].

Significant effort has and continues to be expended in numerous national and international programmes in attempts to understand and assess the potential impact. These efforts have resulted in significantly improved understanding of the relevant processes and produced an important collection of experimental data.

J. Klerkx and B. Imanackunov (eds.), Lake Issyk-Kul: Its Natural Environment, 245–260.

2. Theoretical principles of soil-gas prospecting

The use of deep-seated gases in soil air as fault and fractures tracers has been proposed by several authors since 1930 [19, 30] as the presence of endogenetic gases in soil air can be linked to their ascent along fracture and fault systems which act as preferential migration pathways [11, 24, 18, 20, 41, 42].
Endogenous gases migrate principally by mass transport along zones of high secondary permeability (deep faults and/or fractures) that have high flow rates. This migration process, therefore, is the dominant cause of endogenous gas "anomalies" (anomalous concentrations), generally observed at the surface in proximity to know deep-fracture or fault systems [12, 26, 27]. Radon is a radioactive and a chemically inert element and its atoms generated in crustal materials can entry in fluids present in rocks pore and migrate for very significant distances in respect of their origin site. ^{222}Rn is the most important isotope: it derives from ^{238}U that is 99.28% of all uranium presents in the crust, so the presence of radon is related with the rocks uranium content. The most part of radon in trapped in the same rocks that generated it: a part, however, can entry in the air or in the pore water and be transported by fluids for a long distance. ^{222}Rn is a very good tracers of tectonic discontinuity and buried geological structures as both it can migrate without reacting with any natural substances and it is possible to measure very low radon concentrations in every natural systems.
It is interesting to highlight the roles that CO_2 can play in soil-gases distribution, especially the "carrier" gas role: the association of CO_2 anomalies with radon anomalies suggests that CO_2 could transport trace gas through groundwater up to unsatured soil acting as "carrier" gas. Very high CO_2 concentrations can be found in soil-gases: this depends of many factors, such as organic material in soils, climate variations, carbonate metamorphism, chemical reactions in mineralised deposits and so on. Exhaustive treatments about CO_2 geochemical characteristics and its use in geological prospecting are treated by many authors [37, 25, 5, 17, 3].

3. Mechanisms of gas migration

The two basic processes which are recognised to drive soil gas movement through rocks and sediments are *diffusion* and *advection.*
Diffusion is the process by which gas species move in response to a concentration gradient, from areas of higher concentration to areas of lower concentration. Fick's first law describes the flux of a species in porous media [32].
The diffusion coefficients measured for some gaseous hydrocarbons in a variety of rock types vary widely (10^{-3} to 10^{-8} cm^2/sec), depending on the particular lithology and geologic conditions. Gas diffusion can take place either in gas-filled pores (gas-phase diffusion) or in water-saturated porous media when the gas is in solution (liquid-phase diffusion); the latter is a much slower mechanism of transport than the former. The times required for diffusion through crustal sequences are prohibitively large and often exceed the age of the host rock. As such if diffusion were the dominant process for

migration, then shallow soil gas anomalies would indicate only local, near surface sources. As an example a short-lived radioactive gas such as ^{222}Rn (half-life = 3.8 days) cannot migrate a long distance before it decays; some authors have estimated that it is impossible to distinguish radon coming from a source and radon occurring naturally in the rocks at a distance of 10-15m from that source [29]. However, diffusion can still be considered as an important secondary process, i.e. in small-scale migration of biogenic gases.

Advection is the process whereby mass transport is induced by a pressure gradient that can be caused by tectonic stress, changes of lithostatic and/or hydrostatic load, sudden gas production, atmospheric pumping, etc. Advection can take place with or without groundwater. In the first case we can have liquid-phase (i.e. dissolved gas) or gas-phase (bubbles) advection, whereas in the second case we have simple gas-phase advection. This mechanism is believed to be the dominant mode of moving deep gases to the shallow environment. The sharp location of many anomalies is more consistent with an advective than a diffusive model. Some authors suggest vertical transport rates via advection of ten of meters per day, clearly far greater than diffusive migration distances. The sharp and often linear appearance of the anomalies suggests that faults and fractures play an important role in the movement of deep gases. In seismic areas the soil gas values above active faults increase dramatically shortly after an earthquake.

Exhalation is the process that transfers gas from the soil to the atmosphere via the already-described mechanisms of diffusion and advection. The diffusive exhalation of a gas takes place due to the different concentrations of that gas in the soil and atmosphere. Advective exhalation exists when the gas in the soil has a higher total pressure than atmospheric (eg. barometric pumping). Exhalation can be measured in terms of flux from the soil, i.e. the quantity of gas per unit area per unit time (mass flux = Kg $m^{-2}s^{-1}$; volume flux = $m^3m^{-2}s^{-1}$). Mean values for CO_2 exhalation are 3.7 x 10^{-7} $m^3m^{-2}s^{-1}$ or 0.4 - 4·x 10^{-7} $m^3m^{-2}s^{-1}$, while ^{222}Rn exhalation values are 15 mBq $m^{-2}s^{-1}$ [44] and 22 mBq $m^{-2}s^{-1}$ [38].

Radon flux is partly due to radon coming from the subsoil and partly due to the decay of ^{226}Ra atoms in soil particles and the subsequent *emanation* of radon into the pore space. The fraction of radon emanating in the pore space is the *emanating power* (ε); its value ranges from 5% in rocks to 40% in soils.

4. Gas migration from a repository

Gas migration from a repositoriy can lead directly, through the release of radioactive gases at the surface, to radiological consequences analogous to those that are assessed for the transport of radionuclides from a repository via other pathways. The main identified potential safety issues that may arise from gas migration and two-phase flow effects can be summarised as follows:

a) overpressurisation: if gas cannot escape from a repository as fast as is generated, the pressure will rise fracturing engineered barrier and/or host rock. This would leave

pathways along which radionuclides might be transported more quickly than through the original structures.
b) release of radioactive and flammable gases at the surface: some of the gases produced in radioactive waste repositories may be radioactive because they contain, in particular, ^{14}C, ^{3}H, ^{222}Rn, ^{85}Kr, etc.. these may be transported to the surface by the larger volumes of inactive gases produced in a repository in water-saturated rock. It is necessary also to consider that this flux of gas may provide a vehicle for transporting naturally occurring radon to the surface.
c) effects on movement of contaminated groundwater: it is important to envisage some of the mechanisms by which gas generated in a repository in a water-satured host rock might induced movement of potentially contamined groundwater:1) forcing of water from a repository by the accumulation of a gas cushion within the repository; 2) entrapment of groundwater in streams of gas bubbles; 3) induced movement of groundwater as a consequence of instabilities in gas pathways.
d) transport thorough attachment at gas-water interfaces: a variety of colloidal particles and chemical species are know to attach to or concentrate at gas-water interfaces. In so far as the movement of gas-water interfaces is involved in gas transport from a repository it may provide a mechanism for transporting radionuclides.
e) transport of dissolved contaminants in unsatured rock: transport of dissolved radionuclides back to the biosphere from an unsatured site requires first their transport from the repository, through unsatured rock, to the satured zone below the repository, by water percolating from the surface (or liberated from the matrix of the host rock).

5. Soil gas sampling and analysis

Soil gas samples are collected from shallow point sources and as flux measurements across the ground/atmosphere interface. The soil-gas survey is generally carried out in summer during a period of stable meteorological conditions. Soil-gas sampling was performed using a well tested method [34, 24, 7]. A 1m hollow steel probe with a conic point at the bottom and a sampling port on top is inserted to a depth of 0.5m below the ground surface. Two 50 cc samples of soil gas are extracted from the probe with a syringe; the first sample is analyzed in the field for radon activity by means of an alpha-particle counting EDA Radon Detector RD-200 , whereas the second sample is stored in an evacuated 25 ml steel cylinder for laboratory analysis by means of a Varian Helium Leak Detector mass-spectrometer(^{4}He), and a Fisons 8000 series GC equipped with HWD (CO_2, O_2), FPD (sulphides) and FID (C1-C6 hydrocarbons).

6. Background geology

6.1. ORCIATICO AREA (Tuscany, Central Italy)

The Orciatico area, a few km NW of the town of Volterra, is located on the western edge of the Val d'Era graben. In this area, a small, outcropping subvolcanic intrusion (Selagite) has induced metamorphic effects in the host Pliocene clays (Fig.1).

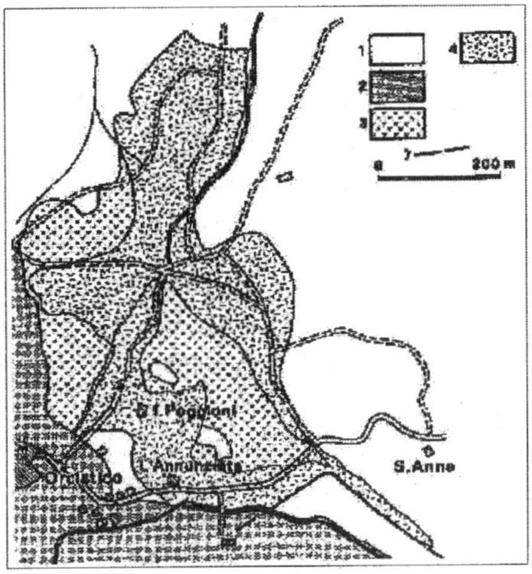

Figure 1. Geological sketch map of the Orciatico igneous body and surrounding rocks.1)Pliocene Quaternary sediments; 2) Miocene lacustrine and marine sedimenyits; 3) thermally methamorphosed rocks ("Termantite"); 4) Orciatico orendite; 5) faults and inferred faults. (from Conticelli et al, 1992)

The emplacement of the Selagite intrusive body was probably controlled by the NNW-SSE trending faults related to the post-tectonic extensional regime. Dating of this unit yields an emplacement age of 4.1 Ma [8], an age which partly overlaps with Tuscan anatectic magmatism but which precedes the Roman magmatism by 3 Ma.
According to Stefanini [40], the Orciatico magma was intruded into shallow Pliocene sediments in the form of a sill-laccolith body fed by a narrow NW-SE trending dike. The intrusion has a chilled margin, varying from glassy to slightly porphyritic, and a massive holocrystalline core. Oxides and K-feldspars are the most abundant phases in the groundmass [10]. The country rocks underwent thermal metamorphism, attaining pyroxene-hornfels facies in the resultant aureole [22]. Late-stage hydrothermal metamorphism has also been recognised which modified the Sr-isotopic ratios of the country rocks, but which did not affect the magmatic body [14].
The simplified geological sketch of the Orciatico structure given in Fig 1 is characterised by:

- *Selagite* [43]: this small, shallow alkali-trachyte intrusion was emplaced in Pliocene clays. It has been described by many authors, including Rodolico [36] and Stefanini [40]. More recently Barberi and Innocenti [4] performed a detailed petrographical and petrological study of the Orciatico laccolith and the closely related Montecatini subvolcanic intrusion.;

• *Metamorphosed clays (Termantiti):* the rocks of the metamorphic aureole can be subdivided into three main groups:

a) hornfels facies: extremely compact, hard, dark, aphanitic rocks,

b) spherulitic rocks: differ from the hornfels facies due to a distinctly lower hardness and the presence of numerous dark spherulites embedded in a yellowish matrix.

c) "pseudo-galestrine" rocks : yellow-grey, underrated fissile shale.

• *Un-metamorphosed clays*: the Pliocene sedimentary formation in which the Orciatico laccolith was emplaced is a thick, prevalently marine, clastic sequence. Locally these sediments are represented by plastic, grey-blue, marly-silty clays.

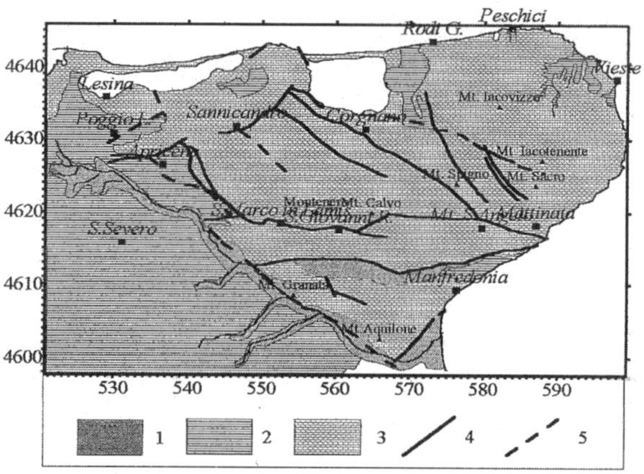

Figure 2. Geological sketch map of the Fucino basin. 1) alluvium; 2) Tertiary sediments; 3) carbonates; 4) faults; 5) buried faults.

6.2. THE FUCINO BASIN (Abruzzo, Central Italy)

The Fucino basin (Fig.2) is an elliptically shaped intramontane tectonic depression (about 250 square kilometers) in the central Apennine chain which has been developing since the Late Pliocene [33]. The basin was originally occupied by a large lake and thus is filled by Pleistocene and Holocene silty lacustrine deposits which are interbedded with coarse alluvial fan deposits at its borders; thin layers of Middle and Upper Pleistocene volcanic fall products originating from volcanoes on the Tyrrhenian coast are also interbedded in the continental sediments [15]. The main faults affecting this basin, the SBGMF and ACF, are characterised by mainly normal or oblique slip movement, indicating that extensional tectonics have been the main mechanism responsible for the evolution of the basin [16]. Remote sensing, subsurface and historical investigations demonstrate that this network of active faults affect the very recent deposits of the basin floor, including the occurence of numerous M>6 events along the SBGMF, TF and ACF over the last 2,000 years. The sedimentation rate in the basin has exceeded the rate of differential movements, resulting in very subtle or no morphological evidence at the surface despite the fact that these structures are capable of surface rupturing (for example up to 0.9 m of vertical coseismic offset was measured along the SBGMF after the magnitude 7.0 Avezzano earthquake of January 13, 1915 [31]. During the 1915 earthquake, gas and water emissions from the fractures were reported, as well as a suite of other typical seismogeological effects like liquefaction and spring anomalies, indicating that some of these structures were also "geochemically active". In general, recent and present-day activity appears to affect a network of active fault segments distributed over the entire basin, although maximum offsets are observed in the eastern sector.

6.3. THE TOLFA MINE DISTRICT (Lazio, Central Italy)

The Tolfa mine district (Fig.3) is located in the Tolfa Mountains area characterised by two major geological units. The Lower Unit (Carboniferous – Upper Cretaceous), only rarely outcropping in the area, is represented by the Tuscan Series terranes and consists of continental sediments, slightly metamorphosed at depth and overlain by a sequence of marine carbonates. Directly overlying the Lower Unit is the Upper Unit (Cretaceous-Paleogene) which is mostly an allochthonous flysh complex with dominant limestones and marly limestones.

During the early Pliocene the region was inundated by the sea [13] and clayey and sandy sediments were deposited. Starting about 4.2 million years ago the area was uplifted by the intrusion of an acidic pluton into the sedimentary rocks [23]. This intrusive event resulted in the development of a system of hydrothermal vein deposits which apparently formed along major faults.

Mineralization in the Tolfa Mountains area can be divided into three stages. The first was the contact metasomatism of the host limestones which occurred upon the intrusion of an acidic pluton. This activity resulted in the formation of garnet-bearing calc-silicate hornfels. Small fragments of such rocks are found randomly but rarely throughout the phanerocrystalline carbonate stocks. During drilling performed for geothermal research, the contact limestone was seen as deep as 330 meters and hornfels was observed at even greater depths. The second stage was the formation of the phanerocrystalline carbonates within major faults. Within the context of large grain size, the carbonates can be distinguished as fine-medium, or coarse-grained, these features being related to sub-stages of formation of the rock. Actually the fine-grained carbonate is often found at the outer walls of the veins, while medium and coarse-grained calcites are generally located in the interior parts. The third stage was the deposition of sparry calcite along veinlets that are as much as 40 wide and which are found widespread throughout the phanerocrystalline carbonate stocks.

Rare quartz and fluorite are found in little veinlets scattered in the stocks. Their occurence is related to the late-stage event [28].

The Tolfa area is one of the most European mine district, active from 1461. Although actually most mines are abandoned, mine resources characterise this area as one of particular geological and mineralogical interest due to the association of different mine products. It is characterised by different mine resources such as:

- products from mercury mines;

- products from alunite, caolino and felspati mines;
- metal sulphides (Galena, Marcasite, Pirite);
- mineral and thermomineral springs (T=56°);
- argillaceous, calcareous and sandy products from caves.

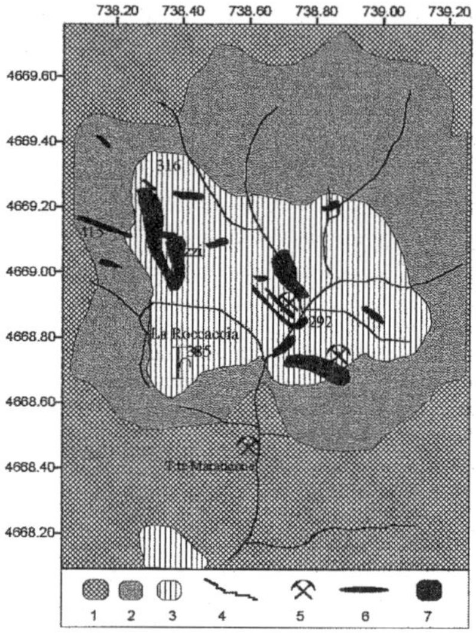

Figure 3. Geological sketch map of the Tolfa mine district. 1)"Palombino" flysh; 2) metasomatised "palombino"; 3)phenocrystalline carbonates stocks; 4) stream; 5) abandoned mines; 6) mineralised veins; 7) mine tailings.

6.4. NEVES CORVO MINE DISTRICT (Baixo-Alentejo region, Portugal)

The Neves-Corvo mine district is in the Baixo-Alentejo region (Portugal) and it is located about 50 km SSW Beja town, between Castro Verde and Almodovar towns. It was discovered in 1977 during a mine prospecting survey. The Neves-Corvo mine district forms a part of the so-called IPB (Iberian Pyrite Belt), a well known metallogenic district that [1] corresponds to an area of Devonian-Carboniferous volcanic and sedimentary rocks containing massive polymetallic sulphide deposits [35, 39]. The IPB is the largest and most important volcanogenic massive sulphide metallogenic province in the world. The original, pre-erosional amount of sulphides concentrated in about 90 known deposits are estimated at more than 1.7 billion tons. Of this amount, about 20% has been mined, and 10-15% lost to erosion.
The Neves Corvo mine is the most important European copper-tin mine and a world class massive sulphide deposit. It is composed of five main ore bodies. These mineral deposits have unique features within the IPB context, namely the high contents of copper and tin. All the deposits exhibit stockwork mineralization in the footwall host rock and are almost always located above a main felsic volcanic episode [21, 6] (Leca et

al, 1983, Barriga & Carvalho, 1983). The Neves Corvo mine is one of the largest deposits in the IPB and certainly the richest ever found with approximately 300 Mt of

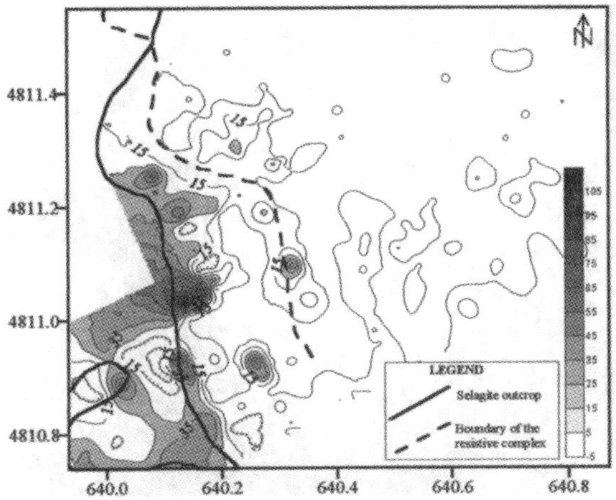

Figure 4. The radon (Bq/l) distribution map shows that highest radon values (> 25 Bq/L) are concentrated in zones where the Selagite intrusion outcrops and where thermally altered clays may be inferred, while further from these fractured zones no appreciable soil gas anomalies (radon concentrations are about 10-15 Bq/L) occur for this gas due to the sealing of these intact clays.

total sulphide-rich rock (ore and sub-ore). Another main feature is the clear lateral and vertical metal zonation patterns exhibited. As a general rule the copper rich sulphides occur at the base of the deposits and are overlain, when present, by the zinc sulphides. The "barren" massive sulphides usually appear towards the upper part of the deposits. Tin occurs in close spatial association with richest parts of the copper mineralization. A peculiar characteristics is the occurrence of banded "rubanè" ore at the hanging wall of the Corvo deposit. Pyrite is the most abundant sulphide, followed by chalcopyrite, spharelite and many other mineral species. Casserite occurs locally in significant economic concentrations.

7. Results and discussion

7.1. ORCIATICO AREA (Tuscany, Central Italy)

The analysis of soil-gas concentration map, were elaborated using the kriging estimation method. The radon contour line map, Fig. 4, shows that higher radon values (> 25 Bq/L) are concentrated in zones where the Selagite intrusion outcrops and where thermally altered clays may be inferred, while further from these fractured zones no appreciable soil gas anomalies (radon concentrations are about 10-15 Bq/L) occur for this gas due to the sealing of these intact clays. This distribution is likely due to intense fracturing of clays along the laccolith boundary, as it is reasonable to suppose that the intrusion of the laccolith caused mechanical stress and contact metamorphism in the clay. This observation indicates that the clay sequence, in spite of its thickness and plasticity, is far more permeable to naturally migrating gas if it is fractured and metamorphosed. In addition, as well as radon distribution, higher carbon dioxide values (>2.0%) are mainly concentrated at the Selagite-clay contact, along the boundary of the resistive complex and, to a lesser extent, in the clays between Selagite and the suggested boundary. The association of a trace gas (radon) with CO_2 suggests that the latter acts as a carrier gas of the former; if the pressure of carbon dioxide is sufficient,it can form a free gas phase (i.e. bubbles) and advect rapidly within the rock, scavenging Rn as it rises through groundwater up to the unsaturated horizon.

7.2. THE FUCINO BASIN (Abruzzo, Central Italy)

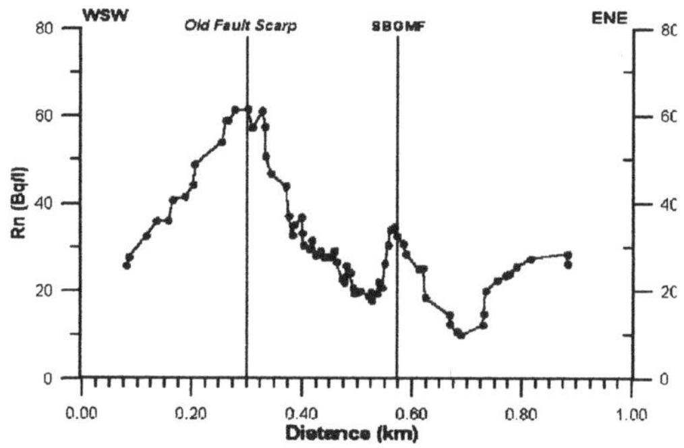

Figure 5. The highest radon values are aligned along the most important faults that border the basin, both in the east (San Benedetto-Gioia dei Marsi Fault - SBGMF) and in the north-west (Avezzano-Celano fault - ACF), as well as along some coseismic ruptures recognised throughout the plain.

The results from the soil-gas survey performed in the Fucino basin yielded anomalies with unique features, reflecting the different gas bearing properties of the seismogenetic faults related to the 1915 earthquake and the hidden structures along the western side of the Fucino plain.

The highest radon values (Fig.5) are aligned along the most important faults that border the basin, both in the east (corresponding to the San Benedetto-Gioia dei Marsi Fault - SBGMF) and in the north-west (corresponding to the Avezzano-Celano fault - ACF), as well as along some coseismic ruptures recognised throughout the plain (e.g. coseismic scarps mapped by *Oddone* [31] where liquefaction phenomena caused pervasive brittle deformation of the surface layers during the 1915 event. The SBGMF and ACF are characterised by extensional kinematics (i.e. dip-slip) and were reactivated during the 1915 earthquake. Moderately anomalous radon values occur along the faults located within the historical lake basin, including the Trasacco (TF) and Ortucchio (OF) Faults. In particular the OF, inferred as the prolongation of the "Alto Sangro" fault which cuts the carbonate rocks south of Gioia dei Marsi, is marked by a continuous, well-defined Rn "ridge", whereas it was recognised only at some discrete points by previous structural surveys [16].

7.3. THE TOLFA MINE DISTRICT (Lazio, Central Italy)

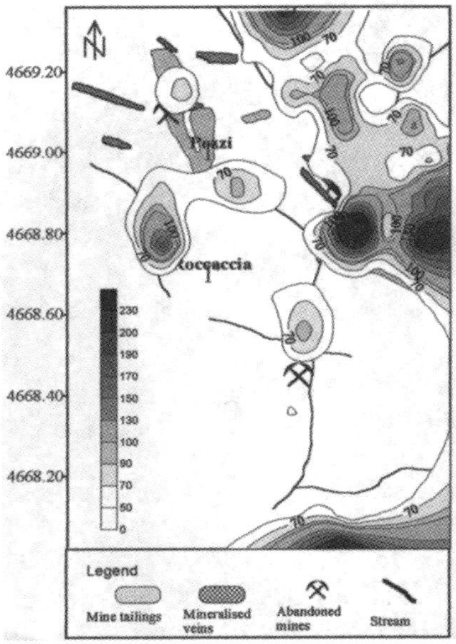

Figure 6. Radon soil gas distribution in the Tolfa mine district. Highest radon values (>70Bq/l) can be found in correspondence of mine tailings, mineralised veins, in proximity of abandoned mines.

A detailed soil gas survey was performed in the Tolfa district, exactly in an area where there was the major extractive activity in the past. This area is characterised by the presence of mineralised veins and by mine tailings. Figure 6 represents the radon concentration distribution: in this map concentration values under 50 Bq/l were not represented as it is the background value. The average value of radon concentration is more than 45 Bq/l and it is well distributed in all the investigated area: for this reason, only anomalous values were represented in the contour line map as the radon anomalies distribution seems to follow particular pathways. In fact, highest radon values (>70Bq/l) can be found in correspondence of mine tailings, mineralised veins, in proximity of abandoned mines and along streams. The maximum radon value, 418.84 Bq/l, is present at the entrance of an abandoned mine (in the eastern sector of the studied area) that was one of the most exploited as it is reasonable to suppose considering the large quantity of tailings that surrounds it. This small area is characterised by the highest anomalous radon values due probably to the presence of mine tailings whilst the high values (>60Bq/l) along streams have to be considered as a result of transport by flowing (or running) superficial waters.

7.4. THE NEVES-CORVO MINE DISTRICTS (Baixo-Alentejo region, Portugal)

In this area two soil-gas profiles were performed following the directions (profile 1: N 36°; profile 2: N 60°) of two previous geological profiles [21]. In this way it was possible to do spatial correlation of anomalous soil-gas values, the main fracturing systems and mineralised bodies. The two profiles intersect the two sulphides deposits of Neves and of Corvo: samples were collected every 50-100 meters in the southern sectors and every 25-50 m in correspondence of mineral ores that are at different depths, from 250 until 700 meters in the north-eastern sector and buried by metamorphic sediments with a low permeability. The results from the two soil-gas profiles are very interesting, especially from the profile 2 where also radon measurements were made. Figure 7 highlights the correspondence between highest radon values (>50Bq/L) and concentration peaks of other soil gases (i.e. CO_2, O_2 and CH_4). As sulphide oxidation produces in turn an acid

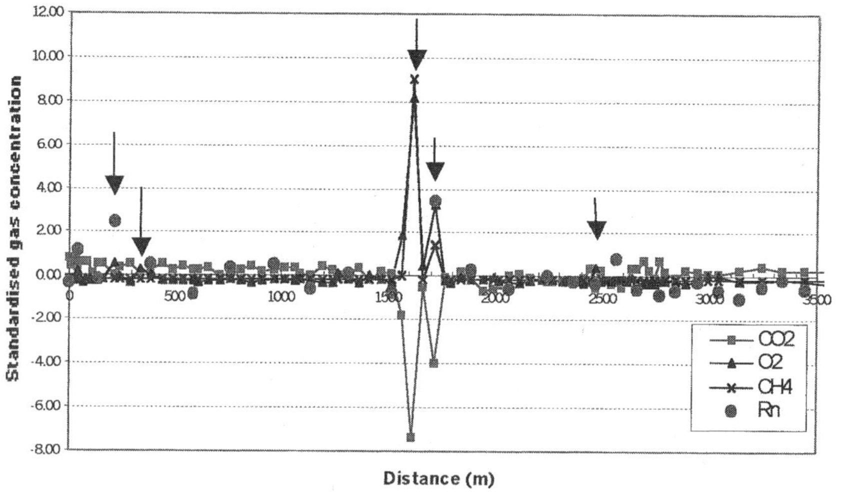

Figure 7. Neves Corvo, profile 2.The graph highlights the correspondence between highest radon values (>50Bq/L) and concentration peaks of other soil gases (i.e. CO_2, O_2 and CH_4).

local environment in which any carbonate minerals present are attached and, in consequence, generate CO_2 some of which migrates into the soil air. As a result sulphide oxidation can modify the oxygen and carbon dioxide concentration gradients in the surrounding rock and soil spaces and so produce anomalies that may be detectable in the soil air near the surface. The radon mean concentration is around 30 Bq/l and it is well distributed along the profile except in those points where the main structural lineaments are crossed and that are connected with deep mineralised bodies.

8. Conclusions

The studied areas represent four different examples of soil gas prospecting application and the obtained results confirm the validity of the method.
In synthesis it is possible consider the Orciatico area and the Tolfa mine district as natural analogue for studying "near-field" migration from a superficial resource. The two areas have geological and structural characteristics completely different but it is possible to note that the clay sequence, in spite of their thickness and plasticity, if fractured and/or metamorphosed, form a lesser impermeable barrier for naturally migrating gas and the presence of transport means (i.e. superficial waters or streams) contributes to the radon migration.
On the other hand, soil gas prospectings performed in the Fucino Basin and in the Neves-Corvo mine district represent natural analogue for studying "far-field" migration from a deep resource: in fact, the Fucino Basin is characterised by the presence of deep faults covered by a significant argillaceous sequence (about 700m) whilst in the Neves-Corvo mine district there are deep ore-bodies (at a depth of 500m) covered by metamorphic material with a low permeability. In both cases it has been possible to investigate the two areas obtaining interesting results. Anomalous radon values are present in the surface indicating the presence of deep faults, in the first case, and ore deposit , in the second one, not easily detectable with traditional methods.
In summary soil gas radon is a valid tool to identify different permeability degrees of clays, planar and geometrically simpler faults, deep ore deposits and preferential pathways of migration.

9. References

1. Albouy, L. (1981) Unités Allochtones à Neves-Corvo. Somincor Unpublished Report.
2. Astorri, F., Lombardi, S. and Paoloni, F. (1999) Soil gas investigations over sulphide ore-bearing fractures – the Fontalcinaldo case study (Boccheggiano and Niccioleta mining district – southern Tuscany, Italy), *Periodico Mineralogia*, **68** (2), 93-108.

3. Ball, T.K. , Crow, M J , Laffoley, N., Piper, and D. and Ridgway, J. (1990) Application of soil-gas geochemistry to mineral exploration in Africa, *J. Geochem. Expl.*, **38**, 103-115.
4. Barberi, F. and Innocenti, F. (1967) Le rocce selagitiche di Orciatico e di Montecatini in Val di Cecina, *Atti Società Toscana Scienze Naturali Memorie*, Serie A, **74**, 139-180.
5. Barnes, I., Irwin, W.P. and White, D.E. (1984) Global distribution of carbon dioxide discharges and major zones of seismicity, *Int. Rep. Dept. Int. U. S. Geol. Survey.*
6. Barriga, F.J.A.S. and Carvalho, D. (1983) Carboniferous volcanogenic sulphide mineralizations in southern Portugal (Iberian Pyrite Belt), in M. J. Lemos de Sousa and J.T. Oliveira (eds), *The Carboniferous of Portugal", Memorias Servicos Geologicos Portugal*, **29**, 99-113.
7. Bertrami, R., Buonasorte, G., Ceccarelli, A., Lombardi, S., Pieri, S. and Scandiffio, G. (1990) Soil Gases in Geothermal Prospecting: Two Case Histories (Sabatini Volcanoes and Alban Hills, Latium, Central Italy), *J. of Geophysical Research,* **95,** 21,475 - 21,481.
8. Borsi, S., Ferrara, G. and Tongiorgi, E. (1967) Determinazioni con il metodo K/Ar delle età delle rocce magmatiche della Toscana, *Boll. Soc. Geol. It.*, **86**, 403-410.
9. Ciotoli, G., Guerra, M., Lombardi, S. and Vittori, E. (1999) The behaviour of rare soil gases in a seismically active area: the Fucino basin (central Itay*), Il Nuovo Cimento*, **22**,567-575.
10. Conticelli, S., Manetti, P. and Menichetti, S. (1992) Mineralogy, geochemistry and Sr-isotopes in orendites from South Tuscany, Italy: constraints on their genesis and evolution, *Eur. J. Mineral.,* **4**, 1359-1375
11. Duddridge, G.A, , Graiger, P. and Durrance, E.M, (1991) Fault detection using soil gas geochemistry *Quatern. J. Engin. Geol.* , **24**, 427-435
12. Etiope, G. and Lombardi, S. (1994) ^{222}Rn soil-gas in sedimentary basins in central Italy: its implications in the radiation protection zoning, *Rad. Protect. Dosim.,* **56/1-4**, 231-233.
13. Fazzini, P., Gelmini, R., Mantovani, M. P. and Pellegrini, M. (1972) Geologia dei Monti della Tolfa, *Mem. Soc. Geol. It.,* 65-144.
14. Ferrara, G. , Leoni, L. , Sartori, F. and Tonarini, S. (1988) Sr-content and Sr-isotopic composition in contact-metamorphosed argillaceous sediments (Orciatico, Tuscany, Central Italy): relation to fluid circulation, *Rendiconti della Società Italiana di Mineralogia e Petrologia*, **43**, 121-128.
15. Follieri, M., Magri, D., Sadori, L. and Villa, I.M. (1991) Palinologia e datazione radiometrica 39Ar/40Ar di un sondaggio nella piana nel Fucino (Abruzzo). *Evoluzione dei Bacini Neogenici e Loro Rapporti con il Magmatismo Plio-Quaternario Nell'rea Tosco-Laziale*, Abstract, 90-92, Workshop, Pisa, June, 12-13.
16. Galadini, F. and Messina, P. (1994) Plio-Quaternary tectonics of the Fucino basin and surrounding areas (Central Italy), *G. Geol.*, **56/2**, 73-99.
17. Gregory, R.G. and Durrance, E.M. (1985) Helium, carbon dioxide and oxygen soil gases: small-scale variations over fractured ground *J. Geochem. Expl.*, **24/1**, 29-49.

18. Hermansson, H.P. , Akerblom, G. , Chyssler, J. and Linden, A. (1991) Geogas, a carrier or a tracer? SKN Report 51.
19. Illing, V.C. (1933) Migration of oil and natural gas, *Inst. Petroleum Technology J.*, **19**, 229-274.
20. Kristiansson, K. and Malmqvist, L. (1987) Trace elements in geogas and their relation to bedrock composition, *Geoexpl*, **24**, 517-534.
21. Leca, X., Ribeiro, A., Oliveira, J.T., Silva, J.B., Albouy, L., Carvalho, P. and Merino, H. (1983) Cadre géologique des minéralisations de Neves-Corvo (Baixo-Alentejo, Portugal). Lithostratigraphie, Paléogéographie et Tectonique, *Bureau de Recherches Géologiques et Miniéres Orléans. Mem B.R.G.M.*, **21**.
22. Leoni, L., Polizzano, C., Sartori, F. and Sensi, L. (1984) Chemical and mineralogical transformation induced in Pliocene clays by a small subvolcanic body and consequence for the storage of radioactive wastes, *N. Jb. Mineral. Mh.*, 155-168.
23. Lombardi, G., Nicoletti, M. and Petrucciani, C. (1974) Età delle vulcaniti acide dei complessi Tolfetano, Cerite e Manziate (Lazio nord-occidentale*), Per. Mineral.*, **43**, 181-204.
24. Lombardi, S. and Reimer, G.M. (1990) Radon and Helium in soil gases in the Phlegrean fields, central Italy, *Geophysical Research Letters,* **17/6**, 849-852.
25. Lovell, J.S. and Hale, M. (1983) Application of soil-air carbon dioxide and oxygen measurement to mineral exploration, *Trans. Inst. Min. Metall.* Sect B, **92**, 28-32.
26. Malmquist, L. and Kristiansson, K. (1984) Experimental evidence for an ascending microflow of geogas in the ground, *Earth Planet Sci. Lett.* , **70**, 407-416.
27. Malmquist, L. and Kristiansson, K. (1985) A physical mechanism for the release of free gases in the lithosphere, *Geoexpl.*, **23**, 447-453.
28. Masi, U., Ferrini, V., O'Neil, J.R. and Batchelder, J.N. (1980) Stable isotope and fluid inclusion studies of carbonate deposits from the Tolfa Mountains Mining district (Latium, Central Italy), *Mineralium Deposita*, **15**, 351-359.
29. Mogro-Campero, A. and Fleischer, R.L. (1979) Search for long-distance migration of subsurface radon. *The Natural Radioation Environment III*, Gesell, Louder, McLaughlin Ed , DOE Conf 780422, Washington.
30. Muskat, M. (1946) *The flow of homogeneous fluids through porous media.* J. W. Edwards Inc. - Ann Arbor, Michingan.
31. Oddone, G. (1915) Gli elementi fisici del grande terremoto marsicano-fucense del 13 gennaio 1915, *Boll. Soc. Sismol. Ital.*, **19**, 71-215.
32. Pandey, G.N., Rasin Tek, M.,and Katz, D.L. (1974) Diffusion of fluids through porous media with implications in petroleum geology,. *The American Association of Petroleum Geologist Bulletin*, **58**, (2), 291-303.
33. Patacca, E., Sartori, R. and Scandone, P. (1992) Tyrrhenian Basin and Apenninic arcs: kinematic relations since Late Tortonian times, *Mem. Soc. Geol. Ital.*, **45**, 425-451.
34. Reimer, G.M. (1980) Use of soil-gas helium concentrations for earthquake prediction: limitations imposed by diurnal variation, *J. Geophys. Res.* , **85**, 3107-3114.

35. Ribeiro, A. and Silva, J.B. (1983) Structure of the South Portuguese Zone, in M. J. Lemos de Sousa and J.T. Oliveira (eds) *The Carboniferous of Portugal, Memorias Servicos Geologicos Portugal*, **29**, 83-89.
36. Rodolico, F. (1934) Ricerche sulle rocce eruttive recenti della Toscana, II. Le rocce di Orciatico e di Montecatini in Val di Cecina, *Atti Soc. Toscana Sci. Nat. Mem.*, **44**, 177-202.
37. Rose, A.W., Hawkes, H.E. and Webb, J.S. (1989) *Geochemistry in mineral exploration,* Academic press, London, 657 pp.
38. Schery, D., Whittlestone, S., Hart, K.P. and Hill, S.E. (1989) The flux of Rn and Th from Australian soils, *Journal of Geophysical Research*, **94/D6**, 8567-8576.
39. Silva, J.B., Ribeiro, A., Fonseca, P., Oliveira, J.T., Pereira, Z., Fernandez, J.P., Barriga, F.J.A.S., Relvas, J.M.R.S., Carvalho, P., Ferreira, A., Beliz, A., Caetano, P., Pacheco, N. and Albernaz, J. (1997) Tectonostratigraphic overview of Neves-Corvo mine in the context of the Variscan orogeny. *SEG-Neves-Corvo Field Conference 1997,. Abstract.*
40. Stefanini, G. (1934) Il complesso eruttivo di Orciatico e Montecatini, provincia di Pisa, *Atti Soc. Toscana Sci. Nat. Mem.* (A), **44**, 224-300.
41. Varheghy, A., Baranyi, I. and Somogyi, G. (1986) A model for the vertical subsurface radon transport in "geogas" microbubbles, *Geoph. Trans.*, **32/3**, 235-253.
42. Varheghy, A., Hakl, J., Monnin, M., Morin, J.P. and Seidel, J.L. (1992) Experimental study of radon transport in water as a test for a transportation microbubble model, *J. Applied Geoph.* , **29**, 37-46.
43. Washington, H.S. (1900) Some analyses of Italian volcanic rocks, *Am J. Sc.*, **9**, 46-48.
44. Wilkening, M.H., Clements, W.E. and Stanley, D. (1972) ^{222}Rn flux measurements in widely separated regions, in *The natural radiation environment 2°, USAEC, VA,* 717-730.

Part III

SOCIAL AND ETHICAL ASPECTS

PHILOSOPHY OF ECOLOGICAL ETHICS EDUCATION, CONSIDERING THE ISSYK-KUL LAKE REMEDIATION MECHANISMS

Umut ASANOVA
National Academy of Sciences of the Kyrghyz Republic
Chui Prospekt, 265a , Bishkek, Kyrgyzstan

It is quite understandable that human factors lie on the basis of the environmental situation decrease alongside with objectively gone global changes factors. Investigations of the environmental state of the Issyk-Kul Lake as any other natural object seem not to be full when limited by drawing the data of only natural sciences.

For the evaluation of the environmental risk related to the Issyk-Kul Lake by collecting the data or integrating the knowledge on this existing so far, it is necessary to turn the attention also to the analysis of data relating to human factors. But first we should define what do we call human factors? Where, when and why are they existing? What is the mechanism of their action? When human factors cross the line to become a negative, destructive force? What are the criteria to evaluate these human factors as real threatening ones? How can we act or modify them to turn out as a positive environmental force? These and many other related questions refer to philosophy of environmental ethics education alongside with the study of natural sciences ethics education.

Let us try to define first the concept of human factors. People adapt their environment and at the same time are themselves adapted by it throughout their historical and individual life. This process of mutual inter adapting takes place in the processes of all the aspects of the economy: industry, agriculture, fishery, breeding, construction, science, even leisure.

Human activity is the centre of this mutual inter-adapting of society and nature. But this phenomenon of inter-adapting is very illusive. What is hidden behind the visible natural occurrences and human reality? When we reveal that hidden something manifested in different diseases, ill-mind, psychosis of a human being from one side, and illness of the environment from the other side we can reveal that "that something" is the very human activity being as a human mal-factor.

Concerning the Issyk-Kul Lake, it is demonstrated in the activities of mal-educated local authorities when without the knowledge of the natural laws they command to turn the streams of rivers not to reach the Lake in order to irrigate the fields around it, for the Lake to become inevitably shallower and shallower, to fertilize the soil with harmful pesticides not being aware that there exists a circulation of underground soil and waters thus poisoning the water of the Lake, where people swim and imperceptibly getting ill,

J. Klerkx and B. Imanackunov (eds.), Lake Issyk-Kul: Its Natural Environment, 263–267.

not saying of eating products grown in this way, when they construct tails just on the places (as an example Kadji Sai uranium tails) from which, should it be even as the result of a small earthquake, there could go all the terrible poisons just to the waters of the Issyk Kul Lake, when build different resort places without consideration all the possible variants of making harm to the environment and thus to peoples' health, when there concentrate more and more waists and no strict rules of their safe transportation and sanitary does exist.

The Issyk-Kul basin is not only the Lake itself, not only the properties of its water, evolution of its changes but rather its coastal life with population reaching more than half a million, with 3 large cities (Kara Kol, Balykchy and Cholpon Ata) and hundreds of settlements and villages, large and small. Some 80 % of its population are Kyrghyzes, about 15% are Russians, 5% are others. All the coastal life is based on the agricultural industry and sheep, cattle breeding.

The transition to the Market Economy has caused tremendous changes in individual life, the economy having turned completely to private, wild forms of the survival market. Rich properties of natural land lead the other years to a very high harvest of fruit - wonderful sorts of apricots, apples, cherries, raspberries etc. But the tragedy of human factors of local authorities displayed in leaving their people without the clever and sustainable organization of these harvested products beneficial marketing is another side of the Issyk-Kul Lake problems.

Saying of human factors it will be not very much as just as to refer it only to the mal-activities of local authorities. Very often it should and must be referred to human activities of the whole community, the absence or insufficiency of their environmental ethics. Throwing the waist and all kind of rubbish just on the shore of the Lake, breaking the glass bottles, making fires on the shore, parking just to the coastal part, alcohol and other drugs drinking, fishing not at the appropriate times and without special permissions and many other mal-activities of people are also called mal-human factors of ordinary people.

So we see that in the question of the evaluation of the environmental state of the Issyk-Kul Lake, we should first evaluate the state of mind of people, their being aware of their own activities threat not only to the environment but to their own health and life quality. It is very much determined by the conviction that we should see the powerful role of mind in the human activity. The quality of mind or awareness or consciousness depends on how deeply, how foreseeingly a human behaves towards the environment, how he co-relates himself with the environment. This state of mind must be the core of the study. As far as it is stated we have the right to propose that one of the very mechanisms of the Issyk-Kul Lake remediation may be the environmental ethics education oriented to humanize, the same as to ecologize the mind of people. Certainly it refers first to the environmental ethics education of local bureaucrats, the policy decision making of whom very much influences the environment quality.

In this philosophy we should come from the most important, to cultivate the principle: how you behave the nature, it will behave you the same way one day. The degree of your health depends on the degree of the environmental health. Obtaining the knowledge of this inexorable law of inter-dependence should lie in the basis of environmental ethics education. The *primary aspect of the humanization or ecologization* of mind is considered to learn (from the side of students) or to teach (from the side of teachers) to be aware (through books, lectures, talks, experience, practice) of natural laws that govern the life, that subordinate the human being, to learn to live in accordance with Nature, in harmony, not in confrontation with this ever powerful creator.

Speculations of this general kind are far from confirmation that may bring to the thought that Kyrghyz people may be considered as an environmentally ignorant or uncultural nation. Their history, traditions, customs, art, the whole culture on the contrary is regarded as uniquely ecologically oriented in all its manifestations culture. Their way of life, economy, thinking, actions were uniquely ecological. The philosophy of environmental ethics education should first of all appeal to this traditional heritage of the Kyrghyz people. The primary goal of such an education is to give a re-birth to that traditional mentality, to appeal to the roots.

It should be said that the Soviet time, though it brought many positive results in the economical, political, educational life of the Kyrghyz people, made a great mistake in having destroyed the institute of religion gone from the time unmemorable thus fully minimizing its role as an inner strongly regulating force of peoples' behavior. In the attempt to unify mentality of all the people of the Soviet Union developing the mentality of the so called one "soviet narod" (people, nation) a great deal of unique traditions, rituals, customs were gradually becoming to be forgotten, especially by younger generations.

Even the native language was being replaced by one more powerful (in its historic, social and political aspects) language - Russian. A nation forgetting its own native language is not surprising to have forgotten its own heritage. It was the Soviet time when Kyrghyz people learned to drink alcohol and some of them to become terrible alcoholics as never they had done and been before. It was this time when they learned to be double-minded, to conduct a disordered life, to do crimes, to divorce, to leave children to grow without ever knowing what love of loving mother or father means … .

It is not surprising then what and how will be the attitude of them to the Nature. A person not able to feel other person's sorrow and pain never is able to feel pain for the Nature. A person not responsible in his own social and family life never is able be responsible for the Nature. So the primary task of any education is to awaken human feelings first. Human's body obeys the same natural laws. The law of nature is to grow, to move spontaneously to happiness, order, harmony, quietness. This law of natural intention of body must be taken as a basis for a human ethics thus environmental ethics education. This may be reached by the meditation practice as one of the necessary and most powerful intellectual tools of harmonizing the human mind first.

The second powerful instrument of ethics education is as it was told above - appealing to our own heritage. Rich folklore, very unique in their symbolic manifestations traditions, rituals, customs may serve as this ever strong tool. Cult of Nature - this is the most characteristics of the Kyrghyz traditional heritage. Cult of mountains, rivers, water, lakes, trees, different animals, sky, sun, moon, earth and other natural phenomena runs all through different legends, fair tales, epical songs, taboo, traditions and rituals.

These traditions of the Nature worshipping have been retained up to present days, displayed in different rituals of sacrifices to the God of Sky, called *Tengri*, which refers also to the Sun *Tengri* or the Moon *Tengri*. The concept of *Tengri* is associated with the highest God - the Sky. Moreover its meaning is very much broader, embracing the meaning of the Universe. Attitude to different places as to sacred ones (*mazars*) and performing the rituals of pilgrimage and worshipping - is another aspect of the cult of Nature.

The Issyk - Kul Lake was the very sacred place (*mazar*) to perform the pilgrimage and worshipping not more than some century ago. Even now one can meet traditional elderly people who do not even swim in the Lake and regards it as a divine (alive) phenomenon. Mountains and rivers running to the Lake were considered sacred, animated ones. There are a lot of places such as Saimaly Tash, Cholpon Ata, located near the known resort town of the same name, Kara Kol Ata etc. considered as mazars. For example, it is believed that a "*saint*" Cholpon Ata (Father Cholpon) was buried on the hill - *dobe*. There one could find two sources (springs), one - with hot and another one - with cold water, there grew poplars and willows. Barren women and sick people came here from the north coast of the Issyk-Kul Lake inhabiting from Chon Oructu till Toru Aigyr villages. Here people maintained rituals of sacrifices - the so called *tuile*. People took with them also sick cattle or sheep, spent the night, pronounced prayers, coming here on Thursdays night and departing on Fridays. Prayers for *Earth - Water (Jer-Suu)* were also done with the aim of requesting rain. Burials of rich people (*manaps*) living not far were also maintained here on the foot of the hill.

There is a belief that the spirit - patroness of the Issyk-Kul Lake is a legendary Fore-Mother of the *Bugu* (Deer) tribe - *Muyiyzduyi Ene* (Horned Granny). One can find a lot of legends concerning this horned mother - deer, who gave the birth to this tribe of Kyrghyzes - *bugu*.

The river Ack Suu (that is the inflow of the river of Jergalan, Issyk Kul basin) was very well known among ancient Kyrghyzes with its warm healing sources, called Alma - Arasan (mineral source in the valley of the river of Ack-Suu, another spring is called Altyn-Arasan), is rounded by dense trees, many apples to grow among them. These trees rounding the spring are considered sacred; one can find many rags of different colour and size hung on the trees: these were and are the sacrifices of Kyrghyzes to the genius of the spring. This mazar is called *Almaluu* (place where many apple trees grow, *alma* means apple trees), people in the past came here, cut the sheep, made bread, drank water from the spring, bathed in the dig filled with spring's water. Some visitors said that they had seen a white young of camel (*ack tailack*). He is considered a "patron" of this spring (*bulacktyn aesi*).

There are a lot of similar beliefs existing in Kyrghyz culture, in the memory of the Kyrghyz people. It would be wrong to consider them as only tales or simple prejudices. It might have taken centuries of experience for the Kyrghyz people in order to establish such a system of the Nature worshipping. We should rationalize them and see the main law in them: unity with Nature; living in harmony with it; strict subordination to natural laws that were not formulated as we are now trying to do, but it was just in practice without any theoretical speculation. It was a practical knowledge of the environmental

laws. And now when we see the Nature being in great danger together with people living within it will be more than silly to think that Kyrghyz people were ignorant in their knowledge of natural laws. The example of unique Kyrghyz natural astral (moon and sun) calendar and activity unexceptionally based on it (when to sow, move from one place to another for Kyrghyzes were nomadic people, when to marry, when to maintain big holidays, how to manage with cattle and sheep breeding, the names of months and years according to the moon and the sun, and many other unique rules of every day life) is the bright evidence of this practical knowledge of the laws of Nature.
So the task of the environmental ethics education regarding the Issyk-Kul Lake remediation is first of all to give a re-birth to this traditional ecological knowledge by more thorough collecting and systemizing, by establishing a Data Base on this Practical, Traditional Ecological Knowledge of the Kyrghyz people. Second - to cultivate this knowledge among people, especially among local authorities, through secondary schools, high institutions, seminars and trainings, radio and television broadcasting. One can assume that this is not at all done now in our practice. Yes, all these tails, legends, traditions are being broadcasted, told, but the task is to turn the attention radically to their essence - environment keeping, loving, perishing meaning, to their ecological wisdom. Being taught on the basis of that wisdom is a primary task for humanization of the mind.
Environmental ethics education alongside with appealing to traditional heritage of the Kyrghyz people, should be based on the modern science data demonstrating how dangerous is the state of the environment, that is of the Issyk-Kul Lake, particularly. The data base collected on the environmental situation of the Lake reported on the NATO Seminar may also serve as one of the effective instruments of such an education.

References

1. Abramzon, S. (1990) *Kyrghyzes and their ethnogenetic and historico-cultural links,* Frunze: Kyrghyzstan (In Russian).
2. Potapov, L. (1990) *Altai Shamanism,* Leningrad: Nauka (In Russian)
3. Muras (Kalygul, Arstanbeck, Moldo Kylych, Aldash Moldo, Jenijok, Tocktogul). - Frunze: Kyrghyzstan (In Kyrghyz)
4. Bernshtam, A. (1998) *Selected works on Archeology and History of Kyrghyzes and Kyrghyzstan.* , Bishkek (In Russian)
5. Bartold, B. (1996) *Selected works on the History of Kyrghyzes and Kyrghyzstan,* Bishkek (In Russian)
6. Asanova, U. (1992) The essence of the environmental education, in *Ecology and Health,* Bishkek, pp. 42 - 51 (In Russian)

ECOSYSTEM MANAGEMENT, ENVIRONMENTAL VALUES AND LAKE ISSYK-KUL

Gregory COOPER
Society and Professions Program on Ethics
Washington and Lee University, Lexington, USA

1. Introduction

The concept of an ecosystem entered ecology when Tansley [1] coined the phrase to emphasize the cohesion of the physical and biological characteristics of the earth's biota. While the term was new, however, the idea was not. Limnologists, in particular, had been treating lakes and rivers from an ecosystem perspective for decades prior to this [2][3]. In more recent times, the ecosystem concept has been making its way from the science of ecology proper into the arena of environmental policy. *Ecosystem management* has emerged as the strategy of choice among a broad range of environmental decision-makers. Here, as well, the biology of fresh waters has led the way. Since at least 1972, when the United States passed the Clean Water Act calling for restoration of the biological integrity of the nation's lakes and streams, environmental managers have self-consciously been managing these water resources from an ecosystem perspective.

In part this ecosystem perspective reflects the recognition that entire watersheds, including the human inhabitants, must be the focus of management activities. However, it is also a reflection of the systemic character of the values that drive the ecosystem approach. The explicit policy goals of ecosystem management – the restoration of biological health and/or integrity and the sustainable delivery of the goods and services that these aquatic systems provide – are grounded in the recognition that it is the entire system, living and non-living, human and non-human, that is the source of value. No doubt the relatively self-contained character of freshwater lakes and streams has facilitated their early treatment from the ecosystem perspective, but that is not the important point here. What is important is the way this rather robust history of ecosystem management in fresh waters gives us relatively taller shoulders to stand on when it comes to asking ecosystem management questions about Lake Issyk-Kul.

But what is ecosystem management? What is significant about it as a management strategy? What are the implications of an ecosystem management approach for the environmental evaluation and remediation of Lake Issyk-Kul? Most important, what sorts of value commitments can and should underwrite an ecosystem management

J. Klerkx and B. Imanackunov (eds.), Lake Issyk-Kul: Its Natural Environment, 269–281.

approach to the lake? This paper will have something to say about each of these questions; however, most of the attention will be devoted to the last one, to the normative standards appropriate to an ecosystem management approach to Lake Issyk-Kul.

2. Ecosystem Management

Although he did not use the phrase, the ideas behind ecosystem management trace back to the environmental writings of Aldo Leopold [4]. Ecosystem management involves identifying the proper entities for management activity in ecological rather than political terms, and evaluating management practices in light of their impact on the entire system rather than on isolated populations and/or locations (see, e.g. [5][6]).

There are a number of distinctive features of ecosystem management, but four stand out as perhaps most fundamental. First, it involves paying special attention to questions of scale in both space and time. Scale is important in individuating the ecosystems that are the objects of management activities; one cannot, for example, hope to effectively manage a stream by focusing on stream reach alone, ignoring the larger watershed that is drained by the stream. Spatial scale is also important for both individuating and understanding the populations of organisms that comprise ecosystems. Colonization and extinction events at one spatial scale may be elements of a more stable metapopulation that is only visible at a larger spatial scale [7]. Temporal scale is important as well. For example, the functional role of many types of ecosystem disturbances, such as fire, can only be appreciated when viewed from the appropriate temporal perspective. These insights have led ecosystem managers to rethink once common management practices such as across-the-board fire suppression.

A second distinctive feature of ecosystem management is the way in which it seeks to bring together the two major perspectives of ecology: the systems ecology approach and the organism ecology approach. The former views biological communities in physico-chemical terms, focusing on flows of energy and materials through the system. Organisms are viewed as transducing nodes that modulate the flow of energy and/or materials through the system, and are classified primarily in terms of broad functional trophic categories (e.g., decomposer, primary producer, etc.). Biogeochemistry is perhaps a descriptively more appropriate name for this field [8][9]. Organismic ecology, on the other hand, is directed towards understanding patterns of distribution and abundance of organisms and, to the extent that this is a separate study, the ways in which those populations aggregate into communities. Ecology in this sense is, in fact, the science that Haeckel named, the science that studies the Darwinian struggle for existence (see [9] for further details). Since the emergence of ecosystem ecology as a strong research tradition after World War II, ecologists have sought a conceptual framework that would unify the two enterprises [10]. While there has been little success unifying these two fields of ecology at a general level, ecosystem management seeks an integrated understanding for specific systems, and some limited progress has been made, especially regarding the relationship between ecosystem properties such as productivity and organismic features such as diversity [11].

A third significant feature of ecosystem management focuses on the incorporation of humans into the system. The most difficult environmental challenges involve managing ecosystems that are heavily shaped by human hands. The ecosystem manager recognizes that successful management must sustain not only valuable biological features of the system, such as the preservation of biodiversity, but also various sorts of desirable cultural practices as well. One of the early shortcomings of the environmental movement, at least in the United States, was a rather myopic focus on wilderness. This has been unfortunate. Though involving numerous practical and logistical difficulties, management strategies for wilderness preservation are relatively straightforward in concept – minimize human impacts as much as possible is a decision principle that will take one a long way in the arena of wilderness management. We confront much more difficult questions when we come to managing the land on which people actually live, when we must decide what kind of human face to give the land. Ecosystem management has taken this process to the most intensively used of human habitats, spawning new fields such as industrial ecology and urban ecology in the process [12][13][14].

Finally, ecosystem management adopts, at least in part, certain ecosystem-level properties as policy goals. Typical goals include the pursuit of ecosystem integrity, ecosystem health, and the sustainable delivery of various ecosystem goods and services. In this paper I want to focus on this last theme, on the policy goals and the values that lie behind them. What gives an ecosystem integrity? What does a healthy ecosystem look like? How do we assign a value to the goods and services that an ecosystem delivers? I want to turn to these questions in the specific context of using ecosystem management strategies in the environmental assessment and remediation of Lake Issyk-Kul, not to suggest specific courses of action – unfortunately the situation at Issyk-Kul is not well enough understood for that – but to suggests the kinds of questions we should be asking. However we must first give some thought to general matters: what sorts of *criteria for assessment* and *targets for remediation* should be used?

We can turn in two directions for help with these difficult matters. On the one hand, we can take advantage of the earlier mentioned fact that, from a biological standpoint, a great deal of effort has been devoted to clarifying the notions of health and integrity for freshwater lakes and streams. This work will help to generate operational criteria for the application of notions like health and integrity, but will not do much to help with questions about the values that ultimately ground these ideas as desirable policy goals. For this we must turn more explicitly to environmental philosophy. I begin with the latter.

3. The Grounding of Ecosystem Values

Environmental philosophy, as a self-conscious domain of inquiry, has existed for about three decades. Perhaps the most striking feature of the field, over this period, has been the incessant and uncompromising clash between two schools of thought – what we might call the moralist and the economist [15]. For the moralist, nature has, in some sense, intrinsic value, and this intrinsic value gives nature a kind of moral standing that must be respected, even when it conflicts with the pursuit of serious human interests [16][17][18] (for a contemporary overview of this line of thought, see [19]). The economist, on the other hand, is concerned with the satisfaction of human preferences. When nature matters, it is because it matters to people [20]. While this is an over-simplification, and while there are many signs that the stark dichotomy is breaking down, it is still useful as a distinction from which to begin.

Historically, those who have focused on ecosystem integrity have tended toward alignment with the moralist camp. They have engaged in something like the following chain of reasoning: ecosystems with the most integrity are those least disturbed by human activity; we should respect and not interfere with these ecosystems because they have intrinsic value; and they have intrinsic value for one of two reasons – either they have a kind of moral standing (see references in previous paragraph) or they have aesthetic value [21]. Those who have focused on ecosystem health, on the other hand, have tended toward alignment with the economic camp. They reason along the following lines: healthy ecosystems are functionally intact; being functionally intact they can continue to deliver the goods and services upon which humans rely; and we care about the delivery of these goods and services because they are essential to human well-being and could be replaced only at significant economic cost [22][13][23].[1]

While this is only the barest of outlines, it does reflect a fundamental fact about attempts to ground the policy objectives of ecosystem management. Such attempts tend to fall into one of two categories: they either base our concern for ecosystems on the intrinsic value of the ecosystem itself, or they base it on instrumental considerations ultimately rooted in a concern for human well-being. We will return to these options, and the differences they make, a bit later, after looking at efforts to articulate the practical import of health and integrity as policy goals.

4. Ecosystem Values in Practice

Environmental philosophy, then, provides very different normative justifications for the twin goals of ecosystem integrity and ecosystem health. When we turn to the biology relevant to operationalizing these goals, there are significant differences as well, at least in theory. James Karr [24] has developed what he calls an "index of biological integrity" to measure the degree to which aquatic ecosystems have integrity. His index involves identifying natural benchmarks that are similar to ecosystem under study, but that have been subject to significantly less human interference. They represent, in effect, what evolution, as opposed to the hand of man, has produced. Ecosystem integrity is

[1] This is not meant to suggest that the economic camp is unconcerned with moral matters – at least to the extent that economic value is ultimately grounded in some form of human flourishing, it is a thoroughly moral position. The main difference is that this camp does not seek to extend morality into the realm of nature.

then evaluated by comparing the target ecosystem with the benchmark along a variety of dimensions (species composition, trophic complexity, etc.). Broadly speaking, the closer a given ecosystem is to the target values in these various dimensions, the more integrity it has.

Ecosystem health is evaluated, in principle at least, by looking more directly at the mechanisms responsible for ecosystem function: are bottlenecks being created that would impair the flow of energy and/or materials through the system, and thus impair the delivery of ecosystem goods and services upon which we have come to reply? A shortage of decomposers to release nutrients back into the system would be an example of such a bottleneck [8]. This is a difference "in principle" because in practice we are often forced to fall back on something like natural benchmarks when it comes to making judgments about ecosystem function; the level of understanding of the mechanistic basis behind ecosystem function is often rather slim. Still there is a real difference between the two approaches [25]. It can be illustrated by the attitude toward introduced species. From the standpoint of ecosystem integrity, introduced species are always to be avoided – they always reduce the integrity of the communities they join. From the perspective of ecosystem health, on the other hand, introduced species may come to play an important functional role and thus contribute to the health of the ecosystem.

There is a further distinction to be drawn here. The ecosystem health view, as sketched above, is itself subject to several interpretations. The crucial notion is ecosystem function. Function is a teleological notion; it describes goal-directed behavior. If health is to be defined in terms of proper function, then the goal of this functional activity must be specified. And here there are several options. First, we must decide whether the functional goals in question have anything to do with human ends. If not, then we have what might be called the *purely ecological approach*, where ecosystem health is rendered exclusively in terms of criteria that are susceptible to ecological measurement. For example, one might view ecosystems as engines designed to capture solar radiation and turn it into biomass. A functional breakdown, then, would be any change that reduces this capacity. Net primary productivity (NPP) might be an appropriate measure of health on this view [8]. Of course, this approach has the unfortunate consequence that some of our most polluted streams and rivers, those subject to excessive anthropogenic eutrophication, would count among our healthiest ecosystems. We may value productivity in a *ceteris paribus* sort of way, but it is not synonymous with ecosystem health.

An alternative is to focus not on productivity itself, but on the manner of productivity. A number of system ecologists who share a thermodynamic approach to the subject matter have capitalized on the fact that ecosystems are dissipative structures [26][27]. The swirling water in the sink that appears when the plug is pulled is an example of a dissipative structure; it develops as an efficient way to dissipate the energy stored in the water contained in the sink. In a similar way, ecosystems dissipate the solar energy that falls on their region of the planet surface. And in a similar way, they can do so more or less efficiently. The "optimum operating point" is the benchmark of optimality in this case. Ecosystem health (or sometimes integrity) increases as an ecosystem approaches its optimum operating point. This approach does not have the same kinds of immediate unhappy consequences as the previous strategy; however, it does have its own set of problems. I will not pause to document them here, save to mention the difficulty that, to me, is most serious. Proponents of this view tend to rely on natural selection as the optimizing agent. The mechanism of selection is somehow responsible for the tendency of ecosystems to move toward their optimum operating point when sheltered from dramatic disturbance. The problem is the "somehow." We have, in my view, no mechanistic understanding of how natural selection is supposed to work in this case (but see [28]). Without such an understanding the teleology floats free of the biology.

It is tempting, at this point, to make the following reply. Why all the fuss about a unitary measure of ecosystem health? If ecosystem function is the key notion then there is no problem; ecologists talk about ecosystem function all the time. It is well understood that primary producers function to capture solar radiation, higher level consumers function to send that energy up through food chains, decomposers function to recycle nutrients, and so forth. Ecologists also understand when these functions are put in jeopardy; the ideas behind ecosystem distress syndrome [29], and ecological risk assessment more generally [30], are grounded in this understanding. But I would argue that this kind of functional understanding is not sufficient to give us a truly workable and comprehensive notion of ecosystem health. Suppose, for example, that the decomposer community in a particular ecosystem is not up to the playing their functional role and a peat bog develops. Can we say that this was an unhealthy development? After all, peat bogs are interesting and, in some ways, valuable places. Somehow the question does not make sense. It makes more sense, I believe, to return to the fork in the road that was presented several paragraphs back – where the question was posed whether ecosystem health was to be defined independently of human ends – and to take that other fork.

If ecosystem health is defined in terms of the delivery of goods and services, then another problem arises. How do we individuate the goods and services that we value? Again, there are, broadly speaking, two approaches. The narrow approach essentially reckons goods and services in economic terms. Ecosystems serve as the source of natural resources and the sink that absorbs our pollution. How do assign value here? The question is: how much would it cost us to get these goods and services in some other way? There is no doubt that this is an enormously significant measure for environmental policy. We are in the process of developing a solid understanding of economic magnitude of services like the water purification function of wetlands; certainly there is a long way to go, but the early returns suggest major impacts (see [31] for examples). The worry is not that it is irrelevant, but that it is too narrow. If we only count goods and services that can be reckoned in this way then perhaps we have not captured the full measure of the value of the various ecosystems.

A broader approach would be more expansive about what counts as a good and/or a service [32][22][13]. Perhaps ecosystems deliver on goods and services that are not easily captured in economic terms. An obvious category here is aesthetic value. Economists use techniques such as contingent valuation in an attempt to measure things like the value of a sunrise over a smog-free landscape, but there are two sorts of problems. How do we attach ontological significance to the measures such studies obtain – what are they measures of? That is the first problem. But suppose it is solved; suppose we have made sense of the notion of a *preference map* that represents the normative stances of various types of individuals in the population. Then a second problem arises. What is the normative significance of preference maps? Why, in other words, does it matter for environmental policy that people prefer what they do?

Here the worries multiply. Sagoff [33][34] has urged a distinction between preferences, the normative attitudes susceptible of economic measure, and what we might call community values – ideas about what we owe to each other, what is good for the group, our responsibilities to future generations, and so forth. On his view, we should set policy on the basis of these non-economic values as revealed ultimately in the democratic political process. It may be that Sagoff is excessively hard on economic value, but this only means that we need to make more room for it as a policy consideration than he does. It does not undermine his worries about a normative arena dominated by economic concerns. There is a closely related problem from political philosophy involved here as well. To what extent does *rational* policy reflect the revealed preferences of the group? We are confronting hoary problems and will not settle them here. Without such a solution, it is best to suppose that not all ecosystem values are susceptible to economic measure; in the discussion to follow, the idea of goods and services will be interpreted broadly. This strategy has its pitfalls as well, not the least of which is assigning operational measures of success in the pursuit of values like sustainability and aesthetic significance. But it seems better to recognize these values, even if we cannot measure them, than to ignore them altogether.

This section began with the question of practical relevance. What, in concrete terms, does it mean to take the pursuit of ecosystem integrity as policy initiative? How do we know when we are enhancing ecosystem health and when we are not? We have identified a number of perspectives from which to address these questions. For integrity we have indices such as Karr's that measure the extent to which an ecosystem has been moved by anthropogenic influence away from its natural state. These indices prompt us to focus on measures such as taxonomic richness, shifts in species composition, health of individual organisms, and food web organization. But Karr's index of biological integrity gives us an operational measure only for the relatively pristine. It aspires to return ecosystems to a state that is as free from human influence as possible. However, this is an aspiration that makes sense for relatively few of this planet's ecosystems, as Karr [25] himself readily admits. As he points out, for the bulk of the earth's ecosystems the most we can hope for is health.

We have seen several approaches to health. The narrowest of these, the purely ecological strategy, capitalizes on an ecological understanding of ecosystem function. Here we face a choice. Either we operate in a rather piecemeal fashion, using the rather unsystematic knowledge of ecosystem function that ecology has developed over the years, or we adopt a unitary measure such as the optimum operating point of the thermodynamics perspective. I have suggested, rather dogmatically I fear, that there are problems with the latter; I will not make further use of that perspective here. But the former, which includes the fairly well-developed methodologies of ecological distress analysis and ecological risk assessment, is clearly worth holding onto. By itself, it is not sufficient to underwrite a very robust notion of health, but it is clearly important as an indicator of when things are likely to go wrong. Attending to ecosystem health in this narrow sense leads us to focus on a number of crucial ecosystem characteristics, including productivity (NPP), levels of species diversity, connectivity, resilience, rates of decomposition, and retrogressions in patterns of successional change (see, for example [29]). We still face the difficult task of normalizing these parameters for the ecosystem in question, but it does give a valuable, even if less than comprehensive, place to start (see [35] for further discussion).

If we include human normative concerns in the definition of health by incorporating the goods and services that ecosystems deliver, then we get something that corresponds more closely to the pre-theoretic notion of health. The clearest demonstrations of the policy implications of this interpretation of health are those situations where the replacement costs for goods and services can be determined. Daily discusses a number of ecosystem goods and services that clearly have economic significance (which is not to say that determining actual replacement costs will be easy). Here are some of the items from her list: "purification of air and water, mitigation of floods and droughts, detoxification and decomposition of wastes, generation and renewal of soil and soil fertility, pollination of crops and natural vegetation, control of the vast majority of potential agricultural pests..." [22](3-4). We can call this third approach, the incorporation of human economic concerns, the *intermediate interpretation of ecosystem health.*

There is, however, more at stake than economics. This brings us to the fourth perspective, the *broad interpretation of ecosystem health*, which incorporates economic value, but in a context that includes other categories of value as well. I will take advantage of the fact that these practical perspectives are, in a way, cumulative. By adopting the last, we get most of the advantages of the earlier alternatives. We can pursue ecosystem integrity where it is appropriate to do so. We can also recognize that attending to ecological health in the narrow ecological sense is likely to be crucial to the sustainable delivery of goods and services, however these get defined. This is consistent with a further effort to define, in economic terms, the goods and services we expect ecosystems to deliver. Finally, we can recognize that there is more than economics involved; we can seek to identify the goods and services involved in the broad conception of ecosystem health.

5. Values and Environmental Policy for Lake Issyk-Kul

What is the relevance of all of this to the environmental assessment and remediation of Lake Issyk-Kul? Let us begin with the philosophical issues discussed two sections back. Should the considerations that ultimately drive environmental policy for Lake Issyk-Kul reflect the biocentrism of the moralist or the anthropocentrism of the economist? It is important, I believe, to avoid getting locked exclusively into either of the extreme positions. The moralist is right that we sometimes value nature for reasons that resist capture in purely economic terms or in the calculus of preference satisfaction (which is not to say that moral standing for nature is the best way to reflect this concern). The economist is right to insist that we recognize the economic value, in terms of goods and services provided, of the ecosystems we inhabit. But the extreme views are the ends of a continuum. In the considerations that drive our environmental decisions however, we are rarely, if ever, at one of the ends. Rather, we are somewhere in the middle; our concerns are a mixture of valuing nature for its own sake and valuing it for its contribution to human welfare. Furthermore, the particular mixture that is appropriate will vary from one context to the next.

Turning specifically to the greater Lake Issyk-Kul ecosystem, it would be surprising to find that one particular mix of the values of the moralist and the economist holds throughout. More likely there will be areas and/or aspects of the larger ecosystem that are closer to natural benchmarks than others, and features that are of long-standing cultural significance. In such cases it may be appropriate to pursue more preservation-oriented goals. There will be other areas and/or aspects of the ecosystem that have been thoroughly integrated into human economic life. In these cases it will be important to take steps to insure that the Lake Issyk-Kul ecosystem continues to sustain these economic activities, and it is important to stress that this may involve some modification of the activities themselves.

One of the first things that is required, therefore, is an assessment of the greater Issyk-Kul ecosystem in terms of the fundamental values that are going to shape human interaction with the land. It appears that most of the areas around the lake, including the first few layers of surrounding mountains, is thoroughly integrated into the agricultural and resource extraction activities of humans (especially fisheries and the mining of precious metals). Tourism is on the ropes, but may recover to previous levels of significance with changes in the political and environmental climate. However, it appears that there is also a tremendous amount of wild and mountainous land as one travels further from the lake. Thus, while ecosystem integrity, understood as valuing ecosystems for their natural and pristine character, is not a live option for the part of the ecosystem immediately surrounding the lake, it may be a reasonable goal for the more remote regions.

For most areas, however, it is a matter of ecosystem health, not integrity. This means attending to those features indicative of ecological health in the narrower, more purely ecological, sense of health, factors such as stability of lake water levels, avoiding excessive salinization, maintaining historical levels of species diversity, avoiding simplification of food web structures, and so forth. With regard to the lake itself, we should be paying special attention to the taxonomic richness of benthic invertebrates and zooplankton and monitoring increases in pelagic phytoplankton biomass.

A concern for ecosystem health, however, also means attending to the ecosystem's capacity for sustainable delivery of the goods and services important to the people of the region. As already indicated, sometimes this can be essentially a process of economic valuation. For example, though we currently lack even a rough measure of the irrigation demands made on the surrounding mountains, by diverting water destined for the lake, such a measure should be relatively straightforward to acquire. Once in hand, its economic impact can be evaluated. A similar thing can be said about the fishery the lake sustains. Ironically, the economic assessment of more directly economic activities, such as the mining of precious metals, may be more difficult. The economic sums for the mining industry, for example, should be easy enough to do at one level: what does it do for local employment? How much does the industry add to the Kyrgyz economy overall?, etc. But at another level, there is much more uncertainty. Mining activities make heavy metals available in the environment, involve the use of potentially destructive processes of extraction, issue in accidents such as the famous Issyk-Kul cyanide spill of a few years back, and in general, yield a diminished aesthetic harvest. It is not so easy to identify the "ecological footprint" of industrial mining in this broader sense. What is this activity doing to human health and to ecological processes in the lake? Does it jeopardize the sustainable delivery of other goods and services delivered by the Lake Issyk-Kul ecosystem?

There are other significant economic activities in the Issyk-Kul basin. I have mentioned agriculture and tourism. The former is currently the dominant form of economic activity; the latter used to be of major significance and, given the natural beauty of the area, certainly has potential for the future. A management plan pursuing the broad health of the Issyk-Kul ecosystem should secure a sustainable basis for both activities. In addition, the lake and its surrounding uplands must be recognized as a sink that absorbs the by-products of human activity. Finally, given the arid nature of this part of Asia and the regional demand for water, it is clear that a management plan must include strategies for handling the snow melt from the surrounding mountains. It is no accident that, on numerous occasions during my visit, I heard the snow that caps the mountains surrounding Lake Issyk-Kul referred to as "white gold."

Beyond purely economic concerns, there are also aesthetic and moral values at stake. Lake Issyk-Kul is widely regarded by the Kyrgyz people as the "pearl of Kyrgyzstan.". It is not hard to see why; the place is blessed with tremendous natural beauty. Direct threats to this beauty could come through physical and biological processes that destroy the lake's water clarity or increase air pollution thus obscuring scenic vistas of the surrounding mountains. In addition, much of the lakeshore is only moderately developed, despite the tremendous attractiveness of potential sites. Depending on the course it might take, such development could have a tremendous impact on aesthetic value. Finally, the Lake Issyk-Kul ecosystem, owing in part to its remoteness and in part to its ecological uniqueness, is at least potentially the home of rare and endemic species. The snow leopard that roams the surrounding mountains is an example of a charismatic megafauna that comes readily to mind. However, the lake itself is also the potential home for unique and endemic species, owing to its isolation from surrounding watersheds; at the present time, the flora and fauna of the lake have not been sufficiently studied to pronounce with confidence on this score.

Turning to moral concerns, we confront an immediate problem. The Issyk-Kul region, like the rest of Kyrgyzstan following the collapse of the Soviet Union, is beset with overwhelming poverty. A central goal of any management plan must be to raise the quality of life for all of the inhabitants. There are also serious inequities in the distribution of the environmental risks that flow from current economic activities in the region. A study underway, and reported in the present volume, shows serious health risks from heavy metal contamination for certain populations living along streams that flow through mine tailings. The demands of environmental justice should prompt a removal of these inequities. Finally, given the central place of Issyk-Kul in the minds of the Kyrgyz people and its role in their conception of who they are, it is crucial to preserve the special virtues of this place as a legacy for future generations.

6. Conclusion

Some pages back, I warned that pursuing broad ecosystem health has the virtue of normative completeness – it is able to accommodate all of the various values that might be relevant to environmental decision – but that it pays for this commodiousness in terms of operational intractability. The lengthy wish-list of values just rehearsed is potent testimony to this problem. It is hard to argue against the development of a sustainable agriculture and fisheries, a mining industry that does not threaten human or ecological health, a regime of water management practices that maintains stability of lake levels, a program of economic development that raises standards of living across the board, an evening out of inequities in the distribution of environmental benefits and burdens, and so on. The problem is that the overwhelming tendency is for these various policy goals to be in tension with one another. Furthermore, since there are several different *kinds* of values at play, and no common currency (either literally or figuratively) in which to reckon them, we face serious incommensurability problems.

Let me give just two examples. At one meeting it was suggested that lake productivity may have fallen as a result of increased salinity of the lake water. It was also suggested that productivity might be increased by introducing plankton adapted to brackish water. Enhancing productivity in this way could lead to a more lucrative fishery and, thus, economic growth in the region. However, such a strategy is likely to directly conflict with at least the following normative concerns: the aesthetic value of water clarity, the ecological values associated with species diversity and endemism, and the economic value of tourism.

A second example concerns the mining industry. Extracting the precious metals from the mountains surrounding Issyk-Kul is already a significant component of the regional economy, and there is apparently the potential for further growth (though a better job could be done at insuring that more of the profits stay at home). However, cost-effective mining runs squarely into conflict, once again, with the economic value of tourism. Relatedly, there is also conflict with the aesthetic value associated with the pristine character of the surrounding mountains, and with the quality of the "white gold" they contain. In addition, at least as currently practiced, mining precious metals appears to pose serious health risks to specific populations in the Issyk-Kul basin.

How are the Kyrgyz people to navigate these difficult waters? How do we improve the economic lot of the population without compromising the environmental quality of one of the most special places on the planet? The decisions are not going to be easy, but holding this conference is an important first step. Recognizing the kinds of issues at stake, and the reasons for caring about them along the lines that I have tried to outline in this paper, is another important step. Since there can be no algorithm for adjudicating the sorts of conflicts of value we have been discussing, making policy must be a matter for the deliberative democratic process. And the deliberative process works best when the participants can clearly express the objects of their concern and the reasons for caring.

7. References

1. Tansley, A. G. (1935) The Use and Abuse of Vegetational Concepts and Terms., *Ecology* **16**, 284-307.
2. Forbes, S. A. (1887) The Lake as a Microcosm., *Bulletin Science Association of Peoria, Illinois* **1887,** 77-87.
3. Thienemann, A. (1918) Lebengemeinschaft und Lebensraum, *Naturwissenschaft Wocheschrift* **17,** 282-290, 297-303.
4. Leopold, A. (1949) *A Sand County Almanac and Sketches Here and There*, New York, Oxford University Press.
5. Costanza, R., B. Norton, and B. Haskell (eds.) (1992*)* *Ecosystem Health: New Goals for Environmental Managemen,* Washington, D.C., Island Press.
6. Grumbine, R. E. (1994) What is Ecosystem Management? *Conservation Biology* **8,** 27-38.
7. Hanski, I. (1996) Metapopulation Ecology, in O.E. Rhodes, R. K. Chesser, and M. H. Smith (eds.), *Population Dynamics in Ecological Space and Time*, Chicago, University of Chicago Press, 13-43.
8. Schlesinger, W. (1994) The Vulnerability of Biotic Diversity, in R.. Socolow et al., 245-260.
9. Cooper, G. (forthcoming). *The Science of the Struggle for Existence: On the Foundations of Ecology.* Cambridge, Cambridge University Press.
10. Allen, T. F. H. and T. W. Hoekstra (1992) *Toward a Unified Ecology.* New York, Columbia University Press.
11. Schulze, E. and H. A. Mooney (eds.) (1994). *Biodiversity and Ecosystem Function,* Berlin, Springer-Verlag.
12. Socolow, R., C. Andrews, F. Berkhout, and V. Thomas (eds.) (1994). *Industrial Ecology and Global Change,* Cambridge, Cambridge University Press.
13. Haila, Y. (1998) Assessing Ecosystem Health Across Spatial Scales, in D. Rapport et al. (eds.), 81-102.
14. Rees, W. E. and M. Wackernagel (1996) Urban Ecological Footprints: Why Cities Cannot Be Sustainable (and Why They Are a Key to Sustainability), *EIA Review* **16,** 223-248.
15. Norton, B. G. (1991) *Toward Unity Among Environmentalists,* New York, Oxford University Press.
16. Katz, E. (1997) *Nature as a Subject: Human Obligation and Natural Community,* New York, Rowman and Littlefield, Publishers, Inc.
17. Rolston, H. (1995) Duties to Endangered Species, in A. Brennan (ed.), 77-86.

18. Westra, L. and J. Lemons (eds.) (1995). *Perspectives on Ecological Integrity.* Dordrecht, Kluwer Academic.
19. Katz, E., A. Light, and D. Rothenberg (eds.) (2000) *Beneath the Surface: Critical Essays in the Philosophy of Deep Ecology.* Cambridge, MA, Cambridge University Press.
20. Baxter, W. (1995) People or Penguins, in C. Pierce and D. VanDeVeer (eds.), *People, Penguins, and Plastic Trees*, 2nd ed., Belmont, CA, Wadsworth Publishing Company.
21. Thompson, J. (1995) Aesthetics and the Value of Nature, *Environmental Ethics* **17,** 291-305.
22. Daily, G. (1997) Introduction, in G. Daily (ed.), 1-10.
23. Costanza, R., R. dArge, R. deGroot, S. Farber, M. Grasso, B. Hannon, K. Limburg, S. Naeem, R. V. O'Neill, J. Paruelo, R. G. Raskin, P. Sutton, and M. vandenBelt (1997) The Value of the World's Ecosystem Services and Natural Capital, *Nature* **387,** 253-260.
24. Karr, J. R. (1991) Biological Integrity: A Long Neglected Aspect of Water Resource Management, *Ecological Applications* **1,** 66-84.
25. Karr, J. R. (1996) Ecological Integrity and Ecological Health Are Not the Same, in P. Schulze (ed.) *Engineering Within Ecological Constraints,* Washington, D.C., National Academy Press, 97-109.
26. Kay, J. J. and E. Schneider (1995) Embracing Complexity: The Challenge of the Ecosystem Approach., in L. Westra and J. Lemons (eds.),48-59.
27. Ulanowicz, R. E. (1995) Ecosystem Integrity: A Causal Necessity, in L. Westra and J. Lemons (eds.), 77-101.
28. Ulanowicz, R. E. (1996) The Propensities of Evolving Systems, in E. L. Khalil and K. E. Boulding (eds.) *Evolution, Order and Complexity,* London, Routledge, 217-233.
29. Rapport, D., R. Costanza, P. R. Epstein, C. Gaudet, and R. Levins (1998) *Ecosystem Health,*. Malden, MA, Blackwell Science.
30. Cairns, J. (1998) Ecological Risk Assessment: A Predictive Approach to Assessing Ecosystem Health, in D. Rapport et al. (eds.), 216-228.
31. Daily, G. (ed.) (1997) *Nature's Services.* Washington, D.C., Island Press.
32. Shrader-Frechette, K. S. (1995) Sustainability and Environmental Ethics, in L. Westra and J. Lemons, ch. 4.
33. Sagoff, M. (1988) *The Economy of the Earth*, Cambridge, Cambridge University Press.
34. Sagoff, M. (1994) Should Preferences Count?, *Land Economics* **70,** 127-144.
35. Woodley S., J. Kay, and G. Francis (eds.) (1993) *Ecological Integrity and the Management of Ecosystems,* St. Lucie Press.
36. Costanza, R. (1992) Toward an Operational Definition of Health, in R. Costanza, B. Norton, and B. Haskell (eds.), 239-256.
37. Westra, L. (1994) *An Environmental Proposal for Ethics: The Principle of Integrity,* Lanham, MD, Rowman and Littlefield Publishers, Inc.

LIST OF CONTRIBUTORS

Prof. N.V. ALADIN
Laboratory of Brackish Water Hydrobiology,
Zoological Institute of RAS
Universitetskaya naberezhnaya, 1,
199034 St Petersburg (Russia)

Dr. Kanatbek ABDRACHMATOV
Institute of Seismology
52/1 Asanbay
720060 BISHKEK (Kyrgyzstan)

Prof. Umut ASANOVA
National Academy of Sciences of the Kyrghyz Republic
Chui Prospekt, 265a
Bishkek (Kyrgyzstan)

Dr. S.E. BEAUBIEN
Earth Science Department, University of Rome "La Sapienza"
Piazzale Aldo Moro, 5, I 00185 Roma (Italy)

Prof. Christian BECK
Laboratoire de Géodynamique des Chaînes Alpines
Université de Savoie
F76766 Le Bourget du Lac (France)

Dr. Meryem BEKLIOGLU
Middle East Technical University, Biology Department,
06531 Ankara (Turkey)

Dr. Laurent BERGONZINI
UMR 8616 "ORSAYTERRE",
Université Paris-Sud, ât 504
F-91405 Orsay (France)

Prof. Gregory COOPER
Society and Professions Program on Ethics
Washington and Lee University,
Lexington (USA)

Prof. Marc DE BATIST
Renard Centre of Marine Geology
University of Gent
Krijgslaan, 281/8
B 900 GENT (Belgium)

Dr. Damien DELVAUX
Royal Museum of Central Africa
Leuvensesteenweg, 13
B3080 TERVUREN (Belgium)

K.D. DJANUZAKOV
Kyrgyz Institute of Seismology, Laboratory of Seismotectonics
720060 Bishkek, Kyrgyzstan

V.N. EGOROV,
The A.O.Kovalevsky Institute of Biology of the Southern Seas (IBSS)
National Academy of Sciences of Ukraine,
Prospekt Nakhimova 2, 99011, Sevastopol (Ukraine)

A.A. FILIPPOV
Laboratory of Brackish Water Hydrobiology,
Zoological Institute of RAS
Universitetskaya naberezhnaya, 1,
199034 St Petersburg (Russia)

Dr. Santiago GIRALT
International Bureau for Environmental Studies
Rue Audrey Hepburn, 9/13
B 1090 BRUSSELS (Belgium)

Prof. Beishen IMANACKUNOV
National Academy of Sciences of the Kyrghyz Republic
Chui Prospekt, 265a ,
Bishkek (Kyrgyzstan)

Yannick IMBO
Renard Centre of Marine Geology
University of Gent
Krijgslaan, 281/8
B 900 GENT (Belgium)

Prof. Janybeck JEYENBAEV
National Academy of Sciences of the Kyrghyz Republic
Chui Prospekt, 265a ,
Bishkek (Kyrgyzstan)

Prof. Ramon JULIA
Institute of Earth Sciences "Jaume Almera" (CSIC)
Lluís Solé i Sabarís s/n,
E-08028 Barcelona (Spain

Dr; Ivan KALUGIN
United Institute of Geology, Geophysics and Mineralogy
Acad. Koptyug Avenue,8
630090 NOVOSIBIRSK (Russia)

Dr. A.S. KARMANCHUK
Institute of Irrigation of Kyrgyzstan
4a, Toktonalieva street,
720055 Bishkek (Kyrgyzstan)

Dr. Rolf KIPFER
Environmental Isotopes Group, Dept.of Water Resources and Drinking Water (W&T),
Swiss Federal Institute of Environmental Science and Technology (EAWAG)
CH-8600 Dübendorf, Switzerland

Prof. Jean KLERKX
International Bureau for Environmental Studies
Rue Audrey Hepburn, 9/13
B 1090 BRUSSELS (Belgium)

Dr. Vincent LIGNIER
Laboratoire de Géodynamique des Chaînes Alpines
Université de Savoie
F76766 Le Bourget du Lac (France)

Prof. Salvatore LOMBARDI
Earth Science Department, University of Rome "La Sapienza"
Piazzale Aldo Moro, 5, I 00185 Roma (Italy)

Dr. A.N. MANDYCHEV
Institute of Water Problems and Hydropower,
533, Frunze str., Bishkek (Kyrgyzstan)

Dr. V.E. MATYCHENKOV
Institute of Water Problems and Hydropower
533, Frunze str., Bishkek (Kyrgyzstan)

PROF. D.N. PACHAJANOV
Nikitin's Institute of Chemistry, Tajik Academy of Sciences
Aini Avenue, 299/2, 734063 Dushanbe (Tajikistan)

D.L. PATINA
Nikitin's Institute of Chemistry, Tajik Academy of Sciences
Aini Avenue, 299/2, 734063 Dushanbe (Tajikistan)

Dr. Frank PEETERS
Environmental Isotopes Group, Dept.of Water Resources and Drinking Water (W&T),
Swiss Federal Institute of Environmental Science and Technology (EAWAG)
CH-8600 Dübendorf, Switzerland

I.S. PLOTNIKOV
Laboratory of Brackish Water Hydrobiology,
Zoological Institute of RAS
Universitetskaya naberezhnaya, 1,
199034 St Petersburg (Russia)

Dr. G.G. POLIKARPOV
The A.O.Kovalevsky Institute of Biology of the Southern Seas (IBSS)
National Academy of Sciences of Ukraine,
Prospekt Nakhimova 2, 99011, Sevastopol (Ukraine)

Dr. S. RIERA
Museu, Laboratorio e Jardim Botanico de Lisboa, Laboratorio de Paleoecologia
Rua da Escola Politecnica 58,
P-1294 Lisboa (Portugal)

Dr. V.V ROMANOVSKY
Institute of Water Problems and Hydropower
533, Frunze str.,
Bishkek (Kyrgyzstan)

A.G. SHABUNIN
Instiute of Water Problems and Hydropower
533, Frunze str., Bishkek (Kyrgyzstan)

G.D. SHABUNIN
Issyk-Kul Lake Observatory
Cholpon Ata (Kyrgyzstan)

Dr. N.N. TERESHCHENKO
The A.O.Kovalevsky Institute of Biology of the Southern Seas (IBSS)
National Academy of Sciences of Ukraine,
Prospekt Nakhimova 2, 99011, Sevastopol (Ukraine)

Pieter VERMEESCH
Renard Centre of Marine Geology
University of Gent
Krijgslaan, 281/8
B 900 GENT (Belgium)

DR. N.I. VOLKOVA
Institute of Geology, Geophysics and Mineralogy, SBRAS
Akademik Koptyug Avenue, 3, 630090 Novosibirsk (Russia)

Dr. Nunzia VOLTATTORNI
Earth Science Department, University of Rome “La Sapienza”
Piazzale Aldo Moro, 5, I 00185 Roma (Italy)